Biology
Fifth Edition

Editor

Phyllis Cellini Braun
Fairfield University

Phyllis Cellini Braun received her Ph.D. from Georgetown University and is presently an associate professor of biology at Fairfield University. As a molecular biologist, her major area of interest is the molecular regulation of morphogenesis. Numerous articles have been published on her research regarding cell wall biosynthesis mechanisms on the dimorphic yeast *Candida albicans*.

A Library of Information from the Public Press

Cover illustration by Mike Eagle

The Dushkin Publishing Group, Inc.
Sluice Dock, Guilford, Connecticut 06437

The Annual Editions Series

Annual Editions is a series of over forty volumes designed to provide the reader with convenient, low-cost access to a wide range of current, carefully selected articles from some of the most important magazines, newspapers, and journals published today. Annual Editions are updated on an annual basis through a continuous monitoring of over 200 periodical sources. All Annual Editions have a number of features designed to make them particularly useful, including topic guides, annotated tables of contents, unit overviews, and indexes. For the teacher using Annual Editions in the classroom, an Instructor's Resource Guide with test questions is available for each volume.

VOLUMES AVAILABLE

Africa
Aging
American Government
American History, Pre-Civil War
American History, Post-Civil War
Anthropology
Biology
Business and Management
China
Comparative Politics
Computers in Education
Computers in Business
Computers in Society
Criminal Justice
Drugs, Society, and Behavior
Early Childhood Education
Economics
Educating Exceptional Children
Education
Educational Psychology
Environment
Geography
Global Issues
Health

Human Development
Human Sexuality
Latin America
Macroeconomics
Marketing
Marriage and Family
Middle East and the Islamic World
Nutrition
Personal Growth and Behavior
Psychology
Social Problems
Sociology
Soviet Union and Eastern Europe
State and Local Government
Third World
Urban Society
Western Civilization,
  Pre-Reformation
Western Civilization,
  Post-Reformation
Western Europe
World History, Pre-Modern
World History, Modern
World Politics

Library of Congress Cataloging in Publication Data
Main entry under title: Annual editions: Biology.
    1. Biology—Periodicals. 2. Bioethics—Periodicals. I. Braun, Phyllis C., *comp.* II. Title:
Biology.
574'.05   84-30680
ISBN: 0-87967-709-0

Fifth Edition

Manufactured by The Banta Company, Harrisonburg, Virginia 22801

Editors/ Advisory Board

To The Reader

In publishing ANNUAL EDITIONS we recognize the enormous role played by the magazines, newspapers, and journals of the *public press* in providing current, first-rate educational information in a broad spectrum of interest areas. Within the articles, the best scientists, practitioners, researchers, and commentators draw issues into new perspective as accepted theories and viewpoints are called into account by new events, recent discoveries change old facts, and fresh debate breaks out over important controversies. Many of the articles resulting from this enormous editorial effort are appropriate for students, researchers, and professionals seeking accurate, current material to help bridge the gap between principles and theories and the real world. These articles, however, become more useful for study when those of lasting value are carefully *collected, organized, indexed,* and *reproduced* in a *low-cost format*, which provides easy and permanent access when the material is needed. That is the role played by *Annual Editions*. Under the direction of each volume's *Editor*, who is an expert in the subject area, and with the guidance of an *Advisory Board*, we seek each year to provide in each *ANNUAL EDITION* a current, well-balanced, carefully selected collection of the best of the public press for your study and enjoyment. We think you'll find this volume useful, and we hope you'll take a moment to let us know what you think.

Education is the instruction of the intellect in the laws of nature, under which name I include not merely things and their forces, but men and their ways.

—T.H. Huxley

Teaching biology is challenging, exciting, and rewarding. The instructor is responsible for presenting the essentials of a discipline that is both dynamic and critically relevant, and for providing a basic understanding of key biological principles, to a class which often consists of students with widely differing backgrounds and needs. Complicating this endeavor is the enormous number of discoveries and fundamental changes in biological science that are occurring at an astounding rate. This revolution in biology is having an impact on the world around us. Announcements appear almost daily about such matters as new techniques to redesign both plant and animal life, production of hormones and vaccines through genetic engineering, and new molecular weapons against cancer and other diseases.

The rapid growth we are witnessing in biology is due to two significant factors: advances in new sophisticated techniques utilizing computer science, chemistry, physics, and engineering; and the trusted assumption of the basic uniformity of life processes. That is, it is believed that the fundamental biological principles that govern the activity of simple organisms, such as bacteria and viruses, must apply to more complex cells; only the details vary. This very logical and productive approach to scientific questions has been the successful driving force that has lead to fundamental breakthroughs on nerve and muscle function, membrane structure, the mode of action of antibiotics, cellular differentiation and development, immunology, and genetics. Current scientific advances are not only affecting genetics and medicine, but also evolution, behavior, physiology, neuroscience, and our environmental interactions. Because of the technological impact these recent advances in the biological sciences have brought to our society, increasing emphasis must also be placed upon scientific responsibility. This responsibility does not rest solely in the hands of the scientist; a knowledge of science is essential for the layperson as well. Scientific understanding is necessary if judicious decisions are to be undertaken by the public. For example, the technology of genetic engineering is currently being debated by both scientists and policymakers. Genetic engineering has dramatically changed the disciplines of microbiology, genetics, plant and animal biology, medicine, agriculture, and physiology. Depending on how we use this technology, the impact upon our environment, natural resources, and society could range from beneficial to devastating. While increased food production and the elimination of diseases could be beneficial, dangerous life forms released by accident or deliberately through biological warfare would be threatening to humanity.

The majority of students who read the selections in *Annual Editions: Biology* probably will not become biologists or even scientists. Nevertheless, they need to become more aware of their role in a complex biological system, and to better understand the world they live in by gaining respect for the intricate balance of people, nature, and the environment. A broad background in biology is needed to fully understand these basic issues. The primary aim of this anthology is to supplement the student's exposure to some of the contemporary topics of biology. Consequently, if the students who read this volume understand that humans are not master manipulators of their world but rather a single species depending on and interacting with thousands of other organisms, perhaps appropriate, intelligent decisions can be made on complex scientific problems that may have wide reaching ethical, religious, and moral implications.

Phyllis Cellini Braun

Editor

Contents

Unit 1

Evolution

Five articles examine the process of evolution and some of the current controversies.

To the Reader   iv
Topic Guide   2
Overview   4

1. **The Evolution Revolution,** Robin Marantz Henig, *SciQuest,* January 1982.   6

   Obscured by the recent *creationism controversy* is the fact that evolutionists are involved in their own debate regarding how *evolution* happened.

2. **Influence of Darwin's Ideas on the Study of Evolution,** William B. Provine, *BioScience,* June 1982.   11

   It took biologists eighty years to accept Darwin's concept of *evolution* by *natural selection.* According to William Provine, it will take them much longer to accept the implications of evolutionary biology.

3. **Earth's Lucky Break,** John Gribbin, *Science Digest,* May 1983.   17

   A clue to the mystery of *life on earth* might be found by looking at the conditions that allowed life to arise here in the first place.

4. **The Search for Life's Origins—and a First ''Synthetic Cell,''** Paul Trachtman, *Smithsonian,* June 1984.   20

   A small group of chemists, biologists, and physicists are attempting to reenact the basic chemical events leading to the *origin of life.* By using nucleic acids, proteins, and lipids, they hope to produce life in a test tube and learn how life evolved four billion years ago.

5. **On the Wings of an Angel,** Michael J. Katz, *Harvard Magazine,* September/October 1985.   25

   There are no biological impossibilities. Within the true province of biology, everything is possible. Michael Katz discusses the *evolution* of biological organisms on the basis of two sequential processes—*ontogeny* and *phylogeny.*

The concepts in bold italics are developed in the article. For further expansion please refer to the Topic Guide and the Index.

Unit 2

Genetics

Nine selections discuss the ever-changing science of genetics, and how genetic engineers can potentially control plant and animal development.

Overview   32

6. **Bits of Life,** Alexander Rich, *The Sciences,* October 1980.   34

   A *living system* is not just a set of simple ongoing chemical reactions, but rather a directed and responsive system in which there is control and regulation of the kinds of reactions that occur.

7. **Genetics: The Edge of Creation,** Albert Rosenfeld, *GEO,* February 1982.   39

   Until now, only *nature* and circumstance controlled the shape and character of living things. That has changed. Scientists are currently learning how to share in nature's work.

8. **Yeast at Work,** Terence Monmaney, *Science 85,* July/ August 1985.   44

   Yeasts, the single cell fungi, are one of the newest biological members to be recruited to biotechnology's workforce. By recombinant DNA techniques, *yeasts* have not only provided new *vaccines* and *medicines*, but have also contributed to a fundamental *understanding of diseases* such as cancer.

9. **Science Debates Using Tools to Redesign Life,** Keith Schneider, *The New York Times,* June 8, 1987.   48

   The recent ability to *parent* higher *life* forms gives biotechnology the go-ahead to genetically alter biological organisms. Keith Schneider discusses both the *benefits* and *ethical objections* of redesigning life.

10. **The Gene Revolution,** Bernard Dixon, *The UNESCO Courier,* March 1987.   52

   The application of *genetic engineering* to *agriculture* is discussed. Through recombinant *DNA* techniques, genes may be introduced to allow plants to live and grow in environments which were previously stressful.

11. **How Genes Work,** Boyce Rensberger, *Science Digest,* June 1986.   55

   What are *proteins*? How are proteins related to *DNA*? Why are biotechnology companies so interested in them? These questions are answered in this report.

12. **Spelling Out a Cancer Gene,** Julie Ann Miller, *Science News,* November 13, 1982.   56

   In a human cell, a single *mutation* makes the difference between a normal gene and a gene that causes tumors. Many *cancers* appear to involve altered forms of basic cellular genes.

13. **The Fate of the Egg,** Stephen S. Hall, *Science 85,* November 1985.   59

   Life starts as a single cell. Scientists are currently trying to understand the mechanism by which *cells differentiate* into tissue and organs.

14. **Life's Recipe,** *The Economist,* May 30-June 5, 1987.   65

   The article reviews the great scientific discoveries about *genes*—findings which have revolutionized biology.

Unit 3

Behavior

Seven articles examine the concept that most living systems are programmed to maintain the biological being in its environment.

Overview   68

15. **Thinking Well: The Chemical Links Between Emotions**   70
    and Health, Nicholas R. Hall and Allan L. Goldstein, *The*
    Sciences, March/April 1986.
    What effect does one's mental state have on *illness*? The authors
    discuss the relationships between *neurological* and *immunolog-*
    ical systems.

16. **Major Personality Study Finds That Traits Are Mostly**   74
    Inherited, Daniel Goleman, *The New York Times,* Decem-
    ber 2, 1986.
    The debate over nature vs. nurture continues. Initial studies on iden-
    tical twins indicate that the *genetic makeup* of a child has a
    stronger influence on *personality* than does child rearing.

17. **The Clock Within,** Philip Hilts, *Science 80,* December 1980.   77
    According to the article, the *body's internal cycles* affect every-
    thing from blood pressure to decision making.

18. **Chemical Cross Talk,** Jesse Roth and Derek LeRoith, *The*   82
    Sciences, May/June 1987.
    For thousands of years, humans have recorded the *medical*
    benefits of various *plants*. The ability of plants to control human
    cellular responses may be due to the constancy of *primordial mes-*
    senger molecules through evolution.

19. **The Instinct to Learn,** James L. Gould and Carol Grant   86
    Gould, *Science 81,* May 1981.
    Bees do it. Birds do it. Perhaps even *humans* are *programmed*
    to acquire critical information at specific times.

20. **Food Affects Human Behavior,** Gina Kolata, *Science,*   91
    December 17, 1982.
    A number of scientists are discovering that people do react to what
    they *eat*, although the effects are subtle.

21. **The Placebo Effect: It's Not All in Your Mind,** Robin   93
    Marantz Henig, *SciQuest,* May/June 1981.
    In recent years, the benefits of the *placebo*, a pharmacologically
    inactive drug, have been documented and studied. These inactive
    drugs have been shown to cure certifiable physical ills. The author
    explains that the positive effects are a response to the psycholog-
    ical conviction on the part of the patient in the ability of the drug
    to work.

The concepts in bold italics are developed in the article. For further expansion please refer to the Topic Guide and the Index.

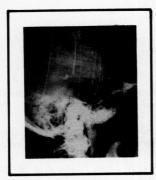

Unit 4

The Brain

Five selections discuss the brain and the nervous system and their relation to the total behavioral patterns of the organism.

Overview   **96**

22. Intelligence Test: Sizing Up a Newcomer, Wray Herbert, *Science News,* October 30, 1982.   **98**

The *IQ test* is obsolete, according to two psychologists. In its place they would substitute a new kind of intelligence test, one derived from recent neuropsychological research on individual thinking styles.

23. What the Brain Builders Have in Mind, *The Economist,* May 2, 1987.   **101**

In an endeavor to understand how the human brain functions in *intelligence*, scientists are attempting to build a *computer* that looks like a brain and learns by itself. This novel approach is called *connectionism* or neurocomputing.

24. Lab Team Enjoys a Cosmic Cruise Inside the Mind, Kathleen Fisher, *APA Monitor,* September 1985.   **104**

Current studies on the *brain* have shifted from the psychological to the biological. *Informational storage, memory*, and brain *mapping* are discussed in this article.

25. Such Stuff as Dreams Are Made On, Robert Kanigel, *Notre Dame Magazine,* Autumn 1985.   **109**

Is the primary motivating force for *dreaming* physiological and not psychological? Dream researchers debate this issue.

26. The Unbalanced Brain, Patrick L. McGeer, *The Sciences,* May/June 1977.   **114**

Patrick McGeer reports that *mood* and *movement disorders* may be caused by different proportions of the same *brain chemicals*.

Unit 5

Physiology

Six articles discuss how organisms age and develop. The articles stress that by understanding what is normal, scientists are better able to deal with what is abnormal.

Overview   **118**

27. The Aging Body, John Tierney, *Esquire,* May 1982.   **120**

After age twenty, the decades take their toll. You may get wiser, but your *memory dims* and your *body parts grow, shrink*, or *disappear*. According to the author, it is a process you can only watch with wonder.

28. Grow, Nerves, Grow, Julie Ann Miller, *Science News,* March 29, 1986.   **126**

Medical researchers are studying *nerve cells* to determine the biological conditions necessary to promote growth. They are hopeful that someday damaged nerves of the central nervous system will be *regenerated* and will return normal functions to millions of people who suffer from head and spinal cord injuries.

The concepts in bold italics are developed in the article. For further expansion please refer to the Topic Guide and the Index.

29. **Immune System: Great Mystery Is Solved After Long Quest,** Harold M. Schmeck, Jr., *The New York Times,* June 19, 1984.   **129**

During this past decade, the mysteries of the **immune system** have been under intense scientific scrutiny. The results of this research have indicated that proper functioning of the immune defenses is dependent on the **cell surface receptors** that aid in recognition.

30. **Weight Regulation May Start in Our Cells, Not Psyches,** Gina Kolata, *Smithsonian,* January 1986.   **131**

Obesity research indicates that the signals to overeat may come from the **fat cells** themselves.

31. **Ergot and the Salem Witchcraft Affair,** Mary K. Matossian, *American Scientist,* July/August 1982.   **135**

An outbreak of a type of **food poisoning** known as convulsive ergotism may have led to the 1692 accusations of **witchcraft**.

32. **Saving Babies,** B.D. Colen, *Health,* August 1986.   **139**

Breakthroughs in **fetal surgery** have made it possible to correct birth defects in the unborn child. The procedure has raised some troubling questions, some of which are addressed here.

Unit
6

Medicine

Six selections examine the relationship between biology and medicine. Particular attention is given to the current status of the war against AIDS.

Overview   **142**

33. **Pain: Many Causes, Fewer Cures,** Stephen Budiansky, *SciQuest,* December 1981.   **144**

Pain is now considered a medical problem in itself, not just a symptom of disease. To block it out may be the best cure, contends the author.

34. **Racing Toward the Last Sneeze,** Dianne Hales, *American Health,* March/April 1982.   **148**

The persistence and elusiveness of **allergic symptoms** have always been with society. The author discusses how immunologists who are intrigued by the enigmas of allergy are now looking beyond the whos, whats, whens, and wheres of allergy to the whys.

35. **The Gene Doctors,** *Business Week,* November 18, 1985.   **154**

Genetic diseases caused by simple errors in the genetic code may soon be cured by replacing the defective gene. The authors report on the various genetic diseases which may be eradicated by such therapy.

36. **Mapping the Genes, Inside and Out,** David Holzman, *Insight,* May 11, 1987.   **159**

Complete **gene identification** promises enormous benefits, including prevention or cure of genetic defects and discovery of new drugs patterned after the body's own chemistry. The magnitude of the task is staggering, explains the author, and so is the cost.

The concepts in bold italics are developed in the article. For further expansion please refer to the Topic Guide and the Index.

37. The Natural History of AIDS, Matthew Allen Gonda, *Natural History,* May 1986.   162

Human T-cell lymphotropic virus type III (HTLV-III) is the causative agent of *AIDS*. Where did it come from, how long has it been on earth, and what are its viral relatives? The author speculates on the *evolution* of the AIDS virus into modern society.

38. What Triggers AIDS? Joanne Silberner, *Science News,* April 4, 1987.   166

Many researchers believe that besides the human immunodeficiency virus, other additional cofactors are necessary to turn a symptomless *AIDS* infection into the disease state.

Unit 7

The Environment

Eight articles examine the world's environment and humanity's interrelationship with it. Topics include the status of energy, the population crisis, acid rain, and the effects of human society on the atmosphere.

Overview   168

39. Energy: A Promise Renewed, Harry Bacas, *Nation's Business,* July 1985.   170

Energy experts anticipate that oil, coal, and nuclear energy will be the primary sources of power far into the next century, but research on *alternative energy* sources continues. Many of these alternative sources such as *biomass energy, hydropower, windpower, geothermal energy*, and *solar energy* no longer seem so exotic or expensive.

40. World Population Crisis, Paul R. Ehrlich and Anne H. Ehrlich, *Bulletin of the Atomic Scientists,* April 1986.   172

Rapid *population growth*, rising *competition for resources*, and increasing *environmental deterioration* are intertwined factors in the human predicament that feed the political tensions and conflicts of the late twentieth century. The authors contend that "the baby race" is as dangerous for humanity as the arms race.

41. 19th Environmental Quality Index: A Nation Troubled by Toxics, *National Wildlife,* February/March 1987.   178

The National Wildlife Federation reviews the United States *environmental record* of 1986. The study focuses on *wildlife, air quality, water pollution, energy use, public forests*, and *soil erosion*.

42. Acid Deposition: Trends, Relationships, and Effects, Arthur H. Johnson, *Environment,* May 1986.   186

Without the 120,000 cubic kilometers of rain which fall each year, the world's continents would be barren. Now the rain in many parts of the world has taken on a new and threatening complexity in the form of *acid rain*, which kills fish and other wildlife, corrodes buildings, and destroys forests.

The concepts in bold italics are developed in the article. For further expansion please refer to the Topic Guide and the Index.

43. **How Much Are the Rainforests Worth?** Philip M. Fearnside, *United Nations University, Work in Progress,* November 1985.   195

The forces that propel *deforestation* are a complex mix of directed self-interests and broad economic processes. Policies, once implemented, can set up vicious cycles leading to very serious consequences.

44. **Beyond the Green Revolution,** Laura Tangley, *BioScience,* March 1987.   198

Creative new strategies of *food production* are needed in the parts of the world not impacted by the green revolution. International agricultural research in creating new crop varieties and studying animal husbandry and livestock disease provides new hope for economic growth for a quarter of the world's people.

45. **Mimicking Nature,** Edward C. Wolf, *Ceres,* January/February 1987.   203

Agricultural scientists have recently begun to recognize that *traditional farming systems* which have persisted for millennia exemplify careful management of soil, water, and nutrients. Further agricultural advancements to raise productivity can be built on the foundation of traditional farming.

46. **Ozone Depletion's New Environmental Threat,** Janet Raloff, *Science News,* December 6, 1986.   207

One of the chief benefits of earth's *stratospheric ozone* layer is its ability to filter out much of the sun's biologically harmful ultraviolet radiation. Until recently, most predictions of the environmental dangers focused on increases of solar ultraradiation and its direct effects (human skin cancer). Atmospheric chemists have added two more hazards resulting from declines in stratospheric ozone: an increase in *smog ozone* and *acid rain*.

Index   210
Article Rating Form   213

The concepts in bold italics are developed in the article. For further expansion please refer to the Topic Guide and the Index.

Topic Guide

This topic guide suggests how the selections in this book relate to topics of traditional concern to biology students and professionals. It is very useful in locating articles which relate to each other for reading and research. The guide is arranged alphabetically according to topic. Articles may, of course, treat topics that do not appear in the topic guide. In turn, entries in the topic guide do not necessarily constitute a comprehensive listing of all the contents of each selection.

TOPIC	TREATED AS AN ISSUE IN:	TOPIC	TREATED AS AN ISSUE IN:
Behavior	Section 3 22. Intelligence Test 24. Lab Team Enjoys Cosmic Cruise Inside the Mind 25. Such Stuff as Dreams Are Made On 26. The Unbalanced Brain 30. Weight Regulation May Start in Our Cells, Not Psyches	**Chemical Basis of Life**	3. Earth's Lucky Break 4. The Search for Life's Origins 6. Bits of Life 7. Genetics 8. Yeast at Work 9. Science Debates Using Tools to Redesign Life 10. The Gene Revolution 11. How Genes Work 12. Spelling Out a Cancer Gene 13. The Fate of the Egg 18. Chemical Cross Talk 26. The Unbalanced Brain 29. Immune System
Bioenergetics and Nutrition	6. Bits of Life 8. Yeast at Work 20. Food Affects Human Behavior 26. The Unbalanced Brain 27. The Aging Body 30. Weight Regulation May Start in Our Cells, Not Psyches 31. Ergot and the Salem Witchcraft Affair	**The Environment**	Section 7 3. Earth's Lucky Break 9. Science Debates Using Tools to Redesign Life 10. The Gene Revolution 31. Ergot and the Salem Witchcraft Affair 34. Racing Toward the Last Sneeze
The Brain: Neuroscience and Integration	Section 4 15. Thinking Well 18. Chemical Cross Talk 20. Food Affects Human Behavior 27. The Aging Body 28. Grow, Nerves, Grow 31. Ergot and the Salem Witchcraft Affair	**Evolution**	Section 1 18. Chemical Cross Talk 19. The Instinct to Learn 35. The Gene Doctors 37. The Natural History of AIDS
Cell and Molecular Biology	4. The Search for Life's Origins 5. On the Wings of an Angel 6. Bits of Life 7. Genetics 8. Yeast at Work 9. Science Debates Using Tools to Redesign Life 10. The Gene Revolution 11. How Genes Work 13. The Fate of the Egg 14. Life's Recipe 15. Thinking Well 18. Chemical Cross Talk 28. Grow, Nerves, Grow 30. Weight Regulation May Start in Our Cells, Not Psyches 35. The Gene Doctors 36. Mapping the Genes, Inside and Out 38. What Triggers AIDS?		

TOPIC	TREATED AS AN ISSUE IN:	TOPIC	TREATED AS AN ISSUE IN:
Genetics	Section 2	**Physiology**	Section 5
	16. Major Personality Study Finds That Traits Are Mostly Inherited		6. Bits of Life
			7. Genetics
	30. Weight Regulation May Start in Our Cells, Not Psyches		8. Yeast at Work
	35. The Gene Doctors		9. Science Debates Using Tools to Redesign Life
	36. Mapping the Genes, Inside and Out		10. The Gene Revolution
	38. What Triggers AIDS?		11. How Genes Work
			12. Spelling Out a Cancer Gene
Medicine	Section 6		13. The Fate of the Egg
	6. Bits of Life		15. Thinking Well
	7. Genetics		18. Chemical Cross Talk
	8. Yeast at Work		24. Lab Team Enjoys a Cosmic Cruise Inside the Mind
	9. Science Debates Using Tools to Redesign Life		26. The Unbalanced Brain
	11. How Genes Work		33. Pain
	12. Spelling Out a Cancer Gene		
	15. Thinking Well	**Population Biology**	5. On the Wings of an Angel
	18. Chemical Cross Talk		6. Bits of Life
	21. The Placebo Effect		7. Genetics
	26. The Unbalanced Brain		9. Science Debates Using Tools to Redesign Life
	28. Grow, Nerves, Grow		40. World Population Crisis
	29. Immune System		
	30. Weight Regulation May Start in Our Cells, Not Psyches	**Reproduction and Development**	6. Bits of Life
	31. Ergot and the Salem Witchcraft Affair		7. Genetics
	32. Saving Babies		9. Science Debates Using Tools to Redesign Life
			13. The Fate of the Egg
			32. Saving Babies

Evolution

When on board H.M.S. Beagle, as a naturalist, I was much struck with . . . the distribution of the inhabitants of South America, and . . . the geological relations of the present to the past inhabitants of that continent. These facts seemed to me to throw some light on the origin of species—that mystery of mysteries, as it has been called by one of our greatest philosophers.

From *On the Origin of Species by Means of Natural Selection* by Charles Darwin

It has been over a century since Charles Darwin published these lines in this introduction to *On the Origin of Species by Means of Natural Selection*, the single most important work in all of biology. In this text, Darwin proposed that over eons of time, species arise from other, preexisting species, through the process of "descent with modification," or evolution. The controversy generated by his 1859 publication still continues. The debate over creationism—the idea that each species was created individually by God—versus evolution still rages. Recently, the US Supreme Court decided that the supernatural theory of creationism was an inappropriate subject in public school curriculums.

Now, evolutionists are involved with their own debate: "punctuated equilibrium" versus gradual evolution. Darwin's theory originally stressed a number of fundamental concepts—gradual changes within populations as one generation replaces another, the idea of reproductive success and natural selection as the guiding force in evolution, and the recognition that differences exist among all members of a population. But today, scientists who study evolution are questioning the speed, patterns, and modes intrinsic to the theory described by Darwin.

"Punctuated equilibrium," the idea that evolution proceeds in discrete leaps rather than gradual change, has recently been presented to explain the lack of transitional forms between many species—a point that has been used by creationists for many years to attack evolutionary thought.

In the article "The Evolution Revolution," Robin M. Henig examines a number of questions now being raised by evolutionists. As the author points out, the facts used to portray the theory of evolution are still being turned inside-out and repositioned.

Scientists are currently not only interested in the biological changes of organisms explained by the theory of evolution, but are also attempting to unravel the mysteries of the origin of life. The question of whether life on the planet Earth was a cosmic fluke or the consequence of the right chemical conditions is discussed in John Gribbin's article, "Earth's Lucky Break." Gribbin maintains that the abundance of water and metals and the earth's distance from the sun allowed life to arise initially. In "The Search for Life's Origins—and a First 'Synthetic Cell,'" Paul Trachtman describes the attempts scientists are making to use their knowledge of both the universe and the cell to recreate the scene on earth, about four billion years ago, when life arose from its chemical constituents.

It took more than eighty years for Darwin's concept of evolution by natural selection to be accepted by biologists. Currently, certain implications intrinsic to Darwin's concept of evolutionary biology are being modified. For biologists to accept these proposed changes will undoubtedly require time and further scientific investigation.

Looking Ahead: Challenge Questions

Using the concept of "punctuated equilibrium," how can one explain the absence of "links" between closely related species?

Why is it postulated that before the first life form on earth, molecules made copies of themselves through a primitive kind of reproduction, growing more complex and evolving through accidental errors?

Living entities are patterns of matter that are at once complex and individual. In terms of the evolutionary process, how do the forces of ontogeny and phylogeny guide the production of these biological organisms?

THE EVOLUTION REVOLUTION

Robin Marantz Henig

Robin Marantz Henig is a science writer based in Washington, D.C.

Charles Darwin's theory of evolution has been under attack ever since it was first offered in 1859. The preposterous idea that humans are descended from the apes sent Victorian England into a tailspin; once the establishment of Mendelian genetics provided scientific underpinnings for the idea, religious fundamentalists began to bristle at the absence of God in all these discussions. From the famous Scopes "monkey trial" in Tennessee in 1925 to the lively debate about "scientific creationism" that flourishes today, evolutionary theory has been the topic of heated controversy.

But obscured by the current creationism flap is the fact that the evolutionists themselves are entangled in their own imbroglio—not over whether evolution occurred (since most scientists take it as a given that it did), but over precisely how it happened. As every schoolchild learns, Darwin's theory of evolution highlights three factors: the randomness of genetic variation, the guiding force of natural selection (a notion later popularized as "survival of the fittest"), and gradual, minute changes accumulated over a long, long time. But scientists who study evolution are now wondering whether evolution occurred in quite the pattern, manner and tempo that Darwin described.

For more than a generation, the prevailing evolutionary orthodoxy has been the so-called modern synthesis. This school of thought, named by Julian Huxley in 1942, blends Darwin's basic theory of evolution (which he called "descent with modification") with the knowledge of basic genetics that blossomed in the 20th century. Experts disagree as to just how closely the modern synthesis parallels Darwin's own writings, but in the current debate, the terms Darwinism and modern synthesis are often used interchangeably.

Three issues seem to be at the heart of the modern synthesis. First, random variation, which is based on barely perceptible genetic mutation, is the raw material for evolution. Second, these variations are differentially spread throughout a population according to the laws of natural selection—that is, mutations that allow their bearers better to adapt to the environment are the ones that will survive, reproduce and prosper. Third, this process occurs at a pace that is slow and continuous, with change occurring so gradually as to be all but undetectable from one century to the next. This change is speciation, the process in which a descendant strain breaks off from its ancestors and differs enough from them, both genetically and physically, that it forms a new species that can no longer mate with its forebears.

The first aspect of the modern synthesis to be questioned is just how random the effects of chance mutations are. Are these mutations in truth genetic accidents or is there some logic to their occurrence or consequences? "Underpinning our view of natural selection is the belief that selection could have pulled things in any direction," says Stuart Kauffman, a developmental biologist at the University of Pennsylvania. "The traditional view is of the organism as a sphere that has an equal likelihood of moving in any direction. But we're beginning to think of the organism as more of a polyhedron, a figure with many faces. Such a figure can tip in several ways, but it usually has preferred directions."

At a recent meeting in Berlin on evolutionary theory and developmental biology, this debate over the randomness of "chance" mutations took center stage. Why, for instance, are there no animals with wheels instead of legs? Why does a camel always give birth to something that is basically a camel, and a human to something that is basically human? Why can't an ant's genome give rise to an elephant? Some conference participants said these apparent consistencies of nature exist only because the genes haven't yet had enough time to express themselves in all possible permutations. But the majority view was that the organism imposes its own constraints and participates in its own evolution.

"Organisms are not pieces of putty, infinitely moldable by infinitesimal degrees in any direction," summarized Stephen Jay Gould, a paleontologist at the Harvard Museum of Comparative Zoology. Rather, organisms are "complex and resilient structures endowed with innumerable constraints and opportunities based upon inheritance and architecture." Evolution has proceeded in the way it has, this view holds, not simply because of the guiding force of natural selection, but also because of the tendency within organisms to change in the direction of greater complexity and greater adaptability.

Critics of the modern synthesis are questioning not only the direction of chance mutations, but their size. According to the modern synthesis, the generation-by-generation changes noted through random variation, and the resulting shift in the gene pool, are infinitesimal; it takes many millions of years for a truly different organism to arise. But now, some heretics are returning to the once-discredited notion of the late Richard Goldschmidt, a German biologist who postulated the existence of "hopeful monsters" that arise from time to time and alter the evolutionary landscape.

"The decisive step in evolution, the first step toward macroevolution, the step from one species to another, requires another evolution-ary method than that of sheer accumulation of micromutations," Goldschmidt wrote in 1940. He theorized, wrongly, that a kind of systemic macromutation must be at work, an enormous error that could either kill its bearer or render him far superior to his brothers.

Today, however, some scientists are looking anew at Goldschmidt's theory, and they postulate a different process. A chromosomal mutation that might seem tiny if it occurred in an adult, they say, could make a profound difference in an organism if it occurred in an embryo—especially if the mutation affects the embryo's developmental clock. If the timing of cell differentiation and growth is thrown out of whack, anything can happen. For example, some scientists calculate that a minor change in the developmental clock of the chimpanzee, one that enables brain growth to continue in the fetus for longer than it should, could give rise to a chimp with a brain the size of a human's.

Natural Selection: How Big a Role?

This brings us to the next point in the evolutionary heterodoxy: the role of natural selection. The doubters are satisfied that natural selection does indeed occur, and that it accounts for a great many changes observed among and within species. Their concern mounts, though, when the modern synthesis applies natural selection to evolution writ large—the "macroevolution" that has taken us from one-celled organisms to complex human beings.

Even the critics admit that, on an individual level, natural selection undoubtedly occurs. They all agree, for instance, that natural selection was at work during the Industrial Revolution when the moths of England changed their coloration from white to black—just as the surrounding tree bark, with which the moths needed to blend as protection from predators, turned black with soot. In addition, the process has been demonstrated empirically: It has been mimicked in the lab and has been the basis for the "artificial selection" that enables domestic animal breeders to create a miniature poodle from the same species gene pool that gives rise to a Great Dane.

But the modern synthesis tends to rely on natural selection to explain a great deal more than individual differences, and as such it has created its own Achilles' heel. Many critics object that the theory of natural selection does not meet some traditional criteria of true science. For one thing, it is essentially a tautology. Those organisms that survive through natural selection are, by definition, the fittest; the fittest are, by definition, those organisms that survive through natural selection.

The theory also fails a standard litmus test of science: the test of falsifiability. It is impossible to conceive of an experiment that could prove the theory false, since natural selection is infinitely malleable and can easily be expanded to embrace even the most contradictory evidence. Since it cannot be proved false, critics say, it cannot in any meaningful sense be true.

"Like the Freudians, the Darwinians get a lot of mileage out of sex," notes Michael Ruse, a philosopher at the University of Guelph in Ontario. Any organic characteristic that seems nonadaptive—a peacock's plumage, for example, or a flamingo's startling color—can be explained as being an aphrodisiac, since what really counts is less the struggle for survival than the struggle for reproductive success. This inclusion of the great unknown of passion at the core of the modern synthesis is a source of some dismay for many scientists.

"Must all the features of organisms be viewed as adaptations?" asks Gould in his book, "The Panda's Thumb." He says Darwin himself would have answered no, but some of Darwin's contemporaries, whom Gould calls the "strict selectionists," would have said yes. Gould blames the strict selectionists, led by Darwin's friend Alfred Russel Wallace, for the modern synthesis theory's deification of natural selection. "In a curious sense," he writes, "they almost reintroduced the creationist notion of natural harmony by substituting an omnipotent force of natural selection for a benevolent deity."

The problem with this view, says Gould, is that natural selection is *not* an omnipotent force. When applied to minute variations, it is excellent at explaining why some features of an organism persist and others die out. But it tends to obscure the forest by its investigation of the trees. As Gould has put it, natural selection can explain webbed feet—but it cannot explain why ducks exist.

Why, then, *do* ducks exist? This grappling for a theory of macroevolution has been at the heart of the rumblings within the evolutionary camp. The existence of complex creatures, each species differing from any other species, makes a belief in the randomness of variation and the rigidity of natural selection hard for some to accept. "The underlying presupposition of most evolutionary theorists is that the origin of life and the increase in its complexity was an enormously improbable event," says Kauffman. "This general belief that the origin of life is profoundly

improbable and mysterious is the essence of how natural selection has been viewed."

But according to Kauffman, life—including complex life—is not so improbable after all. Based on his work with computer modeling, Kauffman concludes that "the crystallization of life and the properties we see in life were initially the inevitable consequences of certain kinds of complex chemicals."

Tempo: Gradual vs. Jerky

The third aspect of the modern synthesis that has been called into question recently is the notion that evolution occurred gradually, continually, as an accumulation over aeons of minute changes from one generation to the next. Central to the modern synthesis is the idea of slow, constant change.

But in 1972, two maverick paleontologists—Gould and Niles Eldredge of the New York Museum of Natural History—proposed a counter-theory that emphasized not change, but stasis. They said the fossils they studied revealed an evolutionary pattern of "punctuated equilibrium"—long periods of relative stability (running into the millions of years) broken by bursts of several thousand years in which entire species suddenly appeared or disappeared. Change, they noted, is the exception, not the rule, in the evolutionary picture. This new version presented a problem for strict Darwinians: The theory of natural selection working on random mutations was excellent at explaining change, but it could not account for nonchange. Kauffman added, "In the paradigm of the modern synthesis, nonchange would be explained by the supposition that the organism was maximally adapted to a stable environment; but neither 'maximally adapted' nor 'stable environment' can in practice be measured."

Before 1972, evolutionists generally ignored stasis. In the modern synthesis, nonchange was nondata. Paleontologists sifting the geological sands for a research topic would discard dozens of static species until they found one that had clearly evolved according to Darwinian theory. "I knew a graduate student at Yale in the 1960s who did something we were all doing at the time," Eldredge recalls. "He looked through 40 million years' worth of rocks in Nova Scotia, and he found fossils for 70 different species. Only three of those species were classic examples of evolution. When he wrote his dissertation, he wrote it about those three species and simply ignored the others."

The debate over the tempo of evolution—was it gradual or jerky?—flares more wildly than any other criticism raised by doubters of the modern synthesis. One reason for the rancor is that the sides cannot agree, even among themselves, just what Darwin and his disciples had to say on the matter. Some say Darwin acknowledged that species might arise rapidly and that evolutionary change might be the exception and not the rule. Others say the fault lies not in the theory itself, but in the fossils examined in attempts to prove or refute the theory—especially since speciating animals are small in number, dispersed from their parent stock and likely to have left behind few signs of their history.

Alan Templeton and Val Giddings are members of the first group. "Darwin clearly pointed out that natural selection more often than not is a force preventing evolution and that only under relatively rare circumstances would it lead to episodes of adaptive change," wrote the Washington University biologists in a letter to *Science.* "He certainly did *not* embrace the view that adaptive changes are continuous over long periods of geological time."

In the second group are Ledyard Stebbins and Francisco Ayala, two geneticists at the University of California at Davis who are considered leading defenders of the modern synthesis. They refer to a classic example of "rapid" evolutionary change, the more than 60 percent increase in brain size observed in humans during the 425,000-year span separating *Homo erectus* from Neanderthal man. Based on the rate at which body size changes have been induced in experimental fruit flies, Stebbins and Ayala compute that a human's modern cranial capacity could have developed in just 540 generations, or about 13,500 years. "Thirteen thousand years are, of course, a geological instant," they write.

One's perspective in the debate over evolutionary tempo depends in large part on one's definition of species. Some biologists believe that a species—a reproductive community that establishes a pattern of interbreeding and genetic consistency—is in fact an artifact of classification, an arbitrary cutoff that separates one creature from the next on a continuum of gradual, adaptive change. This is the belief that has led to the modern synthesis.

But others are now suggesting that species are more like individuals, each with its own beginning, history and ending. This view presents a problem for the orthodox theory of macroevolution. As Niles Eldredge puts it, "Reproductive isolation [the hallmark of a new, distinct species] may be the accidental by-product of adaptive divergence of two isolated populations of a species—but the origin of reproductive isolation is itself not being 'selected for.' Natural selection simply does not, and cannot, create new species," since reproductive isolation cannot be construed to be an adaptive device. In this view, the modern synthesis must be wrong to have ascribed to natural selection a central role in both micro- and macroevolution, since it fails to explain speciation, which is considered the major event of macroevolution.

Some Counter-Theories

Then what *can* explain macroevolution? Eldredge, for one, suggests it might be more appropriate to talk about the natural selection not of individuals but of entire species. This notion of "species selection," further developed by Steven Stanley of Johns Hopkins University, traces the evolutionary history of a particular animal—for instance, the modern horse—as a series of speciating events. Some species lines proceed in directions that are more "advanced" and hence more successful; others proceed in less propitious ways. Over time, those with the more primitive features are gradually driven out through environmental influences, and natural selection, working on entire species lines, moves the animal in the direction of more advanced morphology.

Elizabeth Vrba of the Transvaal Museum in Pretoria offers a different theory: the so-called effect hypothesis. Plotting the history of the modern antelope, Vrba found that the difference between two groups—the Aepycerotini, which includes the impala, and the Alcelaphini, which includes hartebeests and wildebeests—is that the former are generalists and the latter specialists. Impalas, she says, can survive in a wide range of environments and can eat a wide range of food. They are, therefore, relatively immune from extinction, which might result from minor environmental shifts, but they cannot readily tolerate the presence in their range of closely related species. Wildebeests and hartebeests, on the other hand, have a narrow range of food and environmental preferences; as a result, they are quite vulnerable to extinction, but can easily live side by side with other similarly limited species. Based on these differences, Vrba's effect hypothesis states that in a group of related species, such as the antelopes, the environmental history skews heavily toward the specialists—who speciate and become extinct rapidly—rather than the generalists.

A third counter-theory returns to the chromosomal level to seek the mutation that gives rise to genetic variability, but it diverges from the modern synthesis in its interpretation of how such mutations can spread through a population. If any one male in a particular group is equally likely to mate with any one female, a chance mutation, even an adaptive one, would indeed need many, many generations to take hold. But in a social arrangement common to many mammals, including rodents, horses and primates, this process can be accelerated. That arrangement is the harem.

A mutant male in a harem quickly passes his damaged chromosomes to a large number of offspring. If mating then occurs between two offspring who have inherited the mutation—which they carry along with the normal gene in a hybrid genotype—they pass it on to three-fourths of their own offspring, one-half receiving the hybrid form and one-fourth the pure mutant form. Thus, within two generations, the mutation has taken hold in a sizable minority of the mutant's descendants. If it is an adaptive mutation, it can be expected to thrive.

The current scientific maelstrom is, on the whole, an encouraging one to those taking part in it; this is the stuff that scientific revolutions are made of. And the differences among biologists should not obscure their essential agreement. As Niles Eldredge puts it, "It is not so much that the synthesis is wrong as it is in-complete. Indeed, the desire to reduce all evolutionary patterns to the mechanics of generation-by-generation change strikes some of us as the only major flaw in the synthesis."

Creationists have tried, often quite successfully, to magnify the current uproar in evolutionary biology. They point to the scientists' reassessment of neo-Darwinian theory as "proof" that the theory is wrong. But for all the brouhaha, even the most strident critics of the modern synthesis would gladly give Darwin himself the last word: "There is grandeur in this view of life, with its several powers, having been originally breathed into a few forms or into one," he wrote in closing "The Origin of Species," "and that, whilst this planet has gone cycling on according to the fixed law of gravity, from so simple a beginning endless forms most beautiful and most wonderful have been, and are being evolved."

Suggested Readings

(1) Eldredge, Niles, "The Myths of Human Evolution," Columbia University Press, NY, 1981.
(2) Gould, Stephen Jay, "Ever Since Darwin: Reflections in Natural History," W. W. Norton & Company, NY, 1977.
(3) Lewin, Roger, "Evolutionary Theory under Fire," *Science*, **210**, 883–87 (Nov. 21, 1980).
(4) Stebbins, G. Ledyard, and Ayala, Francisco J., "Is a New Evolutionary Synthesis Necessary?" *Science*, **213**, 967–71 (Aug. 28, 1981).

Influence of Darwin's Ideas on the Study of Evolution

William B. Provine

Provine is with the Department of History and the Division of Biological Sciences, Cornell University, Ithaca, NY 14583.

Darwin's *On the Origin of Species* convinced the western world of the fact of evolution. His theory of natural selection, however, was received more hesitantly. Ignorance of the mechanism of heredity, difficulties with understanding the nature of variation in natural populations, and lack of evidence contributed to formation of the opposition. Arguments concerning the continuous or, alternatively, the discontinuous nature of evolution have permeated the history of controversy in evolutionary theory. (*Accepted for publication 5 March 1982*)

CHANCE AND NECESSITY

Charles Darwin was not the first to invent the idea of evolution in nature. Buffon (1707–1788), Lamarck (1744–1829), and Darwin's grandfather Erasmus Darwin (1731–1802) were among the prominent adherents of evolution two or more human generations before the appearance of *On the Origin of Species* in 1859. Darwin's contemporary Robert Chambers (1802–1871) published *The Vestiges of the Natural History of Creation* in 1844, an argument for evolution, and it went through 10 editions and sold at least 15,000 copies in the decade after its appearance. By contrast, John Murray, Darwin's London publisher, sold 9250 copies of the *Origin* in five editions in its first decade.

The evolutionists prior to Darwin shared two fundamental characteristics. Not one of them was able to convince experts in biology or scientists in general that animals and plants had evolved their exquisite adaptations from other different forms of life. The second, crucial to understanding Darwin's impact, was their belief that a guiding force of some kind directed the evolutionary process or set the stage for it. This force might be the Christian God or it might be a wholly unconscious biological urge towards greater complexity. By advocating a directive force in evolution, these thinkers were following one of the deepest traditions in Western civilization.

The Greek and Judeo-Christian heritages share a fundamental assumption: that the physical world and human ethics could not possibly have arisen through the sheer mechanical workings of things. Plato and Aristotle reflected Greek culture when they agreed that nature exhibits reason, purpose, and design inseparable from the foundations of human ethics. In the Judeo-Christian tradition, God created and designed the heavens and earth, animals and plants, and set moral laws for humans. That medieval philosophers could synthesize the Greek and Judeo-Christian heritages into the powerful conceptions that have so deeply affected Western civilization is not surprising.

The deep belief that chance and necessity were insufficient explanations of visible order permeated all aspects of Western culture. Scientists, theologians, social thinkers, and philosophers could all agree. Isaac Newton and William Harvey held views on this issue very similar to those of Martin Luther and Thomas Hobbes. Even Montesquieu, who argued that soil, climate, and social conditions partly determined the ethical rules of a society, declared on the first page of his *Spirit of the Laws:* "They who assert that a blind fatality produced the various effects we behold in this world talk very absurdly; for can any thing be more unreasonable than to pretend that a blind fatality could be productive of intelligent beings?"

Natural theology as it existed in Britain during Darwin's youth was perhaps the most simplistic version of the whole tradition. Bishop Paley argued that the mere existence of adaptations in living organisms proved with certainty God's existence and goodness. As a student at Cambridge University, Darwin found Paley's *Natural Theology* the most rewarding and persuasive book he read during his university education.

Darwin's *Origin of Species* differed fundamentally from the writings of earlier evolutionists. The *Origin* persuaded most scientists that organic evolution had occurred. Using abundant evidence about hybridism and degrees of sterility, geological succession combined with plausible arguments for the imperfection of the geological record, geographical distribution, and embryology, Darwin built a convincing argument for evolution unmatched by any of his predecessors. A quick historical glance at the status of scientific antievolutionists 10 years after the first edition of the *Origin* shows clearly how isolated they were from the evolutionist mainstream. The sad state of Louis Agassiz's prestige at the end of his distinguished career is a typical example.

EVOLUTION, YES BUT NATURAL SELECTION?

Although Darwin successfully made evolutionists of his readers, he was far less successful in convincing them that evolution occurred through his proposed mechanism, natural selection. As Darwin clearly recognized less than a year after he invented the theory in 1837, natural selection was precisely that mechanism, operating by chance and necessity alone, whose existence Western thought denied. Darwin thought natural selection was the primary mechanism leading to apparently intelligent design in organisms. This godless, nondirected, and opportunistic mechanism fashioned the adaptations in which natural theologians discerned the handiwork of God, and in which others found at least some kind of purpose or design. The negative reaction to natural selection was swift and strong. Darwin was able to convince only a few scientists, but those influential ones, that natural selection was the primary mechanism of evolution.

Direct and Indirect Evidence

Those who objected to natural selection had excellent grounds for criticism because Darwin could not produce the direct evidence for natural selection that he had for evolutionary change. He did have indirect evidence and made good use of it. The *Origin* began with a long chapter on the efficacy of artificial selection, presenting ample evidence that the numerous physically diverse breeds of pigeons were created by man's selection and were all descended from the wild rock pigeon. Heritable variation (although less of it) also existed in natural populations. All organisms had geometric powers of increase—the Malthusian parameter—leading to competition for food supplies or the struggle for existence. In turn, the struggle for existence over time became the process of natural selection. The logic was compelling. Darwin gave two examples of natural selection in action—one showing natural selection making wolves swifter and slimmer, and the other a very complex case of natural selection causing the evolution in plants of the ability to produce sweet nectar and sexual differentiation of the flowers, and the physical equipment for bees to suck the nectar. But these examples of natural selection in action were the only ones in the book and were, as Darwin clearly states, "imaginary illustrations."

Adam Sedgwick, Darwin's former professor of geology, expressed the sentiments of many critics when he said that, so far as evidence was concerned, the theory of natural selection was "a vast pyramid resting on its apex, and that apex a mathematical point" (Sedgwick 1860, p. 285) Geneticist E. B. Ford once told me (as he had heard it from E. Ray Lankester, who had heard it from Darwin) that Darwin thought 50 years was the minimum time in which an observer might detect the action of natural selection. So of course Darwin had no examples of natural selection in action, nor

did he have hopes of observing any such examples in his lifetime.

Heredity

Heredity was another problem for Darwin's theory of natural selection. The theory rested upon the assumption that an ample supply of random heritable variation (random, that is, with respect to the direction of selection) existed in natural populations. Natural selection acted upon this variability and was the primary determinant of evolutionary change. But Darwin thought some of the heritable variability was nonrandom, resulting from the inheritance of acquired physical characters or long-continued habit. Inheritance of acquired characters was far less important to Darwin than natural selection as a directive agent in evolution in nature. Darwin's critics responded that, whether by inheritance of acquired characters or purposive forces, most heritable variation was in the direction of adaptational change. Any theory of heredity with directed variation thus diminished the importance of natural selection as a determinant of evolutionary change. Such theories of heredity abounded in the period from 1860 to 1900.

Darwin's "provisional hypothesis of pangenesis" appeared in 1868 in his work *The Variations of Animals and Plants under Domestication*. This theory of heredity was highly influential but unconvincing. It stated that each part of an organism threw off free and minute gemmules that multiplied and aggregated in the reproductive organs, and from there passed on to succeeding generations. Nearly every major theory of heredity proposed in the later nineteenth century began with a discussion and rejection of Darwin's theory. To all it was obvious that the prevailing mechanism of evolution in nature depended upon the mechanism of heredity. Although he defended it vigorously, Darwin's mechanism of heredity was weak and an easy target for opponents of natural selection.

Continuous vs. Discontinuous

Another controversial subject was Darwin's view of evolution in nature. "I fully admit," Darwin said, "that natural selection will always act with extreme slowness . . . often only at long intervals of time, and generally on only a very few of the inhabitants of the same region at

the same time." He added his controversial belief "that this very slow, intermittant action of natural selection accords perfectly well with what geology tells us of the rate and manner at which the inhabitants of this world have changed" (Darwin [1859] 1964, pp. 108–109). Over and over again, Darwin argued that evolution was slow and gradual, although intermittent, just as major geological changes occurred according to the uniformitarian theory of Charles Lyell:

> Natural selection can act only by the preservation and accumulation of infinitesimally small inherited modifications, each profitable to the preserved being; and as modern geology has almost banished such views as the excavation of the great valley by a single diluvial wave, so will natural selection, if it be a true principle, banish the belief of the continued creation of new organic beings, or of any great and sudden modification in their structure. (Darwin [1859] 1964, pp. 95–96)

In Darwin's gradualist view, varieties were potentially incipient species. Natural selection drove organisms into ecological niches like "little wedges," creating new varieties. In time, some, but not all of these varieties gradually diverged in morphology and became more sterile in crossing so that taxonomists would then rank them as species. The heated reaction of antievolutionists and opponents of natural selection was the well-taken point that Darwin, despite his detailed analogy of natural selection with artificial selection, had not one example of the new creation of a species by artificial selection through the whole history of mankind—and thus how could he possibly argue that the more gradual natural selection made species out of varieties?

Even some of the most vociferous supporters of the idea of natural selection disagreed with Darwin about continuity in variation and evolution. Darwin was well aware that large discontinuous variations or sports occurred rarely in natural populations, but he considered them to be too rare to be the primary source of variation; he also thought they were frequently infertile and easily swamped by blending inheritance. Thomas Henry Huxley ("Darwin's Bulldog") and Francis Galton (Darwin's cousin) were both staunch defenders of the idea of natural selection, but both argued that natural selection, to be effective, had to operate upon the occasional large discontinuous variations, leading to discontinuous jumps in the evolutionary process. Huxley and Galton believed their view harmonized far better with the undeniably discontinuous geological record than did Darwin's own view of gradual evolution. The contrast between the continuous versus discontinuous views of the evolutionary process is one of the most pervasive and influential themes in the history of evolutionary biology and remains a very controversial subject.

Other Difficulties

One other problem with natural selection was its difficulty in accounting for the evolution of apparently altruistic social behavior. Darwin acknowledged this difficulty in the *Origin* and proposed that in the evolution of such social behavior natural selection operated upon whole families rather than upon individuals. Critics of natural selection frequently raised cooperative or altruistic social behavior as the ultimate antithesis of individual natural selection and thus a challenge to the whole theory. Not until W. D. Hamilton's theoretical work on the calculus of kin selection in the 1960s (Hamilton 1964) did this problem begin to be solved within the framework of individual selection.

In addition to natural selection and the inheritance of acquired characters, Darwin proposed a third mechanism of evolution in nature. He encountered little difficulty in explaining the origin or maintenance of sexual reproduction (it led to greater physiological vigor, a character of great selective value), but he could not explain gaudy sexual dimorphism or even all the observed differences between human races by natural selection. To explain these, Darwin developed his theory of sexual selection. Female choice, not natural selection, led to brightly ornamented male birds. Indeed, natural selection acted to curb sexual selection in cases where males selected for ornamentation were easily spotted by predators. Alfred Russel Wallace strongly rejected Darwin's theory of sexual selection in the late nineteenth century, as did many other critics.

Nineteenth Century Critics

With his works on evolution between 1859 and 1872 (*On the Origin of Species, Fertilization of Orchids, Variations of Animals and Plants under Domestication, Descent of Man, Expression of the Emotions in Man and Animals*) Darwin stimulated an outpouring of literature on evolution. In 1872 J. W. Spengel published the second edition of his bibliography of literature about Darwinism—36 pages long with 317 different authors, many of whom had several entries (Spengel 1872). Moreover, it was in German and naturally emphasized the literature written in German—no Russian literature was included. Three years later G. Seidlitz published a similar bibliography with more than double the number of entries recorded by Spengel (Seidlitz 1875). By 1900 no one could even hope to present a critical bibliography of all writings related to Darwin. This literature documents Darwin's great achievement in convincing scientists, philosophers, and the theologians that evolution had occurred, but also reveals that every aspect of Darwin's theories of natural and sexual selection had undergone severe criticism.

My own rough list of prominent evolutionists from all countries indicates that at the time of Darwin's death in 1882 those who advocated natural selection as the primary mechanism of evolution were outnumbered by those who believed in alternative mechanisms of evolution by more than two to one. Furthermore, included in my list of natural selectionists are all those who advocated discontinuous evolution in opposition to Darwin's gradualist views. By the turn of the century, Darwin's theory of evolution by gradual natural selection had lost support, and the opposition included the great majority of internationally renowned biologists.

Analysis of the opposition to the idea of natural selection is very complex. I classify opponents of Darwin's views into two categories—those who proposed alternative mechanistic theories (such as Moritz Wagner's (1872) migration theory or Hugo de Vries' (1901–1903) mutation theory), by far the smallest group, and those who attacked Darwin and proposed alternative theories in the hope of retaining a sense of purpose and directedness in evolution. The latter group was attempting to save in some form the deep cultural heritage explained above. Thus Darwin's most active American supporter, Asa Gray, published many attempts to reconcile evolution by natural selection with natural theology. Gray thought that to lose the idea of design in nature was to lose all possibility of meaning in human life. The literature reconciling evolution (generally not by natural selection) with Christianity and religion in general is voluminous in the late nineteenth century.

Darwinism of the 1940s and 1950s resembles Darwin's own views of evolution by gradual natural selection so closely that the depths to which Darwin's views sank at the turn of the century were understandable to many modern evolutionists only by assuming that Darwin's critics simply had not understood what he was saying. Recent historical research proves this surmise wrong. Many of Darwin's nineteenth century critics read him carefully and understood him well, but were unconvinced. In 1900 there was still no example bearing scrutiny of natural selection in action in nature, nor was it intuitively obvious that gradual natural selection had the power to transmute species. Much of the criticism of Darwin was well justified given available evidence. Darwin, despite all his influence, was unable in his century to produce a consensus concerning the mechanisms of evolution in nature.

RISE OF MENDELIAN GENETICS

Ever since the evolutionary synthesis of the 1930s and 1940s biologists have tended to believe that all Darwinism lacked at the turn of the century was an adequate theory of heredity. When this lack was remedied by the newly rediscovered Mendelian theory, the synthesis of Mendelism and Darwinism promptly ensued. Certainly there is some truth to the view that the synthesis of Mendelism and Darwinism created the possibility of a powerful neo-Darwinian view, but the synthesis occurred neither immediately upon the rediscovery of Mendelism nor easily.

By 1900 the argument between the advocates of gradual Darwinian evolution and the advocates of evolution by discontinous leaps had already begun to intensify. In England, William Bateson was vociferously defending the discontinuous view with the support of Francis Galton and Hugo de Vries against the arguments of the othodox Darwinians E. B. Poulton, E. Ray Lankester, and the biometricians Karl Pearson and W. F. R. Weldon. When Mendelian heredity was injected into the debate, Bateson and de Vries took it for their discontinuous view of evolution. Mendel had utilized sharply discontinuous characters for his experiments with peas, and the theory did appear at first to fit well with discontinuous evolution. The orthodox Darwinians, particularly the biometricians, thus proceeded to attack Mendelism. The battle between the Mendelians and biometricians during the years 1900–1906 was heated and highly publicized. The net effect was to exacerbate the argument over continuity in evolution and initially, as Mendelism gained, Darwinism lost (Provine 1971).

The most influential proponent of discontinuous evolution was Hugo de Vries. By 1899 de Vries had observed many examples of discontinuous "mutations" in his stocks of the evening primrose, *Oenothera*. Some of these bred true and were sterile with parental stocks, so de Vries assumed that he had observed the creation of new species. In 1901–1903 de Vries published throughout his two-volume *Mutations theorie*, an argument for discontinuous evolution. Part of de Vries' popularity resulted from the rise of experimental biology in the late nineteenth century; unlike Darwin, de Vries at least had some direct experimental evidence favoring his theory.

In 1906 Mendelism and the mutation theory appeared closely allied and on their way to victory over gradual Darwinian evolution. But the great problem with all the non-Darwinian theories was the same as for natural selection; positive evidence was meager. When the positive evidence for the mutation theory weakened, as did the belief that gradual selection was ineffective in changing a population, Darwin's views again began to appeal to biologists.

SYNTHESIS OF MENDELISM AND DARWINISM

The dominant view in the first decade of the century was that selection of small differences could only cause a population to exhibit characters already present in the original range of variability. In other words, selection could not progress beyond the range of phenotypes already present in the population. Francis Galton had popularized the view that regression would soon balance selection unless a new mutation appeared. Wilhelm Johannsen's "pure line theory" of 1903 supported Galton's argument. In a genetically pure line, selection must be ineffective and regression complete.

In a curious twist often exemplified in science, geneticists motivated by the mutation theory to conduct experiments soon produced evidence to show that, far from being contradictory, Mendelism and Darwinism fit together nicely. The American geneticist William Castle demonstrated (to his own surprise) that artifi-

cial selection for increased and diminished size of pigmented stripe on the backs of "hooded" rats soon produced strains with average pigmentation well beyond the original range of variability. The selected strains bred true; selection of small differences was effective.

Castle's colleague at the Bussey Institution of Harvard University, Edward M. East, proposed, along with the Swedish botanist H. Nilsson-Ehle, the multiple-factor theory to account for apparently continuous variability in terms of Mendelian heredity (Nilsson-Ehle 1909, East 1910). By 1918 A. H. Sturtevant had not only shown that similar selection was effective in *Drosophila melanogaster*, but that the modifiers were inherited according to Mendelian inheritance. He had even associated some of the modifiers with particular chromosomes (Sturtevant 1918). Mendelian inheritance was then seen to be a mechanism for recombining chromosomes and, through crossing-over, individual genes. With only 10 independent loci, each with two alleles, 59,049 different genotypes were possible. Thus any population with a modicum of genetic variability could exhibit only a tiny fraction of the possible genotypes at any one time. Selection could make less likely genotypes become more likely. Thus Mendelian heredity and Darwinian selection were complementary rather than contradictory. By 1918, when de Vries' "mutations" in *Oenothera* were known to be balanced lethal hybrids, most experimental geneticists believed in a general way that Mendelism and Darwinism could be synthesized.

Because Mendelian heredity is a regular system of inheritance, under a given set of assumptions the quantitative consequences of Mendelism can be computed. A simple consequence of Mendelian heredity, for example, is that for a single locus with two alleles, a stable equilibrium of allelic frequency (in the absence of selection) results after one generation of random breeding. This is the famous Hardy-Weinberg equilibrium principle taught to all beginning students in genetics. Beginning in the late 1910s, R. A. Fisher, Sewall Wright, and J. B. S. Haldane all developed quantitative models for the distributions of gene frequencies in Mendelian populations. All three wished to show that Mendelism complemented the theory of evolution by natural selection.

Fisher, Haldane, and Wright each began by assuming a generalized Hardy-

Weinberg equilibrium in a population; they then examined the effects upon gene frequencies of factors including selection, dominance, mutation, epistasis, population structure, breeding structure, linkage, balanced polymorphisms, random processes, and group selection. Although they used different quantitative methods, all three came to agree upon almost all purely quantitative questions. Thus Fisher and Wright thought in 1931 that all their quantitative results were in complete agreement.

Agreement on the numerical consequences for a given set of assumptions by no means guaranteed agreement upon the mechanism of evolution in nature. Although they corresponded cordially and substantively in the period 1928–1931, Fisher and Wright broke communications when the deep extent of their disagreement about evolution in nature became evident. Fisher believed that evolution in nature was dominated by very small selection rates acting upon equally small modifiers in very large populations. This was twentieth century Darwinism. Wright, on the other hand, believed that natural selection acted primarily, not upon single genes, but upon gene interaction systems created in part by random genetic drift in relatively small, partially isolated populations. These different conceptions of the evolutionary process led to a series of specific disagreements, perhaps the most notable concerning the evolution of dominance.

The quantitative models of Fisher, Haldane, and Wright strongly influenced evolutionary thinking in at least four ways (Provine 1978). First, the models demonstrated that Mendelism and natural selection, in combination with known processes in natural populations, were sufficient to account for observed evolution in nature. Darwin's belief that a very small selection rate could alter the hereditary constitution of a population, a far from intuitively obvious proposition, was verified by the models. Second, the models indicated that some earlier views were untenable. One popular conception of laboratory geneticists had been that mutation pressure was the dominant factor in evolutionary change. By elucidating the relationships between mutation rates, selection pressures, and changes of gene frequencies, the quantitative models showed unmistakably that selection was vastly more effective than mutation as an agent of evolutionary change. Third, the models clarified and complemented field researches already completed or in progress, thus giving the field research greater significance. One prominent example was Haldane's use in 1924 of available data on the frequency of melanic and nonmelanic forms of the moth *Biston betularia* in the area of Manchester, England. Haldane calculated that the melanic form was twice as likely to survive as the previously prevalent nonmelanic form. This is perhaps the first example of natural selection in action in nature documented beyond reasonable question (65 years after Darwin published the idea!). Fourth, the models stimulated and provided the intellectual framework for later field research. The most impressive example of this is the great influence of Sewall Wright's models in Theodosius Dobzhansky's monumental series of 43 papers on field researches with *Drosophila pseudoobscura* and relatives under the title, *The Genetics of Natural Populations* (Lewontin et al. 1981).

THE EVOLUTIONARY SYNTHESIS

The work of the mathematical population geneticists did not by itself create the synthetic view of evolution in nature that emerged in the 1930s and 1940s. Fisher, Haldane, and Wright had very little direct knowledge of genetics and evolution in natural populations. Wright's theory of evolution in nature was primarily an extension of his theory of evolution in shorthorn cattle. The fields of biological research that had become specialized and cut off from one another in the late nineteenth and early twentieth centuries came together in a synthetic view of evolution. Biologists could then see the fields of genetics, systematics, cytology, paleontology, and embryology in both botany and zoology as complementary parts of one large view of the evolutionary process. The synthetic view was exemplified by such works as Dobzhansky's *Genetics and the Origin of Species* (1937, 1941, 1951), Julian Huxley's edited *The New Systematics* (1940) and his *Evolution: The Modern Synthesis* (1942), Ernst Mayr's *Systematics and the Origin of Species* (1942), George G. Simpson's *Tempo and Mode in Evolution* (1944), and Ledyard Stebbins' *Variation and Evolution in Plants* (1950) (see Mayr and Provine 1980 for a comprehensive view of the evolutionary synthesis).

Relatively few prominent biologists offered fundamental challenges to the syn-

thetic theory after 1940. Richard Goldschmidt argued in his 1940 book *The Material Basis of Evolution* that speciation was abrupt and required macromutations in addition to the usual more gradual heritable variation, but the reaction of other biologists to this discontinuous theory was almost uniformly negative. By the early 1950s the new synthetic view of evolution had become well delineated and commanded greater respect and agreement than any previous theory of evolution. Evolution in nature was dominated by relatively small selection pressures upon natural populations with considerable genetic variability. The process was gradual and adaptive. These were, of course, characteristics of Darwin's own view of evolution in nature, and at the time of the centennial of the *Origin* in 1959, modern evolutionary theory was most frequently and appropriately called "twentieth century Darwinism" or "neo-Darwinism." To the young student of that time, the new Darwinism appeared to differ from the original chiefly by the addition of Mendelian heredity, field research, and an overwhelming vote of confidence from biologists.

DARWINISM BETWEEN THE TWO CENTENNIALS 1959–1982

No sooner had the neo-Darwinian synthesis become the new orthodoxy among evolutionists and in textbooks than challenges began to appear. Pioneered by the work of R. C. Lewontin and J. L. Hubby on allozyme variability in *Drosophila,* biologists discovered immense amounts of unsuspected genetic variability in natural populations. Were the genetic variations adaptive? Motoo Kimura and others proposed that much of the variability was selectively neutral and the particular constellation to be found in a population might be due more to chance than to adaptive selection (see Kimura and Ohta 1971). Stephen Jay Gould's and Niles Eldredge's proposal of their theory of punctuated equilibria in 1972 challenged the neo-Darwinian view of macroevolution as the natural extension of gradual microevolution (Eldredge and Gould 1971). The geological record with its sharp discontinuities between species found in different layers was not, according to Gould and Eldredge, the imperfect record Darwin imagined it to be. Instead, new species really did appear rather suddenly after long periods of stasis, and such changes were reflected in the geo-

logical record. Adherents of the neo-Darwinian view have countered that gradual microevolution is perfectly consistent with the known geological record (Stebbins and Ayla 1981). The intensity of the disagreement was evident at the Macroevolution Conference at Chicago's Field Museum of Natural History late in 1980.

This new controversy on the old theme of continuity or discontinuity of the evolutionary process stems in part from a desire shared by both camps (and scientists in general) to devise the simplest and most economical explanations for observed phenomena. Gould and his supporters focus upon the geological record. It is undeniably discontinuous in general and exhibits punctuated equilibria. The simplest explanation for this discontinuous record is a theory of discontinuous speciation. The neo-Darwinians, on the other hand, focus upon evolution in natural populations, and they wish to have the simplest possible explanation for observed phenomena of change in natural populations. To them, the neo-Darwinian synthesis is the compelling explanation.

The problem with Gould's view is that the hypotheses about mechanisms of evolution in natural populations must be multiplied (he now advocates a "species" selection)—hypotheses that appear superfluous to most neo-Darwinians. And for their view, neo-Darwinians must, as Darwin did, multiply hypotheses to explain the appearance of the geological record, hypotheses that Gould considers superfluous. Evidence is strong that rates of geological deposition far exceed (by a factor of 10) the depths of observed geological layers, indicating most of the record is lost. But this loss does not warrant the conclusion of evolutionary continuity or absence of true punctuated equilibria.

This particular incarnation of the controversy may soon subside as others have, but is unlikely to be settled by incontrovertible evidence. Selection pressures far too small to be accurately measured by field workers can greatly change the genetic constitution of a population in geological instants. Little is known about the variables governing evolutionary change in natural populations. All quantitative models of stochastic distributions of gene frequencies prominently feature the quantity Nm, effective population size times the migration rate. The whole conception of the process of microevolution rests upon assumptions about effective population size and migration rates. At the present, biologists know little about effective population sizes in natural populations and even less about exact migration rates. Thus both Gould's view of discontinuous species change and the neo-Darwinian view of gradual species change are consistent with the geological evidence, and nothing less than a massive and precise understanding of evolution in nature will settle the issue. The contrast between the continuous and discontinuous views of evolution, so prominent in Darwin's own time and ours, may still be controversial at the second centenary of Darwin's death in 2082.

IMPLICATIONS OF MODERN EVOLUTIONARY BIOLOGY

Evolutionists still disagree about the precise mechanisms of evolution in nature, but they have nevertheless given overwhelming support to Darwin's belief that design in nature results from purely mechanistic causes. As Jacques Monod, E. O. Wilson, and many other biologists have pointed out, modern evolutionary biology has shattered the hope that some kind of designing or purposive force guided human evolution and established the basis of moral rules. Instead, biology leads to a wholly mechanistic view of life, as Darwin suspected.

Here are the implications of this mechanistic view of life as I see them. First, except for purely mechanistic ones, no organizing or purposive principles exist in the world. There are no gods and no designing forces. The frequently made assertion that modern biology and the assumptions of the Judeo-Christian tradition are fully compatible is false. Second, there exist no inherent moral or ethical laws, no absolute guiding principles for human society. Third, humans are marvelously complex machines. The individual human becomes an ethical person by two primary mechanisms: heredity and environmental influence. Some sociobiologists (and Darwin himself) emphasize heredity, but they have remarkably little evidence for their assertions, except for the obvious conclusion that humans have the hereditary capacity to be moral beings. A person is morally what his culture educates him to be. The educational process is complex, but purely mechanical. Moral systems reflect the perceived realities of human social experience, which of course includes cultural traditions. When perceived social realities change, so do moral rules. Recent changes in moral rules are easily cited, with birth control and abortion being prime examples.

Fourth, free will, as usually conceived, does not exist. What choices a person makes can already be predicted with high accuracy by a close friend with no scientific training in analysis of human behavior. Any attempts to find the basis of free will in quantum mechanical indeterminacy (already damped out at the level of DNA molecules) or ambiguities in complex computing systems (rapidly disappearing) are doomed to fail. "Free will" generally means that one is "free" to choose some options over others—and this freedom is precisely what the mechanistic view denies.

Eighty years were required for Darwin's concept of evolution by natural selection to be widely accepted by biologists. Acceptance of the implications of evolutionary biology and rethinking the bases of moral rules will require a much longer time. The totality of Darwin's influence has scarcely begun to be accepted.

REFERENCES CITED

Darwin, C. [1859] 1964. *On the Origin of Species*. John Murray, London. Facsimile edition, Harvard Univ. Press, Cambridge, MA.

DeVries, H. 1901–1903. *Die Mutationstheorie*. 2 vols. Von Veit, Leipzig.

East, E. M. 1910. A Mendelian interpretation of variation that is apparently continuous. *Am. Nat.* 44: 65–82.

Eldredge, N. and S. J. Gould. 1972. Punctuated equilibria: an alternative to phyletic gradualism. Pages 82–115 in T. J. M. Schopf, ed., *Models in Paleobiology*. Freeman, Cooper and Co., San Francisco, CA.

Hamilton, W. D. 1964. The genetical evolution of social behavior. *J. Theoret. Biol.* 7: 1–16.

Kimura, M. and T. Ohta. 1971. *Theoretical Aspects of Population Genetics*. Princeton Univ. Press, Princeton, NJ.

Lewontin, R. C., J. A. Moore, W. B. Provine, and B. Wallace, eds. 1981. *Dobzhansky's Genetics of Natural Populations I-XLIII*. Columbia Univ. Press, New York.

Mayr, E. and W. B. Provine, eds. 1980. *The Evolutionary Synthesis*. Harvard Univ. Press, Cambridge, MA.

Nilsson-Ehle, H. 1909. Kreuzungsuntersuchungen an Hafer und Weizen. *Lunds Universitets Årsskrift*. N. F. Afd. 2. Bd. 5, Nr. 2.

Provine, W. B. 1971. *The Origins of Theoretical Population Genetics*. Univ. of Chicago Press, Chicago, IL.

———. 1978. The role of mathematical population geneticists in the evolutionary synthesis of the 1930's and 1940's. *Studies in the History of Biology* 2: 167–192.

Sedgwick, A. 1860. Objections to Mr. Darwin's theory of the origin of species. *The Spectator* 32: 285–286.

Seidlitz, G. 1875. Literatur zur Descendenztheorie. An appendix to *Die Darwinsche Theorie*. Second edition. Breitkopf and Härtel, Leipzig.

Spengel, J. W. 1872. *Die Darwinsche Theorie*. Verlag von Wiegant und Hempel, Berlin.

Stebbins, G. L. and F. G. Ayala. 1981. Is a new evolutionary synthesis necessary? *Science* 213: 967–971.

Sturtevant, A. H. 1918. *An Analysis of the Effects of Selection*. Carnegie *Inst. Wash. Publ.* 264.

Wagner, M. 1868. *Die Darwinsche Theorie und das Migrationsgesetz der Organismen*. Duncker and Humblot, Leipzig.

EARTH'S LUCKY BREAK

JOHN GRIBBIN

John Gribbin, a Ph.D. and astrophysicist, is the physics consultant to New Scientist. *His most recent book is entitled* Future Weather.

One of the great mysteries of the universe is the existence of life on the planet Earth. Is life here a cosmic fluke that has arisen nowhere else in the universe? Or is life as we know it the inevitable consequence of a happy combination: just the right temperature, just the right amount of radiation, just the right chemical ingredients? A clue to this mystery can be found by looking at the conditions that allowed life to arise here in the first place.

When planets formed out of the dusty material of the nebula that contracted to become the sun and the solar system, they did so in the face of conflicting forces. Gravity tended to pull aggregates of matter together as primordial planets, but the increasing radiation from the young star at the center of the nebula tended to blow these aggregates of matter apart, with the lightest, most volatile material boiling away into space.

The critical variable was the distance of the primordial planet from the sun. Those planets nearest the sun were subject to the greatest heat and consequently lost most of their lighter elements. Those four planets—Mercury, Venus, Earth and Mars—are sometimes called the terrestrial planets because they are all solid, rocky spheres surrounded by thin layers of gas. Farther on, where things were cooler, lightweight elements could condense out in gaseous compounds that could be held together by gravity.

The result was the four giant planets—Jupiter, Saturn, Uranus and Neptune. While the terrestrial planets are solid with a fringe of atmosphere, the giants are made almost entirely of gas—primarily hydrogen, helium, methane and ammonia—with perhaps a small rocky core. (Pluto, the outermost planet, is solid, but it is almost certainly an escaped moon that did not originate as a planet.)

Although hydrogen is the most common element in the universe (as well as in the sun and the pre-solar nebula), almost all of it blew away from the inner solar system, and the terrestrial planets are made up almost entirely of leftovers.

The Earth itself is composed of elements that made up less than one percent of those found in the original nebula, and it has retained only a fraction of its original hydrogen. Most of that hydrogen can be found combined with oxygen to produce the wet seas of our planet.

As well as being wet, the Earth is a small, rocky planet with an oxygen-rich atmosphere. All of these features fit together and relate to our position in the solar system: The general proximity of the Earth to the sun makes the planet rocky; the *exact* distance of the Earth's orbit from the sun has determined the nature of the atmosphere and the oceans that cover the planet's surface.

It is the abundance of liquid water that seems to be the key to the formation of life on Earth. No other planet in the solar system has liquid water. Mercury, the innermost planet, is rather like the core of the Earth stripped of even the elements that might have formed a thick rocky crust. It is a very dense, small planet that is very rich in metals compared with the other terrestrial planets. Mercury is so hot that it has scarcely a trace of atmosphere. It could never have had oceans of running water and can be ruled out as a home for life.

Venus and Mars, however, are more promising. Although they are at different distances from the sun, Venus and Earth are almost the same size and have very similar compositions. Venus has a thick atmosphere, rich in carbon dioxide. This gaseous blanket traps solar heat in the same way that the glass walls of a greenhouse do and raises the temperature at the surface to nearly 932 degrees Fahrenheit—much too hot to allow liquid water.

Mars, farther out from the sun than is the Earth, is a lighter planet (following the general rule that lighter material dispersed outward across the solar system as it formed) with a thin but respectable atmosphere. Mars, however, is too cold for water to exist as a liquid.

So thanks largely to its distance from the sun, the Earth alone has an atmospheric blanket that is just right to keep the surface of the planet hotter than the freezing point of water and cooler than its boiling point. The result is a wet planet, where water continuously evaporates from the oceans and is recycled as rain—conditions that are ideal for life as we know it. How did these perfect conditions come about?

Let's start with the Earth as a rocky ball, stripped of its atmosphere. This seems a reasonable assumption about the conditions on this planet as our story begins. Whatever leftover scraps of light gases were still associated with the terrestrial planets as they formed were probably blown away during the erratic activity of the young sun before it settled down into the steady glow we know today. This occurred about 4.6 billion years ago. The present atmospheres of the inner planets came from gases seeping out from the planets' interiors—"outgassing" from the rocks and volcanic activity, as well as from vaporization produced when large meteorites hit the surface at high speed.

It used to be thought that this primordial atmosphere was rich in gases such as methane and ammonia, similar to the atmospheres of the gaseous giants. This idea was tied to the search for the origin of life. Laboratory experiments had shown that by mixing gases such as methane and ammonia with molecular hydrogen and water vapor in a sealed flask, then passing electric sparks or ultraviolet radiation through the mixture, molecules regarded as the precursors of life could be formed. The early Earth was bathed in ultraviolet radiation from the sun, and the early atmosphere must have provided plenty of sparks in the form of lightning. So scientists guessed that methane and ammonia were also present to start life on its way.

But more recent experiments have shown that precursor molecules can also be built up in test tube "atmospheres" rich in carbon dioxide—and astronomers Fred Hoyle and Chandra Wickramasinghe argue that precursors of life are even present in interstellar gas clouds and the material of comets! Today, some atmospheric scientists argue that the kind of atmosphere produced by the original outgassing must have been rich not in methane and ammonia, as originally thought, but in the same gases that now escape from the interior of the Earth. This conclusion is very strongly supported by the discovery that the thick atmosphere of Venus and the thin atmosphere of Mars are both rich in carbon dioxide. These planets, however, seem to have lost their life-giving water, while we have lost our carbon dioxide. Why? And how?

Again the answer can be found in the orbital distances of the three planets from the sun. Just about the only simple thing physics can tell us about conditions at the surface of a rocky planet a known distance from the sun is its surface temperature—and that, it turns out, is all we need to know!

For Venus, the stable temperature at which heat coming in from the sun is balanced by heat being radiated into space is 188 degrees (in the absence of an atmosphere). So, as soon as gas escaped from the rocks and began to build up in the atmosphere, it stayed as gas. Not just carbon dioxide but the water also would have stayed in its vapor state. Both water vapor and carbon dioxide permit the short-wavelength radiation from the sun to reach the planet's surface, but they trap the longer wavelength infrared waves that are radiated by hot rocks. The result of this so-called greenhouse effect was an initial surface temperature of 188 degrees,

which rapidly rose as an atmosphere developed. Soon it went above the boiling point of water and kept climbing until it reached its present oven-hot state—and all chance for life was gone.

THE GREENHOUSE EFFECT

On Mars, where the stable surface temperature was about 67 degrees *below* zero before outgassing got under way, things went very differently. Water could not even melt, let alone evaporate. Although the thin carbon dioxide atmosphere does produce a greenhouse effect, it is not enough to melt the frozen water today. It is just possible that sometime in the past, the atmosphere was thick enough to do the job, and that water did flow on Mars, carving out the canyons and lineated systems that look so much like dried-up river beds. But judging from the number of meteorite craters that scar the "rivers" of Mars, it has been at least 500 million years since water, or whatever liquid it was, flowed on the surface of that planet.

Earth remains the odd planet out, the wet one between the extremes of heat and cold. Here, the initial surface temperature before outgassing was about –13 degrees but later became warmer, high enough for liquid water to flow but not so high that enormous quantities of water vapor got into the atmosphere to produce a *runaway* greenhouse effect. Quite the reverse—the warm waters dissolved carbon dioxide out of the atmosphere, checking the greenhouse effect. The temperature rose somewhat and then settled down to an average of around 59 degrees. It has remained there ever since, thanks partly to the natural thermostat that the clouds provide.

Suppose the sun warmed up a little, as it might during its lifetime. Instead of the Earth getting hotter and perhaps developing a runaway greenhouse effect, the slight increase in temperature might evaporate more water from the oceans and thus produce more white clouds to reflect away the sun's new heat. Or imagine the sun cooled slightly. Less heat means less water evaporating, producing fewer clouds. With a larger fraction of the diminished solar heat getting to the ground, cooling would be less severe than it would otherwise have been. In other words, once the temperature of the Earth got into that nice, life-preserving zone, it stayed there, thanks to the protective action of the clouds.

But conditions were not yet right for the beginning of life on Earth—at least, not on the surface of the Earth. Lethal ultraviolet radiation from the sun could penetrate to the surface of the Earth and kill any primitive forms of life that might arise there. The oceans were different, however; they filtered out the harmful ultraviolet rays and provided a safe haven

for life to develop. Life arose, and soon it began to play its own part in molding the surrounding environment.

The first life forms found oxygen poisonous, a dangerous waste product of their life processes. But by a couple of billion years ago oxygen produced by those primitive creatures was beginning to build up in the atmosphere. There, chemical reactions stimulated by the sun's radiation (photochemical reactions) led to the production high in the atmosphere of ozone, a triatomic form of oxygen. (Normally, two oxygen atoms form an oxygen molecule, abbreviated as O_2. Ozone is O_3.) To this day, the ozone layer acts to filter out much of the sterilizing ultraviolet radiation from the sun. Under this protective filter, life began to move out of the sea and onto the land, while an abundance of atmospheric oxygen allowed new life forms—the first animals—that used the oxygen as an energy source to thrive.

SPECIAL CONDITIONS

How does this understanding of the origin of the air that we breathe help us to assess the likelihood of life existing elsewhere in the universe? A surprising number of people seem to take the view that the story I have just outlined hints that life may be very rare. They argue that the story shows the conditions for life are very special. They point to the fact that life does not seem to exist on either of our near neighbors, Venus and Mars.

The whole argument, however, can be turned on its head. If a star like the sun has a family of rocky, terrestrial planets, then it is very hard to see how one or more of them could fail to orbit the star in the temperature band where a wet planet like the Earth can exist. Look at the situation from a different viewpoint. Suppose our sun were just a little bit hotter. Then our planet would perhaps be a little warmer than it is now but would still be in the viable zone for wet, life-bearing planets. Venus would be just as undesirable as before. But now Mars too would be a wet planet with a reasonably thick atmosphere! And if the sun were just a little cooler, then Venus and Earth could indeed be sister planets. From this point of view, it seems that we are actually *unlucky* in this solar system, because we have a sun that is just at the right temperature for only one of its terrestrial planets to be wet, but not quite right for there to be two livable planets within our solar system.

This is a very important shift of viewpoint in terms of the prospects of finding life elsewhere in the universe. The pessimists who say that if the Earth were a little closer to the sun we would fry, while if it were a little farther away we would freeze, are being excessively gloomy. Almost anywhere from just this side of Venus to just this side of Mars a planet like Earth will inevitably end up wet and livable! The air that we breathe, and the oceans that surround us, are not freaks of nature. They are the inevitable accompaniments of a rocky planet at such a distance from its parent star that the equilibrium surface temperature, before the atmosphere formed, was at least a few degrees above the freezing point of water, but cool enough for most of the water to stay as liquid rather than vapor. That is a small enough range to make the chance of any particular rocky planet being wet fairly small, but if the spread of planets in our solar system is typical, the livable

band is wide enough to make the chances of finding at least one wet planet in each such system fairly good.

Stars like our sun seem to be ideal breeding grounds for life. The sun is a long-lived, stable star, with a fairly broad "life zone" in which a wet planet can exist comfortably for thousands of millions of years while life evolves. Smaller, cooler stars live even longer—but their life zones are correspondingly narrower, decreasing the chance of finding a wet planet in just the right orbit. Bigger, hotter stars can have very broad life zones—but they go through their own life cycles more quickly than cooler stars, leaving insufficient time for the process of evolution, which, on Earth, has taken so long to produce us and all the life around us.

And this is, perhaps, the happiest realization of all. Wherever stars like our sun exist, there ought to be planets like the Earth, with the key feature of free water running on their surfaces. So it may, indeed, be possible not just that life exists elsewhere in the universe, but that it lives on watery planets with blue skies and white clouds, rivers and trees, and surface temperatures in the range we are used to here on Earth. This, rather than the science fiction of methane-breathing monsters or intelligent crystalline life forms, inhabiting worlds with purple or green skies and lakes of liquid ammonia, seems to be the simplest interpretation of the evidence. If we are not alone in the universe, the air we breathe may be very similar to the air "they" breathe.

The search for life's origins— and a first 'synthetic cell'

*Scientists use precursors of genes and enzymes
to learn how life evolved four billion years
ago; they hope to repeat the process in the lab*

Paul Trachtman

*Paul Trachtman, a member of the Board of Editors,
wrote on children "without language," February 1984.*

In David White's chemistry laboratory at the NASA-Ames Research Center in Mountain View, California, there are no spectacular electric sparks or lightning flashes, no bubbling cauldrons or vaporous clouds rising from the racks of test tubes. "My experiments involve fairly boring test tubes with a little bit of dried-out clay in the bottom," White says. What is going on inside those test tubes, however, could turn out to be among the most exciting chemical interactions in the past four billion years. White is one of a small group of chemists, biologists, physicists, geologists and other scientists who are edging closer to a reenactment of the basic chemical events leading to what could be called "the origin of life."

These scientists, who now command the technology to peer into the minute molecular structure of cells and to gather data from the atmospheres of remote planets, are using their new knowledge of both the cosmos and the cell to re-create the scene on Earth, about four billion years ago, when life first arose from its chemical constituents. Their research is stripping away biology's most ancient mysteries, but its most extraordinary implications lie in the future. For, if scientists can repeat in the microcosm of the test tube the chemical processes by which nature first brought forth life, they will have found the recipe that allows them to make synthetic life for the first time.

While the recipe is not yet clear, the main ingredients amount to only a short list: nucleic acids, proteins and lipids. These provide the cell with its basic necessities: "information molecules" that carry its genetic code, "enzyme molecules" that keep its chemical reactions going, and "membrane molecules" that determine what flows in and out of the cell.

Nucleic acids (the DNA and RNA of the genetic code) are generally regarded as information molecules, while proteins provide the enzymes and lipids the membranes. In nature, genes and enzymes are very large, complex molecules, assembled in the cell from smaller precursors—nucleic acids from nucleotides, proteins from amino acids. In the laboratory, scientists can't yet synthesize complete genes and enzymes, but they have made a start by synthesizing intermediate molecules from the small precursors.

In the past few decades, researchers have demonstrated clearly that the necessary building blocks of the cell, including amino acids and nucleotides, can be made out of simple, ordinary compounds found in any chemistry lab and likely to have been present on the ancient Earth. Now, as their research goes beyond the building blocks, scientists are trying to demonstrate how the basic components could have evolved and organized themselves into the first cell. They are trying to show how, before there was life, molecules could have made copies of themselves through a primitive kind of reproduction, grown more complex and evolved through accidental errors like mutations—and also how conditions in their environment could have enforced a kind of "natural selection" to direct their evolution toward life.

In David White's test tubes small precursors of proteins and nucleic acids are gently heated and dried out on a clay surface for up to a month; then the products of any chemical reactions that may have occurred are analyzed. In the new molecules that may be formed, White is looking for evidence to support, or disprove, his theory of how life first arose. He has proposed that living systems were formed far more rapidly on the primitive Earth than most scientists have supposed, and from much simpler compounds than are usually thought to be necessary.

To test his theory, White is searching for the simplest combination of chemicals that would include both an information molecule and a primitive enzyme, thus providing a prototype of the living cell's chemistry. But White is experimenting with molecules so small and simple that they have seemed outside the realm of biology. "To use an analogy," White explains, "if the simplest living cell has the amount of information of the collected works of Shakespeare, we are talking about a couple of short words like cat and dog, and maybe one of them is misspelled."

Scattered around the country are other laboratories like David White's. At Columbia University's medical center in New York City, Donald Mills and Fred Kramer watch short RNA strands "reproduce" and "evolve" in solutions where one kind of enzyme assists their synthesis while another preys on them by chewing them up—resulting in a kind of molecular natural selection. At Cornell University's Department of Chemistry, David Usher is trying to simulate conditions in ancient tide pools where the simple molecules that make up RNA would have been subjected to repeated cycles of dry hot days and cold wet nights, a process he thinks could have provided both the energy for chemical evolution and another sort of natural selection, favoring some chemical bonds over others.

At the Salk Institute in La Jolla, California, Leslie Orgel's experiments are raising expectations that RNA molecules might have been all that were needed to begin a process of replication and evolution that led to life. Still another image of life's origins comes from David Deamer's laboratory at the University of California's Davis campus. In Deamer's experiments, using fluctuating "tide pool" conditions, the kinds of molecules that make up cell membranes, called lipids, assemble themselves from simple precursors. Then, when the membrane molecules are dried out and flushed again in the company of large molecules of DNA, the liplids engulf and capture the DNA in primitive cellular structures.

"This is as far as we have gotten toward producing a cell-like system under plausible prebiotic conditions," Deamer says. "It shows that big molecules could be encapsulated, very simply. It just takes one incoming tide, one drying cycle, and then the next tide washes away these 'cells.' So it must have been happening. It's just too simple not to have happened."

Only in the past few years have experiments such as these become practical. "We couldn't begin to think about the original processes," says Deamer, "without advances in understanding membranes, nucleic acids and proteins in contemporary cells."

That understanding points up the extraordinary nature of living cells. University of Iowa biochemist William Day, author of a comprehensive new text, *Genesis on Planet Earth*, points out that even the simplest bacterium has 2,000 to 3,000 different enzymes, and the activity of these enzyme molecules is almost beyond imagining: a single enzyme molecule can typically participate in chemical reactions with more than 500,000 other molecules per minute.

It is the increasing ability of scientists to measure and manipulate these basic molecules in the laboratory that is producing a new wave of experiments, and theories about the origin of life. "It's nice that so many people are starting to do real experiments," says David Usher. "There's been so much hypothesis and philosophy surrounding this entire field of science."

The Russian biochemist Alexander Oparin laid the basis for these experiments in 1924 with the suggestion that evolution was at work even before life began, "selecting" increasingly complex organic molecules (those containing carbon and hydrogen) with new intrinsic properties from the background of mixed organic compounds on the early Earth. Four years later, Englishman J. B. S. Haldane predicted the source of the organics. Harsh ultraviolet light would have energized the gases in Earth's first atmosphere, he said, causing organic molecules to form and collect in the oceans like a massive primordial broth—a chemical soup from which life would emerge.

Reconstructing a picture of life's earliest habitat has been a considerable challenge because crucial evidence has been entirely erased. The mystery is what happened on Earth *between* 4.6 billion years ago, when our planet formed from a vast cloud of gas and dust, and 3.8 billion years ago, the age of the oldest terrestrial rocks. The geological evidence has vanished, leaving what British geologist Stephen Moorbath calls "an embarrassing little gap of more than 700 million years." Significantly, it was toward the end of this enormous blank period that the very first inhabitants, single-celled and microscopic, probably emerged. Paleontologists have found traces of primitive algae and bacteria in rocks, from a lonely outpost in western Australia, dated at 3.5 billion years. Considerably simpler organisms must have preceded them, according to the current consensus in the field.

Clearly, it is necessary to fill in the "embarrassing

little gap." And, to a surprising extent, scientists have reconstructed a dramatic picture of life's first habitat based on data from other planets, and on back extrapolations from information about today's atmosphere, oceans and landmasses.

Four billion years ago, the Earth was a study in black and blue. Within the first 600 or 700 million years after the planet coalesced from the solar nebula, a black rocky "skin" appeared and started to thicken. This original crust, speculates Stephen Moorbath, was covered in most places by a blue ocean. All over the globe, dark volcanic islands jutted upward and strong waves beat against hundreds of miles of curving black beaches, wide lava flows and rocky shorelines.

The rims of giant impact craters probably also loomed from the oceans, says planetary scientist Brian Toon of NASA-Ames Research Center. Like the moon, the Earth was heavily bombarded by massive asteroids for the first billion years or so of its history. Thus the sky, which was probably pale blue due to volcanic ash, would have been emblazoned regularly by brilliant meteors and asteroids.

The bombardment was so extensive that eventually, scientists believe, all the rocks of the initial crust would have melted in impact events. "This," says geologist Steve Squires of NASA-Ames, "would set a fundamental limit on the oldest rocks" that is several hundred million years less than the actual age of the Earth. And it gives an explanation of the mysterious gap in the geological record.

But there is much that remains uncertain about the world in which life arose. "The day was shorter about four billion years ago, but no one really agrees on how short," says David Usher, who is trying to simulate that prebiotic world. "In my experiments, I'm using somewhere between five and ten hours. And there's a fair bit of controversy, but not wildly emotional, about what the average temperature was." Usher uses current estimates from places like the Namib Desert and Death Valley, where surface temperatures range from six to 75 degrees C. "Those are fine," he says. "You can do a lot of chemistry with that."

The turbulence of the setting, while it may have erased any record of the first organisms, may also have been a significant contributor to their emergence. There is a growing consensus that fluctuating environments with cycles of wet and dry, hot and cold, were needed for the kind of chemical evolution that could lead to life. "At every beach these natural experiments were going on," Deamer surmises, "where dilute components were washed into tide pools and dried out, allowing the simple sorts of chemistry that cause small molecules to become big molecules, until one of these suddenly found a way to replicate itself. That probably occurred at many places, and we don't necessarily have to think of one molecule starting up at one place and beginning this whole process."

One of the most hotly debated questions about this primitive environment has been the composition of the atmosphere, which would have determined the kinds of molecules available for those first "natural experiments." The early biochemists, including Oparin and Haldane, had assumed an atmosphere lacking oxygen but rich in the original gases of the solar system: methane, ammonia and hydrogen. This mixture was the basis of a historic experiment, carried out in Harold Urey's laboratory at the University of Chicago in 1953, which resolved a paradox that had stymied previous students of life's origins.

Life's precursors in a spark chamber

The problem was that the kinds of organic molecules essential to life could not be synthesized from the inorganic molecules of the nonliving world. But if only living cells could make the organic compounds needed to create life, how could the process ever have gotten started? A student of Urey's, Stanley Miller, designed an apparatus to test his professor's conviction that in the Earth's original atmosphere, organic compounds could have formed before there was any life. Its upper chamber was filled with the gases thought to represent the ancient atmosphere, and a lower flask was filled with water to simulate the oceans. The scientists shot sparks through the system for a week to simulate lightning, while heating the water so that it would evaporate and rain down again in repeated cycles. Soon their in-vitro ocean was brimming with organic compounds—including amino acids, the precursors of proteins.

This experiment, carried out in the same year that Watson and Crick deciphered the genetic code of DNA's double helix, and Frederick Sanger first figured out the basic structure of a protein molecule, opened a new era of experimentation as one laboratory after another now demonstrated the synthesis of life's basic molecules. Cyril Ponnamperuma at the University of Maryland and Joan Oró at the University of Houston showed how precursors of nucleic acids could have formed under primitive conditions, while Sidney Fox at the University of Miami showed how amino acids could be combined into more complex molecules, and could form proteinlike spheres which he called "protocells." More recently, Oró and Deamer have both shown how lipids in the prebiotic soup could have formed more lifelike protocells.

The first successes of these chemists were questioned by other researchers who challenged their assumptions about the atmosphere, some claiming that the Earth's original atmosphere was blown away and replaced by outgassing from the Earth's core during eons of volcanic eruptions. The proposed second atmosphere was less favorable for forming organics, but the chemists continued producing new batches of the essential organic precursors of life even from the latter mixtures of gases. Moreover, there was new evidence from outer space to support the hypothetical experiments.

In the late 1960s, astronomers detected enormous clouds of organic molecules in the spiral arms of our Galaxy. By now, more than 50 compounds have been identified, including formaldehyde and hydrogen cyanide. Molecules of the former can react to yield amino acids, and of the latter to produce the precursors of nucleic acids.

In addition, the meteorites called carbonaceous chondrites—which splashed into the early oceans by the millions—still fall to Earth regularly, and Ponnamperuma and others have claimed that these jagged, charred-black rocks do contain a palette of biological precursors. Recently, in fact, Ponnamperuma reported finding all five of the nitrogenous bases that occur in DNA and RNA in a single rock from outer space, the Murchison meteorite, which fell in Australia in 1969, although other workers claim that the meteorite was contaminated by organic compounds after landing on Earth.

With the wealth of biological precursors apparently well established, scientists have turned to the question of how life could have organized itself out of that original organic soup. And here the controversies are as thick as the evidence is thin.

Some researchers have simply removed the problem to someplace else in the universe, proposing that life arrived on Earth as an interstellar microbe. Nobelist Francis Crick and Leslie Orgel of the Salk Institute proposed a spacecraft from a more advanced civilization as the vehicle, while British astronomers Fred Hoyle and Chandra Wickramasinghe have suggested the heads of comets as life's first home.

Taking a more Earthly tack, oceanographer John Baross and colleagues at Oregon State University are looking for evidence that life first formed, and may still be forming, from organic precursors in hot ocean-bottom vents. Baross was aboard the research submarine *Alvin* in 1979 when it descended to the deep seafloor off the Gulf of California. In those black, frigid waters a mile and a half below the surface, the explorers discovered a field of hot springs inhabited by strange marine life and scattered with tall chimneys spewing jets of superheated, mineral-laden water. Samples taken from the heated water of these "smokers," Baross reported, revealed bacteria that can thrive at 250 degrees C or higher—cells alive in a veritable cauldron.

This astonishing claim has provoked skepticism from some of Baross' colleagues, who believe that his water samples from the depths, like many meteorites from outer space, were contaminated. But Baross maintains that these thermal bacteria may be modern counterparts of the first living cells, noting that many fossil remains of ancient microorganisms have been found in ancient marine volcanic sediments.

Baross is now constructing a complex chamber in his laboratory to simulate conditions in these natural seafloor pressure cookers, and one of his colleagues has returned to the smokers to collect more samples.

Probing the chicken-or-egg paradox

Meanwhile, a host of researchers without oceans to dive into have begun plunging into the ultimate paradox of life's origins. Even the simplest living cell needs nucleic acids in order to make its proteins, and proteins in order to make its nucleic acids. Thus, to ask whether the first cell evolved from nucleic acids or proteins is like asking Which came first, the chicken or the egg? Scientists, studying these essential components in different laboratories, have tended to divide into "nucleic acids first" and "proteins first" camps.

Recent experiments in Leslie Orgel's laboratory and elsewhere have stirred increasing interest in the RNA-first theory. With a short strand of RNA and a mixture of simple molecules that serve as letters in its mixture of simple molecules that serve as letters in its code, Orgel has shown that the strand can "instruct" the proper molecules to pair up with it in a double helix, even without any enzyme present. However, Orgel has not figured out how to separate the paired strand into new copies, as would be necessary to demonstrate self-replication. "Only *cells* are capable of replicating themselves," observes Queens College biologist Barry Commoner, a critic of both theoretical camps. "If something is self-replicating, it's got to be able to do it by itself, with no clever biochemist standing on the sidelines helping it along."

Confronted by the chicken-or-egg paradox, some scientists are trying to look at the process of life's origins in new ways, seeing the cell as something more than the sum of its parts. The paradox, they suggest, does not really exist in nature but rather in the scientists' way of explaining life.

It is not easy to explain what is going on here, but a metaphor may help. One theorist, Chilean neurobiologist Humberto Maturana, uses the analogy of an automobile. To understand what a car is, and how it runs, it is useful to know about all of the basic components such as cylinders and spark plugs, gears and wheels. But if these are laid out on the floor, one still has no idea of what a car is—until the parts are connected and the vehicle begins to move. Researchers on the origins of life are at the stage where they are shifting from asking What are the parts? to How are they organized to function as cells? The difference between a car and a cell, Maturana points out, is that a car keeps going by cyclic changes in the *positions* of its parts, while a cell keeps going by cyclic changes in the *productions* of its own parts.

New ways of thinking, and new words

"There are whole areas in ferment now," says David

Deamer, "because not only are we trying to find new ways of thinking about the phenomenon, but new words to describe it."

Manfred Eigen in Germany, Maturana in Chile and David White in California are three of those who have drawn up blueprints for a system that involves a "feedback loop" or cyclic relationship between nucleic acids and proteins. Eigen has described his system in terms of synchronized "hypercycles" involving both proteins and nucleic acids—"neither one could precede the other," he says; "they had to evolve together." Maturana has coined the term "autopoiesis" (from the Greek for self-producing) to define a system in which "the various kinds of molecules that make up the cell participate in producing the same kinds of molecules." And David White, working with his "cat" and "dog" molecular components, has developed a similar concept he calls the "autogen."

"The idea of autopoiesis has direct consequences for the laboratory," says Lynn Margulis at Boston University, a leading researcher on cell evolution whose own work has been guided by the concept. "It means that looking at nucleic-acid replication by itself does not necessarily lead to self-*maintaining* systems." David White has equally practical concerns. "The autogen theory has one purpose above all else, to suggest what is worth doing in the laboratory," he says.

In his test tubes, White has found molecules made of only two amino acids that act like a primitive enzyme, and he is trying to understand how the information of nucleic acids is translated into the amino-acid language of the proteins. In its simplest terms, his autogen would be a system in which information molecules direct the production of enzyme molecules they need to make more copies of themselves.

This new direction may bring science closer to an explanation of life's origins—or it may make clearer how far there is yet to go. "What is occurring now is the interplay between the sorts of questions we have been asking," says David Deamer. "What were the mechanisms that permitted the first replicating systems and the first catalytic systems to talk to each other, and to get together in a cyclic process?" Deamer's own research adds yet another level of complexity, looking at nucleic acids and membranes together.

"It may be that just working with nucleic acids, or proteins or lipids is not enough," agrees David Usher. "Probably the real advance is going to come from someone who is brave enough to put those three together at the same time. But that is *horribly* complex to investigate. Especially for someone who has been trained to investigate the details of a mechanism, it's very difficult to suddenly say, 'Well, I'm going to throw all these things in, and we don't know quite what's happening, but it might produce something interesting. Then we'll go in and look at what happened.' It's so complicated that to do anything experimental at this level is a real challenge."

How great a challenge was made clear by the pioneering biochemist J. B. S. Haldane when he turned to his colleagues at an origins-of-life conference 20 years ago and predicted that "some of us, or of the next generation, will try to make a living organism." Was Haldane too optimistic in his prediction? "I wouldn't be too sure," Usher says.

On the Wings of an Angel

An exploration of the limits of biological enterprise

Michael J. Katz

Michael J. Katz '72, M.D.-Ph.D., teaches neuroanatomy in the medical school of Case Western Reserve University. An extended version of this essay appears in No Way. On the Nature of the Impossible, *edited by P.J. Davis and D. Park (W.H. Freeman).*

> When you have eliminated the impossible, whatever remains, however improbable, must be the truth.
> —SHERLOCK HOLMES in *The Sign of the Four*

Biologists rarely use the word "impossible." To a biologist, the range of the possible is so large, the potential biological entities so overwhelmingly numerous, that the impossible is only a tiny issue.

To a biologist, impossibilities are the wall at the edge of the physical universe—real and formidable constraints, but constraints lying somewhere far away, somewhere in the realm of the physicist.

The biologist sees these physical constraints as if through the wrong end of a telescope. Demagnified and miniaturized, the limits of the physical universe form toy fences in someone else's province. With more than 1,500 species of daisies in Europe alone and more than 2,000 species of crickets world-wide, with 30,000 different proteins specific to the brain of the rat, biologists have little room on their desks for perpetual motion machines or for rockets that travel faster than the speed of light.

The mainstream of biological tradition is natural history, the record of Nature's accomplishments, the careful charting of the possible. Moreover, biologists are artisans at heart; they want to do things and make things, and they actively seek out the possible. Beyond this, however, I suspect that there is a more fundamental reason that biologists rarely speak in terms of the impossible. Perhaps there are no biological impossibilities. Perhaps, deep within the true province of biology, everything is possible.

The province of biology

The wide range of the possible in biology depends on the unique and peculiar province of the living. Living beings form a special realm of science, filled with eye-popping collages of butterflies, sheets of grasses, mildewing molds, people, bears, whales, and bats, eggs and embryos, grandmothers and grandfathers. Biology is about life, and life is organisms.

We know them well, these organisms. We pass them every day as we walk on the grass, under the trees, past the birds and squirrels. Gardening, we run our hands through the cool earth, crumbles of plant detritus, worms, larvae, hundreds of thousands of microorganisms. We know them by touch, smell, and sight; and those organisms that never come within our grasps, such as the gulls at the shore, we know by sight and sound.

But most of all, we know the human organisms. We know our parents and our friends; we know strangers on the bus, on the street, and in the stores; and, of course, we know ourselves. How many times a day do we look at our hands? A hundred? A thousand? We feel our toes from outside and from inside. We hear our heartbeats at night. We smell our sweat, we taste our blood. We are not all physicists or economists or mathematicians, but we are all biologists. We know organisms, and most of us know them quite well.

Biological organisms fall into a few general types but into innumerable specific varieties. In fact, each organism is essentially unique. Human "identical" twins differ in many ways; for example, their fingerprints are different. Even a pair of cloned organisms are sufficiently complex so that, during their creation, the stochasticism of the world can insinuate itself and change a molecule here or an organelle there and in this way produce two slightly different individuals. The complexity and the individuality of its particular items of study—that is, organisms—are certainly characteristic of the biological realm, but these features do not distinguish biology from physics or metallurgy or economics. Each of these disciplines faces items of study that have many interactive parts, with those parts forming unique wholes. What is it, then, that distinguishes biology from other sciences?

Ontogeny

Organisms are patterns of matter that are at once complex and individual, and the features that distinguish these patterns from other complex and individual patterns of the natural world are the two sequential processes that produce organisms: ontogeny and phylogeny.

Ontogeny is the history of a living entity from conception through birth to maturity and death, the laying out of all the stages in its transformation from an unspecialized embryonic form to a particular and idiomatic machine. Ontogenies come in all shapes and sizes. At one extreme, bacteria go through an ontogeny that is entirely internal: the transformations from a single parent cell to two daughter cells are a series of changes of molecules *inside* the cell. On the other hand, multicellular organisms, such as squid, butterflies, and people, begin as single fertile cells—zygotes—and transform into unified collections of millions of cells. The ontogeny of a multicellular organism is a cascade of intracellular, cellular, and extracellular changes that establish whole cities of specialized cells. During the ontogeny of a multicellular organism, interactive pockets of cells are geographically segregated into organs and tissues connected by highways of nerves and vessels. The construction of these cities is continuously dynamic, and it proceeds inexorably in a particular sequence, the characteristic ontogeny of that organism.

An ontogeny is stereotyped and highly reproducible. It is like a phonograph record: when conception sets the needle in the first groove (and given that the basic machinery has an appropriate supply of energy), it plays out the full music of a life. Random dust will always change the notes a bit along the way. Sometimes the environment intervenes to turn the volume up or down. Occasionally, the needle gets caught by a scratch and falls into the endless loops of a cancer. Usually, however, a scratch or even a jarring of the turntable causes only a skip in the sequence as the needle falls into a different groove and proceeds once again resolutely on its inevitable path.

Ontogenies are dogged things, and organisms are those highly complex patterns produced by ontogenies. It is ontogeny—the repeated generation of stereotyped yet complex patterns—that first sets the biological realm apart from the other spheres of natural science.

Phylogeny

This is the book of the generations of Adam. . . . And Adam lived an hundred and thirty years, and begat a son in his own likeness, after his image; and called his name Seth. . . . And Seth lived an hundred and five years and begat Enos. . . . And Enos lived ninety years, and begat Cainan. . . . And Cainan lived seventy years and begat Mahalaleel. . . . And Lamech lived an hundred eighty and two years and begat a son: And he called his name Noah . . . and Noah was five hundred years old: and Noah begat Shem, Ham, and Japheth.
                                                  —GENESIS, Chapter 5

Generation after generation, organisms beget like organisms. This is phylogeny, the ancestral lineages of organ-

> Traced back into the dim reaches of time, the human phylogeny contains all manner of creatures: apes, small bright-eyed mammals, dog-sized sharp-toothed reptiles, flat snub-nosed amphibians, even fish. Exuberant variety is a constant in biology.

isms. Ontogenies are the life histories of individual organisms, and phylogenies are the repeated unfoldings of ontogenies. Our ancestors are our phylogeny.

Biological time is different from physical time, and the biological clock of phylogenies ticks in generations. In one hundred years, a human phylogeny contains five generations, a buttercup phylogeny one hundred generations, a fruit fly phylogeny 2,500 generations, and a bacterial phylogeny can contain 2.5 million generations.

A human phylogeny of five generations is, on the whole, a very short time, and it represents an almost unchanging set of transformations. The striking similarities between ancestors and descendants transcend a handful of generations—how often have we heard: "He certainly reminds me of his grandfather." Those differences that do show up during a few generations are really rather subtle, and each child is inordinately more like his parents than he is different from them. A few generations of phylogeny is a biologically stable time interval, but a million or ten million generations is quite another story.

In the course of millions of generations, the difference between members of a phylogeny can become so marked that we say the original organism has evolved into a new organism. "Evolution" means "change," and long phylogenies tend to change. If traced back into the dim reaches of time, the human phylogeny contains all manner of different creatures: apes, small bright-eyed mammals, dog-sized sharp-toothed reptiles, flat snub-nosed amphibians, and even fish. If we take a patient astronomical view, we can see clearly that we have evolved from our ancestors. Phylogenies slowly evolve, and evolving phylogenies characterize the biological realm.

One distinguishing characteristic of the province of biology is ontogeny, the recurrent stereotyped re-creation of a very complex pattern. The other distinguishing characteristic of the biological realm is phylogeny, the ancestral lineage of these ontogenies. While an ontogeny is a relatively stable sequence, a phylogeny is a slowly evolving sequence of ontogenies. When we ask what is impossible in biology—in biology specifically, above and beyond physics and chemistry, besides psychology and economics, separate from mathematics and the arts—we are asking what is impossible in ontogeny and in phylogeny.

The richness of natural ontogenies and phylogenies

If there is a constant in biology, it is its exuberant variety. A major contributor to this rich unpredictability of living things is unnecessary complexity: biological systems often contain more machinery than is necessary to make them work properly. Excess complexities permeate life. At the molecular level, there is the DNA that does not code for any proteins—noncoding DNA can outnumber coding DNA a hundredfold in some cells. Then, there are the "futile metabolic cycles" in cells, circular chemical reactions that go back and forth producing and unproducing the same molecules and depleting energy stores to no apparent purpose.

Another example of unnecessary complexity is the blood-clotting cascade. When you cut your finger, blood proteins immediately begin to clump together, the wound is soon dammed up, and the cut stops bleeding within five to ten minutes. To staunch the blood flow, the initial injury sets off a waterfall of from eight to thirteen separate chemical reactions in two chains, with each chemical transformation giving rise to the next in an orderly sequence. At least thirteen different proteins—coagulation factors—form the normal clotting cascades in humans, and if one of these factors is missing the person can have a bleeding disorder such as hemophilia.

The complete blood-clotting cascade is quite complex, and a theoretical biologist would be hard-pressed to predict its actual details from a priori considerations, from first principles, or from the requirements of blood-clotting systems. One of the factors—Hageman Factor or Factor XII—even appears to be unnecessary. Those people who, through genetic disorders, develop without any Factor XII do not have bleeding problems; and whales, dolphins, and porpoises, all of which survive injuries quite normally, do not have any Factor XII.

Complex and elusive intricacies also characterize the tissue level of biological organization. Consider the corpus callosum, one of the largest bundles of axons in the human brain. Although it interconnects most areas of the cerebral hemispheres, its function is so subtle that for years no one understood exactly what it does. The five out of a thousand individuals born without a corpus callosum cannot normally be distinguished from those with one. The corpus callosum is found only in placental mammals: other mammals (such as opossums and kangaroos) and all nonmammals live quite happily without it. Only through an ingenious series of psychological experiments did Roger Sperry finally show how the two halves of the brain normally use the corpus callosum as their most intimate route of self-communication.

From his desk, the theoretical biologist could not determine the role of the corpus callosum with certainty, and he could not predict its appearance or its use in those animals that have acquired one during the last 200 million years. Who could have imagined that the human brain contains two separate minds, a right mind and a left mind, each localized in one of the major cerebral hemispheres? Normally, the two minds are in such close touch that they think alike, they trade thoughts instanta-neously, they share the same sensations and emotions, and they act as one. All this intimacy flows through the corpus callosum, and the intercommunication is smooth and efficient. At the same time, each separate brain is a powerful and complete mind. Amazingly, without a corpus callosum, the nervous system still functions as a smooth and efficient unit—one brain, to almost all outward appearance. Normally two brains make each human, and two brains are a wonderful but unnecessary complexity.

The corpus callosum is not a necessity, but is it just a frill? Two minds are not a necessity, but is the second one a frill? Such questions, with words like "frill" or even "necessity," are slightly askew. They are difficult to answer because they are built from peculiarly human judgments. As Richard Bentley wrote (in the late seventeenth century):

> All pulchritude is relative. . . . We ought not . . . to believe that the banks of the ocean are really deformed, because they have not the form of a regular bulwark; nor that the mountains are out of shape, because they are not exact pyramids or cones; nor that the stars are unskillfully placed, because they are not situated at uniform distance. These are not natural irregularities, but with respect to our fancies only; nor are they incommodious to the true uses of life and the designs of man's being on earth.

Nature need not adhere to human standards, and she need not follow human principles. Nature does as she does, and we can only be secure in our science when we act as natural historians, conscientiously describing the natural record retrospectively. We walk a precipitous course when we attempt a priori evaluations based on anthropocentric standards.

Ontogenies and phylogenies are not limited to the simplest or the most efficient paths. In the natural realm, organisms are not built by engineers who, with an overall plan in mind, use only the most appropriate materials, the most effective design, and the most reliable construction techniques. Instead, organisms are patchworks containing appendixes, uvulas, earlobes, dewclaws, adenoids, warts, eyebrows, underarm hair, wisdom teeth, and toenails. They are a meld of ancestral parts integrated step by step during their development through a set of tried and true ontogenetic mechanisms. These mechanisms ensure matching between disparate elements such as nerves and muscles, but they have no overall vision. Natural ontogenies and natural phylogenies are not limited by principles of parsimony, and they have no teleology. Possible organisms can be overdetermined, unnecessarily complex, or inefficiently designed.

Many roads

Two roads diverged in a yellow wood,
And sorry I could not travel both
And be one traveler, long I stood
And looked down one as far as I could
To where it bent in the undergrowth;

Then took the other, as just as fair,
And having perhaps the better claim,
Because it was grassy and wanted wear;
Though as for that the passing there
Had worn them really about the same. . . .

—ROBERT FROST

The constraints in building organisms are usually insufficient to limit Nature to only one blueprint, and a wide range of alternate constructions have evolved. There is no one "right" way to build an eye. The octopus and the human both have eyes that appear quite similar, but the human eye is built exactly inside out when compared to the octopus. In an octopus, light passing through the lens falls directly on the photoreceptors, while in a person, light must travel through many layers of cells and axons before reaching the photoreceptors, which are themselves pointing the wrong way—that is, toward the back of the eye. Likewise, Nature has used a number of radically different designs for building wings: bat wings, for instance, are modified hands, while insect wings are entirely separate appendages.

Even molecules themselves can have architectural latitude. Although certain parts of a biological molecule are fairly immutable, there is often no one right overall molecule. For instance, insulin is an essential protein hormone that is built of about fifty subunits (amino acids). Three to five of these subunits differ between the insulin molecules of pigs, cows, and humans. Nonetheless, the insulins from pigs and cows are perfectly acceptable substitutes for human insulin, and both pig and cow insulin are commonly used to treat human diabetes.

In terms of many roads, the capricious courses of phylogenies are most telling. Evolution has followed the exigencies of the times under the whims of chance and the accidents of history. Had the continents not drifted apart, Australian fauna and flora would undoubtedly be less peculiar—Australia would probably have had indigenous hoofed animals and indigenous apes, animal groups that never developed on that island. Had Alexander Fleming not discovered penicillin, penicillin-resistant bacteria would be a freakish oddity rather than ubiquitous inhabitants of our planet. Extant organisms are legacies of habits acquired by their ancestors, but these habits coalesced from a plethora of possibilities.

What Nature cannot do in ontogeny and phylogeny

I am Rose my eyes are blue
I am Rose and who are you
I am Rose and when I sing
I am Rose like anything

—GERTRUDE STEIN

Science fiction comes in two varieties. On one hand are the tales that explore worlds harboring phenomena that scientists think are impossible. These stories ask the questions of dreams: How would people spend their evenings if everyone had a perpetual motion machine in the basement? How soon would you get bored if you lived forever?

Then there are those tales that explore worlds that just might exist. They ask the questions of science: In what language could we talk to an extraterrestrial creature? What could we do with self-reproducing automata? What will people do when the sun goes out?

The standard science fiction of biology certainly falls into both of these categories, but which biological tall

In science fiction, the mushroom people of the planet Basidium-X subsisted on chicken eggs. But carnivorous mushrooms do exist. Certain species trap and eat worms, and from worms it may be a small step to chickens.

tales are the stories of dreams and which the stories of science?

Consider, for a moment, the mushroom—"the elf of plants," Emily Dickinson called it. Actually mushrooms are only distantly related to plants. They are many-celled fungi, relatively advanced organisms with cell walls but with no ability to manufacture their own food (e.g., no photosynthetic machinery), no ability to move, and no nervous system. We are all aware that many animals are fungivorous, but it took science fiction to popularize the idea that mushrooms could be carnivorous. In *The Wonderful Flight to the Mushroom Planet,* for instance, Eleanor Cameron invented the mushroom people of the mysterious planet Basidium-X, who must eat chicken eggs to remain healthy. The stories of dreams? Surprisingly not. Carnivorous mushrooms actually exist here on earth: certain species of woodland toadstools trap and eat worms—and from worms it may be only a small step to chickens. Carnivorous mushrooms, once in the realm of science fiction, are now unequivocally science fact.

Or, consider the square organism, once a creature confined to E.A. Abbott's *Flatland,* where all "Professional Men and Gentlemen are Squares." Today, the square organism has found a home on our well-worn earth, swimming in the brine pools of the Middle East. There tiny, flat, transparent bacteria in the form of thin square sheets float like ghostly salt crystals, mimicking the perfect planar polygons and belying the notion that—to reduce their surface-to-volume ratios—cells must be spheres.

Carnivorous mushrooms and square bacteria bring a smile to the biologist, but they do not stretch the bounds of biology because they can be explained by mechanisms that sit somewhere on Nature's cluttered shelf of standard organismic machinery. True, the biologist may have to hunt around a bit among the everyday mitochondria, the familiar Krebs cycles, and the mundane cyclic AMPs to retrieve all of the appropriate mechanisms. He will certainly have to spend some time in serious study to find how these mechanisms have been stuck together in each of their peculiar combinations. Nonetheless, somewhere in a corner of her cupboard Nature is sure to have just the right bits and pieces to construct these natural oddities.

Nature regularly builds baby carnivorous mushrooms from spores of parent carnivorous mushrooms and replicates daughter square bacteria from parent square bacteria, and Nature derived the parent carnivorous mushrooms from other preexisting mushrooms and the parent square bacteria from other preexisting bacteria. Strange as they are, carnivorous mushrooms and square bacteria—the incarnations of biological tall tales—are neither ontogenetically nor phylogenetically impossible.

Angels, on the other hand, are somewhat different. Although they adorn the spiritual world, our natural world has no angels. Why is this? While it is not absolutely impossible, it is nonetheless difficult for Nature to construct an angel from an extant phylogeny. Besides the arms and the legs of a human, an angel has a set of wings along its back. Wings are complex structures sculpted of muscles, bones, and nerves; and angels' wings are covered with feathers. To introduce wings or any other complex appendage into an existing organismal lineage, Nature needs the appropriate raw materials and organizational blueprint—preexisting structures that can be transformed—because complex biological forms cannot be created *ex nihilo*.

For angels' wings, the preexisting structures are simply not available. The wings of the natural extant flying vertebrates—the birds and bats—are direct modifications of preexisting front limbs. The muscles, bones, and nerves were already there in ancestral organisms, and Nature proceeded to evolve wings by stretching, shrinking, folding, and bending those elements. Furthermore, through all of the transmogrifications, the overall organization of the front limb has remained the same during evolution. For example, the upper limb always has a single long bone, the humerus, and the lower limb always has a pair of parallel long bones, the radius and the ulna.

The back of a mammal has no structures that can be stretched or shrunk, folded or bent into a wing. To make an angel, the ground plan of the existing elements must be tampered with and new structures must be generated without precedent. This Nature cannot easily do.

Angels are probably destined to remain spiritual, and winged horses like Pegasus are likely to be forever myths; but biologists do not consider them impossible. Instead, they are put into another realm: they are highly improbable biological phenomena. Improbable biological phenomena cannot easily be pieced together by Nature from any of the mechanisms in her crowded cupboard of organismic machinery. Without the coincidence of a number of highly improbable events, Nature cannot generate a winged horse or an angel in an existing phylogeny.

What a biologist cannot do

He said "I look for butterflies
That sleep among the wheat;
I make them into mutton pies,
And sell them in the street.
I sell them unto men," he said
"Who sail on stormy seas;
And that's the way I get my bread—
A trifle, if you please."

—LEWIS CARROLL

Embryonic eyes from baby newts can be transplanted to tail buds. Adult newts, with two eyes fore and one aft, swim in laboratory aquaria such as Nature has never seen.

With the development of his complex and specialized brain, man has taken a place beside Nature as a biological creator. Nature creates through ontogenies and phylogenies, but man is an engineer and can construct biological forms from other beginnings and through other ad hoc processes. Man is not limited to the natural routes of creation, and in the laboratory he can generate biological phenomena that would be highly improbable in nature.

A protein, an organelle, a cell, a tissue, an organism—each is made of a great many different parts, and in each case these parts are organized in a particular and characteristic design. The many parts and the unique designs are found at every level—we see them whether we look at an elephant from a distance or we examine its gall bladder under a microscope. These biological items are truly complex in all ways. Not only are they composed of many different parts interrelated in unique designs, their fabrication is complex. In most cases, the parts of a biological item will not fully self-assemble. You cannot shake a beaker of salts and amino acids and make insulin, and you cannot stir a soup of cells and make a mouse. To build a protein or an animal, one must carefully put all of the parts together in their single proper order. One must impose detailed external information—templets—on the raw materials.

Biological items form only a small subset of all the possible items that one might construct from the same raw materials. This means that, as an architect, the biologist cannot merely choose the right bricks and mortar, but he must also draw up the right plan and then contrive to interweave the building blocks into just the right design. This is a difficult set of tasks. Often the parts are tiny and cannot be easily moved about or stored by themselves, isolated from their natural settings. Moreover, gluing these tiny parts together in such a way as to create the proper order at all levels, from macroscopic to microscopic to molecular, takes extreme patience, steady hands, X-ray vision, and highly specialized, Rube Goldberg contraptions.

In the face of these problems, biologists have been undauntedly optimistic, twiddling and fiddling, tinkering with bones and nerves, gingerly reconstructing hormones, and rearranging genes. The job of building a biological item entirely from scratch is usually too overwhelming to be practical. The fabrication of a biological item from extant biological scraps, however, and the sculpting of new biological items from preexisting ones, these have become everyday operations in the laboratory. In baby newts, embryonic eyes are transplanted to tail buds, where they eventually mature and send out nerves. The adult newts, three-eyed oddities, swim in laboratory aquaria as Nature has never seen, with two eyes fore and one aft. In embryonic chicks, extra limb buds are grafted alongside the normal ones; later, the adult chickens run through the laboratory flailing supernumerary appendages. Mouse and human cells are fused to form hybrid mammalian cells, biological items never found in nature but powerful laboratory tools for mapping human chromosomes. Copies of human insulin genes are inserted into yeast or bacterial cells, and these tiny and primitive creatures, as different from us as any organisms on earth, will now manufacture human proteins. In the lab, the biologist is busily creating highly improbable biological phenomena.

Building ontogenies and phylogenies

Given the range, the power, and the detailed precision of modern technologies, is there anything that the biologist cannot do within the bounds of the physical constraints of the universe? Today's biologists have the faith that it is possible to construct almost any biological item from precursor materials.

A gene, a protein, a cell, a tissue, an organism—these all seem to be in the realm of possibility. It may not always be practical to create these items from the most elemental materials. Nonetheless, new genes can be manufactured by mutating and rearranging existing genes, and short proteins can be made to order from their constituent amino acids. Primal protocells can be formed in appropriate man-made molecular soups, and new and complex cell types can be pieced together by fusing whole cells or by combining parts of cells. New tissues can be designed by growing cells on artificial templets. New organisms can be constructed by mutations, by genetic engineering, and by embryonic reconstructions such as grafts and transplants.

Biologists can build improbable biological items in the laboratory, but the hallmarks of the biological realm are more than individual biological items. Life is not a DNA molecule or a nerve cell or a kumquat or a wolf spider. Life is a special set of sequences; it is the autonomous and recurrent stereotyped re-creation of certain very complex patterns. Life is a child growing and becoming a mother and eventually a grandmother. Can the biologist create new grandmothers, that is, can he generate ontogenies and phylogenies never before seen in nature?

The answer is yes, although the new ontogenies and phylogenies take advantage of natural cascades of developmental events that are normally found in preexisting ontogenies. To begin, the biologist makes an improbable change in some developmental event; for example, he grafts a frog eye primordium into the side of a newt embryo. In nature, the two sets of tissues would never interact, but in the laboratory the hybrid organism undergoes an ontogeny. The frog cells form an eye and send an optic nerve into the newt nervous system; concurrently, the newt skin cells form a lens and the newt nervous system accommodates the aberrant nerves. Frog cells integrate with newt cells, newt cells mesh with frog cells, and the strange three-eyed chimera that develops unfolds through a truly new ontogeny.

Similarly, new phylogenies—ancestral lineages never before seen in nature—take advantage of natural cascades of ontogenies that are normally found in preexisting phylogenies. Here, the biologist makes an improbable change in the stuff of inheritance; for example, he grafts a sequence of human DNA into the DNA of a bacterium. In nature, the two sets of genes would never interact, but in the laboratory the hybrid bacterium divides and quickly becomes a grandmother. All of her children and grandchildren will manufacture certain human proteins, and a truly new phylogeny has been founded.

Such ontogenies and phylogenies are new, but they are not unnatural. Once triggered, they unfold spontaneously and thereby enter the natural realm, producing surprising wonders like three-eyed frog-newts and insulin-secreting germs or carnivorous mushrooms and square bacteria. Man-made ontogenies and phylogenies are autonomous and recurrent stereotyped re-creations of certain very complex patterns, just as are naturally initiated ontogenies and phylogenies. In the laboratory, it is only the initiating event that may have been "unnatural."

When the biologist founds a new ontogeny or phylogeny, he can, of course, understand the initial improbable event. Many times, he can also understand the initial event when Nature founds a new ontogeny or phylogeny. But in nature initial events can sometimes remain arcane. As Dr. Seuss wrote in *The 500 Hats of Bartholomew Cubbins*:

But neither Bartholomew Cubbins, nor King Derwin himself, nor anyone else in the Kingdom of Didd could ever explain how the strange thing had happened. They only could say it just "happened to happen" and was not very likely to happen again.

In the province of biology, buttermilk-thick with life and under the patience of millions of generations, Nature sometimes stumbles on the extremely improbable, and arcane initial events can indeed be quite natural. They just "happen to happen" and are "not very likely to happen again." Moreover, when they happen in a natural ontogeny or phylogeny, arcane events can trigger sequences that are as natural as apples. It is the initiating event in the generation of an angel that would undoubtedly be arcane, but the autonomous development of an angel from a tiny wisp of an angelic embryo or the spontaneous unfolding of a lineage of angels, once set on their way, become natural phenomena no more impossible than the development of an oak from an acorn or the spontaneous unfolding of the ancestral lineage of the great Bach family.

Biological possibilities

What is an impossibility? I think that for the biological realm, the requirements are rather special. Not only must we be able to write science fiction about it, we must also be able to imagine it as a part of a natural ontogeny and phylogeny. Truly biological entities are always enmeshed in a developmental and an evolutionary sequence: they are dynamic, they have a lifespan, they have ancestors, and they beget progeny. In this way, a biological impossibility would be something—be it an organelle, a cell, or a creature—that we could imagine in an ontogeny or in a phylogeny but that cannot ever exist in the real world.

Physically impossible organisms, such as hedgehogs that can run faster than the speed of light and perpetual motion bees, can be dreamt by the physicist, but I cannot easily imagine a biologically impossible organism. When we have eliminated the physically impossible, when we remain within the constraints set by the phys-ical limits of the universe, whatever remains—no matter how improbable—must be considered biologically possible. With the biologist as creator, the improbable has ofttimes become probable. But, Nature herself is wild and rich and her splendor is unconstrained. Afternoons poking about the Woods Hole seashore among the horseshoe crabs, the seaweed, and the tunicates or munching blue-eyed scallops and beach peas on a rocky island in Penobscot Bay or chipping ornate brachiopods from the shale of the Chagrin River make me hesitate to think that I could ever dream of a creature that might not creep out from among the cattails one windy spring morning.

Genetics

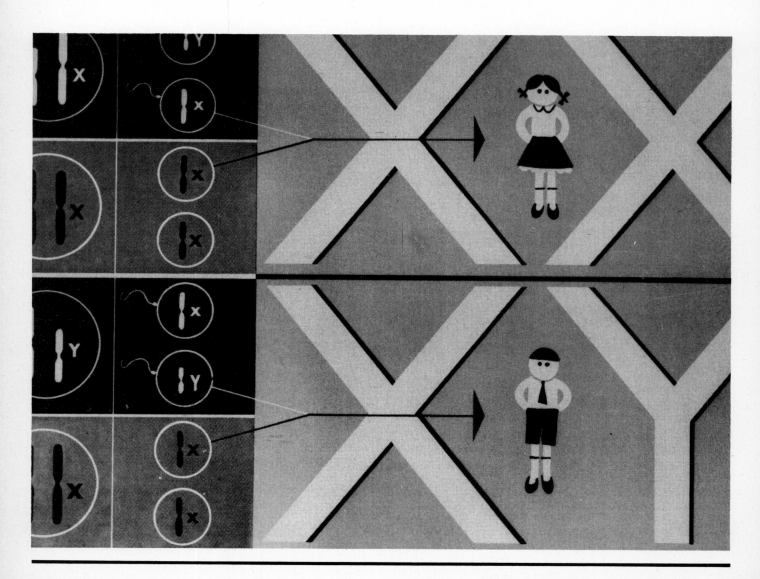

In the mingling of the seed, sometimes the woman, with sudden force, overpowers the man, and then the children, born of maternal seed, will resemble more the mother; but if from paternal seed the father. The children you see resembling both their parents, having the features of both, have been created from father's body and mother's blood, when the seeds course through the bodies excited by Venus, in harmony of mutual passion, breathing as one, with neither conquering and neither being conquered.

Lucretius

Human fascination with the mechanisms of reproduction and genetics has existed for centuries. The opening quotation from the Roman poet Lucretius (c. 96-55 B.C.) is evidence that this interest was often combined with a sometimes amusing lack of understanding. During this past century, however, scientists probing into the areas of genetics and inheritance mechanism have replaced these fanciful explanations with scientific fact. Knowledge of the biochemical basis for heredity as seen through the structure of DNA and the significance of RNA and proteins has been elucidated. While it may seem that the entire discipline has been revealed, geneticists have barely begun to scratch the surface.

Molecular genetics is one of the most exciting, most rapidly moving fields of biology. There are reports almost daily of geneticists doing things that would have been considered impossible a decade ago. Since newspapers, radio, and television are saturated with this new information, terms like genetic engineering, genetic counseling, recombinant DNA, and genetic diseases are becoming more familiar to the layperson.

Scientists are already using genetic engineering in industry and agriculture and the results are impressive. Future use of these techniques in the field of human medicine offer outstanding promise. At the present time, over 2,500 diseases are classified as having a genetic cause. Although most of these are not common, scientists are nevertheless excited over the prospect of "gene repair," using molecular techniques such as gene splicing and gene transfer to correct human deficiencies. Couples who worry about passing genetic defects to their children are able to seek counseling about the chances of their offspring acquiring the parental disease.

As Albert Rosenfeld states in "Genetics: The Edge of Creation," "Until now, only nature and circumstance controlled the shape and character of every living thing." This is changing as scientists learn how to manipulate the basic unit of heredity—the gene.

The question of potential ethical problems associated with genetic manipulation is addressed in a number of articles appearing in this section. Keith Schneider's, "Science Debates Using Tools to Redesign Life" states that, "There are severe limits to the extent of the modifications we can make. If you mix genes from genetically distant organisms that don't fit each other well, you will not have an organism. We're not going to have Frankensteins crawling around."

The modern molecular technology described in this section can be used to change the blueprint of life and the course of evolution. To many, this ability holds both great promises and potential perils. This technology will force society to make choices—choices that are likely to be more profound than any other civilization ever faced.

Looking Ahead: Challenge Questions

The code for the protein that determines a particular cell's form and function originates within DNA. How is this DNA code expressed in the synthesis of new cellular components?

What ethical problems have been associated with the concept of genetic manipulation? Are these problems warranted?

Why are bacteria not able to be utilized in the production of human proteins?

Bits of Life

Alexander Rich

Alexander Rich is Sedgwick Professor of Biophysics at the Massachusetts Institute of Technology. He has carried out extensive research on the structures and functions of nucleic acids.

A living system can be characterized in terms of the flux or flow of various things in it. These are in three general classes. One can describe a living cell as having a flow of matter: Chemicals come into the cell, they are broken down or transformed into other chemicals which then leave the cell. One can also describe a living system as having a flow of energy. Energy comes into a living system either in the form of chemical energy associated with chemical bonds, or in the form of radiant energy from the sun which is trapped and utilized. However, having described a living system in terms of both matter and energy, the description is still not complete. There is a third element which is crucial—information. A living system is not just a set of simple ongoing chemical reactions, but rather, a directed and responsive system in which there is control and regulation of the kinds of chemical reactions that occur.

In order to understand how living systems are maintained, one has to understand how this information is contained, replicated, and finally, expressed. The answers to these questions lie in the study of the large molecules which constitute the cell's informational system—the nucleic acids and the proteins.

The information strategy that has evolved in living systems is probably of ancient origin, perhaps three to four billion years old. Two basic types of informational polymers are used and both are nucleic acids: DNA (deoxyribonucleic acid) and RNA (ribonucleic acid) contain and transmit the genetic information of the cell. In terms of informational content, they are the most important elements in present-day living organisms.

The instructions contained by the nucleic acids are finally expressed by another group of polymer molecules called proteins. Proteins form many of the structural and mechanical components of living systems. Some of them, the enzymes, play very important roles as chemical catalysts. The chemical activities of the cell are thus controlled indirectly by DNA and RNA, which direct the synthesis of proteins. These in turn regulate the metabolic activities of the cell.

The Genetic Alphabet

The nucleic acids consist of a backbone chain containing phosphates and sugars, and the sugars have side-chains of nitrogen containing molecules called bases. There are four bases in DNA—adenine, guanine, thymine and cytosine. RNA has a different sugar in its backbone and one base is slightly different. It is fair to say that information in biological systems is contained in exactly the same way that information is contained in the English language. When you read a book, it has meaning for you because of the order of the letters. In biological systems, the same is true; it is the order or the sequence of these four different bases which provides the meaning of the genetic material.

How do you replicate such a polymer? It turns out to be helpful to have things which are complementary to each other, that fit together in such a unique way that if you have one element, it assures you that you will have the other, its complement. The double helix uniquely answers the question of how a molecule replicates itself by using complementary interactions: guanine pairs with cytosine and adenine pairs with thymine.

The molecular basis for our understanding of this system are these famous base pairs of DNA, discovered by James Watson and Francis Crick some 27 years ago. These bases combine with each other in a highly specific fashion—not like with like, but complementarily. When DNA replicates, the helix opens up and the strands separate. The single strands then act as templates for building up a polymer chain which is complementary to the original strands.

The DNA is organized to form a three-dimensional molecule made of two chains, with base pairs holding them together. The common analogy is that of a spiraling ladder, with base pairs forming the "rungs" and the sugar and phosphate backbone forming the "sides" of the ladder. These sides are wound around each other in the form of a right-handed double helix with the bases tucked in the middle. It wouldn't be obvious that it is an information-containing helix, except that there are four different kinds of materials which make up the base pairs, and they have an order to them in much the same way that letters in the English language have an order to them. However, this is a four-letter alphabet.

From *The Sciences,* October 1980. ©1980 The New York Academy of Sciences.

Master Blueprint

Carrying and replicating information is one feature, but expressing information is something different and it is a far more complex project. Furthermore, there is a great deal of information to be expressed.

If you took the DNA from the common colon bacteria *E. coli*, and put it in a straight line, it would be about two millimeters long. These two millimeters contain about 10 million base pairs. If you wanted to write out in a book the genetic code of *E. coli*, you would need a volume of about 2,000 pages, with 5,000 letters to the page. Eventually we will have such a book. We now have methods for determining this sequence in its entirety.

What about a higher organism, such as *Homo sapiens*? In a sense, a true measure of a man (or woman) is DNA. The DNA in one human cell is about 1000 times the DNA content of a single *E. coli;* the DNA of a human cell is about two meters long, or just over six feet. It would thus take 1000 of these huge 2000-page volumes to describe a single human cell in the same four-letter code. Such an encyclopedia would contain all the information needed to make a human being.

How has nature organized the system for expressing genetic information? In the early evolutionary phase of development, various systems probably competed with each other to see which would be the most effective in expressing information. This evolutionary lottery was "won" by the amino acids which can polymerize to make proteins. The amino acids are fairly simple molecules. However, they have a number of different side-chains and the side-chains introduce important variations. The side-chains can be hydrophilic or hydrophobic, that is, water-loving or water-repelling. They can be positively or negatively charged. They can be saturated or unsaturated. There are 20 different kinds of amino acids that are used in contemporary biological systems, and with them one can create a great variety of microenvironments on the surface of protein molecules which can perform the kind of catalysis that is needed to regulate the chemical reactions inside living cells.

Although it was believed during the late 1950s that the genetic information in DNA had to be used somehow to direct the synthesis of proteins, the mechanism for this was by no means clear. It was widely believed that RNA was involved in this activity, but the situation was confused by the fact that the organelle or machine of protein synthesis, the ribosome, contained a great deal of RNA. It took a significant time period before the relationship between information-containing RNA and the structural RNA of the ribosome could be separated. Around 1960 a number of workers developed the concept that there was a rapidly synthesized and rapidly turning-over segment of RNA called messenger RNA which was copied off DNA and then rapidly made its way to the organelle of protein synthesis, the ribosome, where it directed the synthesis of proteins.

DNA may be regarded as the master blueprint. The cell must make copies of that blueprint for use in the workshop. These will be destroyed after they have been used. During the copying process, the same opening up or separation of the two strands of the DNA double helix occurs which is used in replication; however, only one of the strands is copied.

Messenger RNA (mRNA) contains a set of information for assembling amino acids in the correct sequence for the formation of different proteins. The sequence of amino acids in a protein tells you what kind of protein is being formed and defines the nature of its chemical reactivity.

In order to understand this process, we had to solve a cryptographic problem. The nucleic acid code is a four-letter code (four bases), which we want to translate into the 20-letter protein code (20 amino acids). How can you translate from one language to another if you have different numbers of letters in the respective alphabets? That problem was solved long ago. Basically, the idea is to use many more symbols in the "symbol-poor" language to equal a number in the "symbol-rich" language. What this means is that we must use groups of nucleotides to define individual amino acids. A simple calculation will tell you that if two nucleotides define one amino acid, then there are 4×4, or 16, combinations, clearly inadequate because there are 20 amino acids. Therefore you must use three nucleotides to define one amino acid. This gives you $4 \times 4 \times 4$, or 64, combinations to choose 20 different amino acids; this is called the "genetic code." What it does is allow us to define the relationship between a group of three nucleotides and each amino acid. It also means that there are several groups which define the same amino acid.

Figure 1: A diagrammatic sketch of the flow of ribosomes over messenger DNA.

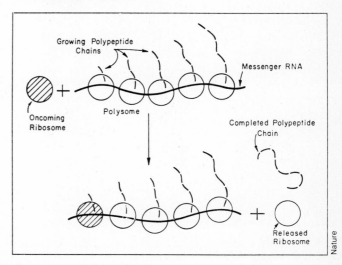

"Tape-Driven Assembly"

Proteins are made in the ribosomes, cellular organelles about 250 Å (angstroms) in diameter (one-millionth of an inch). They are fairly large and complicated, consisting of over 50 different proteins and several different types of RNA.

The mechanism for making protein is such that the ribosome attaches to the messenger RNA, and then starts moving along the RNA chain. As it moves along, it decodes the message the mRNA is carrying. With this is associated the gradual lengthening of the polypeptide chain. Finally, the ribosome is released and the completed polypeptide or protein chain is ready to fold up and carry out its activities. It is a kind of tape-driven assembly system, complicated, and yet compact.

Proteins can contain from 50 to 1500 amino acids in one chain. Thus the messenger RNA can be longer than one ribosome, and can, in fact, stretch across several of them. In 1962 these considerations led us to look for a somewhat larger structure which might be the site of protein synthesis. That led directly to the discovery of clusters of ribosomes, which we called polysomes, which were all acting simultaneously on that same messenger RNA to translate the information into proteins. This system is shown diagrammatically in Figure 1. The ribosome engages the message near one end and continues moving along while the growing polypeptide chain is elongated steadily. At the end, the ribosome is released as well as the completed polypeptide chain. Figure 2 shows electron micrographs of polysomes from the rabbit reticulocytes which are active in the synthesis of the protein hemoglobin. These generally consist of five or so ribosomes attached to the same message (top). A negative stain electron micrograph reveals both the ribosomes and the messenger RNA stretching between them (bottom).

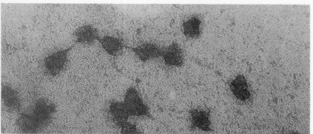

Figure 2: Electron micrographs of rabbit reticulocyte polysomes synthesizing the protein hemoglobin.

Each triplet of nucleotides specifies a particular amino acid but, of course, you need to know where to begin. There is a special nucleotide triplet that says "start" and is recognized by the ribosome. It uses a particular amino acid, methionine. There is also another group of nucleotide triplets called the termination codons which stop the process and release the protein.

But how does the right amino acid find a particular triplet of nucleotides? About 25 years ago, when we were thinking about this problem, we had all kinds of strange thoughts that perhaps there would be special holes in the surface of an RNA molecule that would define sites for individual amino acids. But then Francis Crick had a rather simple idea. Crick remarked on the fact that what nucleic acids do rather well is grab on to other nucleic acids by taking advantage of the same complementarity that is used in the replication of DNA and the formation of RNA. He suggested that there might be some molecule (which he called an adapter molecule) which would interact with mRNA at one end and with a particular amino acid at the other. We have indeed discovered such molecules, which are now known as transfer RNA (tRNA).

Turning a Corner

Transfer RNAs play a central role in the transfer of genetic information. They stand at the crossroads of information encoded in nucleic acids and information encoded in the polypeptide chain. At one end they interact with messenger RNA through hydrogen bonding of three bases, called the anticodon, to three bases on the messenger RNA codon. This hydrogen bonding is approximately the same type as is found in the Watson-Crick base pairs in DNA. At the other end of the molecule, they are conveniently linked to a polypeptide chain which grows continuously as the ribosome works its way down the messenger RNA strand. These molecules are called transfer RNAs because they are active in transferring free amino acids into polypeptide chains.

The transfer RNA molecule is a small polynucleotide chain containing from 75 to 90 nucleotides. Almost 200 of these have been sequenced, and they all fall into a pattern which was first described by Robert Holley in 1965, the so-called "cloverleaf sequence." From the sequence it was inferred that the transfer RNA molecule exists in the form of a number of stems and loops which fold back on themselves, the stems consisting of double helical RNA segments. The relationship of the cloverleaf to the actual three-dimensional structure was not clear at that time. In 1971 we were able to obtain crystals of yeast phenylalanine transfer RNA which yielded diffraction patterns with almost 2 Å resolution. By 1973, we had solved the crystal structure and traced the backbone of the transfer RNA molecule and were able to show that it exists in the form of a compact folded structure, more or less L-shaped. The anticodon bases are at one end of the L and the end of the molecule which binds to the amino acid is at the other end of the L. The overall shape of the molecule is shown in Figure 3. The darker spheres in the

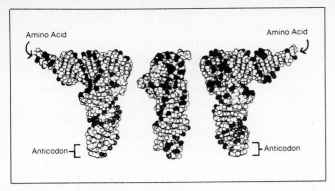

Figure 3: The molecular architecture of the yeast transfer RNA which codes for the amino acid phenylalanine.

diagram represent the negatively charged phosphate groups; positively charged magnesium ions are also shown in a shaded form. The molecule has a number of interesting features to it. What one sees is a highly engineered molecule in which the parts fit together precisely in order to allow the double helix to turn a corner. In the course of doing this there are a number of hydrogen bonding interactions between bases in addition to the hydrogen bonding interactions in the double helical stems. Whereas the double helical stem interactions are virtually all of the Watson-Crick type, the same as those found in double helical DNA, the additional so-called tertiary interactions involving the various loops of transfer RNA with each other are largely of a different type.

How Proteins Are Made

In present-day biological systems, there are 20 different amino acids which are used in forming proteins. Associated with this are 20 different families of transfer RNA molecules, each of which will accept a particular amino acid. This joining process is carried out by a synthetase enzyme which specifically joins a particular amino acid to a particular family or transfer RNA molecules. Once the amino acid is joined to the transfer RNA, the aminoacyl transfer RNA goes into the ribosomal machinery.

In order for the process of protein synthesis to occur, two transfer RNA molecules interact side-by-side, occupying two different sites in the ribosome. The purpose of this is to permit the sequential addition of transfer RNAs with amino acids attached to adjacent codons on the messenger RNA strand. The machinery of the ribosome transfers the growing polypeptide chain to the adjacent aminoacyl tRNA, thereby lengthening that chain by one residue. Further events in the ribosome result in evicting the transfer RNA from which the peptidyl chain has been transferred. This is followed by a process of translocation, whereby the transfer RNA, with its newly joined polypeptide chain, moves over into the adjacent

ribosomal peptidyl site. The vacated aminoacyl site can then be used to add the next amino acid. The specificity in this process largely resides with the interaction of codon and anticodon bases. Although this process is described in outline here, it should be understood that many of the finer details of the mechanism are not clear at present and a great deal of additional work has to be done before we will have a complete understanding of the molecular dynamics of the ribosome.

The transfer RNA molecule may have an L shape in order to allow two of these molecules to lie closely together, side-by-side, on adjacent codons. The L shape may make it possible to bring both ends together so that the growing polypeptide chain can be transferred from one molecule to the other.

In a rapidly growing bacterial cell, it is possible to add 20 amino acids to a rapidly growing polypeptide chain in about one second. Thus, it takes about 50 milliseconds to complete one cycle of the process. Considering the fact that a very large number of processes have to go on during this time period, the mechanism is actually fairly rapid. Nature finds that it is very important that this process be carried out with great accuracy. Accordingly, a great deal of effort and biological energy is taken in making sure that this process occurs with a relatively small number of errors. A variety of error-correction mechanisms are involved in the ribosomal machinery.

New Twists and Pieces

Molecular biology is still young enough to treat us to a great many surprises. For example, we've been living

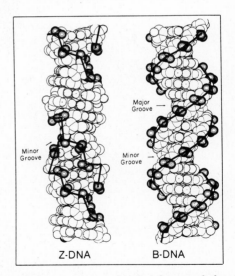

Figure 4: Two molecular forms of the DNA double helix, right-handed B-DNA and left-handed Z-DNA. The heavy black line in each goes from phosphate group to phosphate group, indicating a smooth right-handed helix in B-DNA, but an irregular left-handed helix in Z-DNA.

quite comfortably with the right-handed double helix called B-DNA for many years. Recently, however, we discovered that there are also left-handed double helices. We called this form of the molecule Z-DNA because of the zigzag array of the backbone (Fig. 4). We discovered this in a crystallized segment of DNA, which was solved to atomic resolution. Even though the guanine-cytosine base pairs are the normal Watson-Crick type, the helical ladder is folded the other way.

The left-handed double helix seems to be the most stable when there are specialized sequences of bases. Although we believe that most of the DNA in a cell is right-handed, some of it may be able to assume a left-handed form. In that form quite different parts of the molecule are exposed and that may have some interesting biological consequences. The reversed twist cells us that the double helix has much more configurational flexibility than we were aware of, suggesting that DNA can adopt a number of different forms.

This altered configuration exposes segments of DNA which are otherwise hidden in the normal right-handed form. Some of these segments are sites where carcinogens bind. It is entirely possible that carcinogens interact with DNA when it is in this left-handed form. There are also some suggestions that high rates of mutation may occur in sequences which can adopt this left-handed configuration. It is possible that this may be a physical basis for a high rate of mutation.

Finally, we have had a comfortable feeling about the way that genetic information is transmitted and expressed from DNA to RNA to proteins. It was such a nice and simple scheme. But three years ago something quite startling was discovered. The message which appears in mRNA during protein synthesis in higher organisms is not organized in the same manner in the DNA. In the DNA, the message is split up into pieces, divided by segments which are called intervening sequences. These sequences are present in DNA, but not in RNA. The discovery of intervening sequences is causing a mini-revolution in molecular biology. It tells us that the DNA containing information for the same protein strand comes in segments, with other pieces stuck in between. We don't know why this is occurring. It is possible that this allows evolution to go on more rapidly, because you can think of mixing together different pieces of DNA, cutting out larger pieces, and making new combinations. It is also possible that these intervening sequences have some special utility that we have not yet recognized.

Basically, what we have found is that the nice compact picture we had of information transfer in biological systems is inadequate; the mechanisms involved are much more complicated than we had originally understood them to be, and we will have to work hard to figure out what's going on.

The major difference between living systems and non-living systems is related to the informational organization in the living cell. That organization directs the chemical activities of the cell, responds to external stimuli and is, indeed, the physical repository of the evolutionary history of the organism. We have learned much about this in recent years but a full understanding is still left for the future.

GENETICS
THE EDGE OF CREATION

Albert Rosenfeld

With its golden-yellow color and its slight froth of silent bubbles, the liquid in the flask seems to generate an eerie glow. The young man holding the flask has about him the air of a latter-day alchemist. Perhaps it is only the way the lab lights illuminate the scene, or merely my own boggled mind conjuring up a medieval atmosphere based on my knowledge of what the flask contains: a thick soup of *pure genetic material,* the raw chemical information that dictates what all living organisms, including people, are and do. As I reach to touch the liquid's surface, I am surprised to discover that it is in fact not liquid at all: it is crisp to the touch, readily scraped into a powder. The bubbles are tiny empty pockets left over from the chloroform used in the purification process. I remember reading a book in the late 1950s in which the great French biologist Jean Rostand calculated that the amount of DNA (deoxyribonucleic acid) required to transform the heredity of the entire human population would fit into a cube measuring only one twenty-fifth of an inch on each side—

something like a thimbleful. And here I am in Rockville, Maryland, scarcely two decades later, standing in the laboratory of the Genex Corporation next to a biochemist named Stephen Lombardi, who calmly holds in his hands not a thimbleful but a *10-liter flaskful* of the very stuff of life!

What the flask contains is not yet the same fully constituted, working DNA that resides in the cells of your body and mine, spelling out our hereditary traits as well as the detailed instructions for the minute-by-minute functioning of our cellular processes. This solid broth does, however—in concert with the contents of three similar flasks—hold the makings of every possible variety of DNA molecule. The genetic alphabet consists of only four chemical "letters," called bases or nucleotides, and each of the flasks holds 220 grams of one of these nucleotides in pure, raw form with one of its ends chemically "open," ready to join with the others in whatever combinations might please the new breed of genetic engineer exemplified by Lombardi and his colleagues at Genex.

It is basically these four genetic letters, in the diver-

sity of their arrangement, that determine whether a given creature turns out to be a rattlesnake or a grizzly bear, a mollusk or a Michelangelo. The same four-letter genetic code specifies the characteristics of every organism that now lives, or has ever lived, on earth—and could specify those of creatures that have never before existed anywhere in the universe. Until now, genetic change or mutation came about only through the forces of nature—blind or purposeful, evolutionary or divine, but in any case outside the control of mere women and men. But now that we have learned how to manipulate genes, we have suddenly become the trustees of our own further evolution, if any—as well as the mediators of all future life on our planet.

I am not suggesting that the Genex scientists, or that any scientists anywhere, are yet capable of creating living creatures out of their new chemistry sets. But so breathtakingly rapid has progress been in this field that geneticists find themselves matter-of-factly doing things that 10 years ago they would have deemed close to impossible. Though it was theoretically

plausible to think about recombining genes, for example—that is, snipping genes out of one organism and transplanting them into the cells of another organism, even of another species—no one had the foggiest idea of how this might actually be done. Yet today not only has the feat been attained in the laboratory but it has been attained so readily and so easily that recombinant DNA, or "gene splicing," has become the basis of a rapidly growing new industry. Genex, with operations already spreading to Europe and the Far East, is only one of more than a hundred new bioengineering firms that have sprung up in the United States alone over the past few years. So far the companies are based more on promise than on product—but what promise! The ability to splice genes and to implant them with exquisite precision into the genetic apparatus of bacteria means that the host organisms are endowed with the capacity to turn out whatever protein product is dictated by a given gene. The bacterium thus implanted can then be cloned—that is, pure strains can be grown from the original, each one inheriting the new gene. In

From Cell . . . to Nucleus . . . to Chromosome . . . to DNA

All life is based on the cell—a nucleus surrounded by cytoplasm. The human body is comprised of 60 trillion of these living units in a variety of shapes and sizes.

The nucleus is the director of all activity in the cell. The genetic instructions that determine what each cell's particular function will be are sent from the nucleus.

Every living thing has a specific number of threadlike chromosomes in its nucleus. The chromosomes always exist in pairs: human beings have 23 pairs, or 46 chromosomes.

The DNA coiled within each chromosome contains our hereditary plan. It consists of four chemical bases—thymine, adenine, guanine and cytosine—wound in a double helix.

fact, multiple copies of the gene can be spliced into the same microbe. Suppose it takes a bacterium 20 minutes to divide; when you consider that in 24 hours you will have billions of bacteria manufacturing your product, this offers an inkling of the vast new possibilities that have opened up, virtually overnight, for industry, agriculture and medicine.

There have been extensive debates about the ethics of conducting some of this research. It is to the scientists' credit that they were the first to bring the issue to public attention. With the first successful experiments in gene transfer and gene splicing in 1973 and 1974, it was clear to everyone that we had suddenly entered a vast, unknown territory. Might we endow infectious microbes with the power to resist antibiotics? Or inadvertently insert genes into organisms that could get into our bodies and increase our cancer risks?

Small groups of molecular biologists began to meet under the auspices of the National Academy of Sciences, and in July 1974, an unprecedented event took place: a letter signed by Stanford University's Paul Berg and nine other scientists appeared simultaneously in *Science, Nature* and *Proceedings of the National Academy of Sciences,* perhaps the three most prestigious scientific journals in the English-speaking world. The letter asked scientists everywhere to institute a voluntary moratorium on certain kinds of genetic experiments until an international conference could be held to discuss possible dangers and necessary safeguards. To ensure that the public would know what was going on, a news conference was arranged to coincide with the letter's publication.

In February 1975, a group of 139 researchers from 17 nations, along with a smattering of lay people and a large corps of journalists, assembled at Asilomar, a conference center in Pacific Grove, California, to spend several long and frenetic days hammering out guidelines and recommendations for conducting research in recombinant DNA.

Even before the conference, a storm of controversy far beyond the scientists' expectations erupted—and it was destined to accelerate in the months following Asilomar. Scare scenarios were freely composed in which bizarre new microbes escaped from the lab to decimate the earth's population like nothing since the black plague. In many communities around the United States, people worried about research being conducted at their local universities, and in some cases they demanded that the work be monitored if not altogether halted.

The scientists' task at Asilomar was made somewhat easier by new reports assuring them that bacteria could be specifically bred and "disarmed" so that they could not survive outside the laboratory environment. Experiments were classified according to their estimated degree of hazard, and strict safety and containment measures were recommended for each level. Thus, only properly equipped labs would be permitted to carry out the riskier experiments. The irony was that by the time the National Institutes of Health had appointed its Recombinant DNA Advisory Committee, held meetings and issued its own stringent guidelines, many of the Asilomar scientists had come to believe that much of their earlier concern had been unwarranted. Apart from the disarming of the bacteria employed, it became clear that recombinant DNA had been going on in nature over the millennia—transferred by viruses, for example. Moreover, it is no easy matter for a microbe to learn to become infectious: it requires long periods of evolving with the organism that eventually becomes susceptible to infection. So a brand-new organism would have a hard time finding a hospitable host—unlike the celebrated "Andromeda strain" of science fiction. These were some of the considerations that led to a gradual relaxation of gene-splicing regulations and even to some sentiment in favor of dropping them altogether.

But the implications for humans are even broader than these debates imply in what we must begin to recognize as a new Age of Genetics. To transfer naturally occurring genes is impressive enough, but scientists are now learning how to fabricate genes out of the basic building blocks—as the Genex people were preparing to do with their flasks. In fact, there already exist "gene machines" so automated that almost anyone can, with minimal training, learn to turn out gene fragments in a few hours, a task that formerly would have taken a skilled chemist several months of assiduous effort. Scientists have also begun to "map" genes—that is, to pinpoint the locations of specific genes on specific chromosomes; and the first attempts have been made to replace missing or faulty genes with the intent of curing, or at least of treating, human genetic disease. It is becoming increasingly evident, too, that genes contribute substantially not only to our physiological makeup but to our personalities and behavior as well.

In studying and manipulating genes, scientists naturally turned to the simpler organisms first. When Paul Berg of Stanford inserted the first foreign gene into an organism, he chose the SV 40 virus, which has only five genes. (Berg and his asso-

ciates were later able to transfer a bacterial gene to a human cell in tissue culture, where it was able to produce the missing enzyme that caused a human genetic disease.) The genome, or full set of genetic instructions, of the common intestinal bacterium *Escherichia coli*, or *E. coli* for short, with its single chromosome and ready accessibility, has been investigated diligently; and *E. coli* remains the favorite vehicle for both research and production in gene splicing. But the human genome has barely begun to be explored, and its exploration is of course incomparably more complex and difficult than that of *E. coli*. At one time, locating the insulin gene was beyond anyone's capacity. Fortunately, the protein chemists

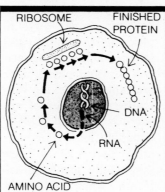

What Happens

The code for the protein that determines a cell's particular form and function originates with the DNA. This code is copied in the nucleus by the RNA, which then goes out into the cell body and collects the necessary amino acids. At the ribosome, the amino acids are assembled in the order that was originally dictated by the genetic information contained within the DNA. The amino acids are then linked in a protein chain and are ready to work.

did know how to take apart the insulin molecule itself. With that information, it became possible to reconstruct the gene. Genentech, Inc., the best-known and so far the most successful of the new bioengineering companies, farmed out the task to scientists at Cal Tech and City of Hope National Medical Center in Duarte, California. As a result, Eli Lilly and Company, under an agreement with Genentech, is already able to produce human insulin in experimental quantities. In fact, the first trials undertaken by Lilly with British and American volunteers indicate that the product is both safe and effective.

Gene splicing must surely be rated as one of the seminal biotechnological breakthroughs of this century—and therefore of all time. It suddenly became feasible in the early 1970s with two discoveries. One was that a class of bacterial proteins called restriction enzymes had the capacity to sever DNA molecules at predictable sites (other enzymes had already been found that could "suture" genes back in place). The other was that loose, circular pieces of DNA called plasmids could pass genetic information along from one generation of *E. coli* to the next. The plasmid that has now become historic is the one called pSC101; the *p* is for "plasmid," and *SC* are the initials of Stanley Cohen, who is one of Berg's colleagues at Stanford.

Cohen's pioneering feat was achieved in collaboration with biochemist Herbert W. Boyer of the University of California at San Francisco and two associates. They first trans-

ferred genes to *E. coli* from another *E. coli;* then from another bacterium, *Staphylococcus aureus*—the first gene transfer between species; and finally, working with other scientists, from an animal—the frog *Xenopus laevis*. These experiments represented the practical beginning of genetic engineering.

As I sat with Cohen recently in his lab at Stanford, he was willing enough to reminisce a bit about these exciting events out of a past that already seems remote. But he was more interested in talking about the scientific questions now engaging his attention—and that is what I was more interested in hearing about, too.

The irony is that in the period just before the gene-splicing breakthroughs occurred, many molecular geneticists were feeling glum because they thought there was really not much more to discover about DNA. Oh, yes, more details to be sure, but the basics were all in place. They could not have been more mistaken. The current paradox is that while we can manipulate genes with an ease and a dexterity not even contemplated a decade ago, we have at the same time discovered how much less we know about genes than we thought we did. New mysteries arise, it seems, with each new set of experiments. Such a state of affairs would produce despair in most other spheres of human endeavor, but scientists find it exhilarating. It is nevertheless also disquieting to learn that a stable, reliable, orderly molecule such as DNA is really full of quirks and caprices.

One of the aberrations of

DNA that has particularly preoccupied Stanley Cohen's lab involves a class of "transposable elements," or transposons—segments of DNA that can combine and recombine in various ways, switching around whole groups of genes among plasmids and viruses as well as within the genomes of living organisms. Earlier in the 1970s, in West Germany, Peter Starlinger and his associates at the University of Cologne had come upon curious DNA fragments they called "insertion sequences," which, though they had no discernible function of their own, were mobile units that readily intruded on other genes. Meanwhile, experiments by Cohen and other investigators in England offered clues about how movable elements were able to transfer antibiotic resistance from one bacterium to another. Transposons appear to be important in causing some types of disease to occur—and they also help to set up bacterial defenses against man-made remedies designed to combat those diseases.

Apart from the obvious health implications, another reason for Cohen's fascination with transposons and their ability to bring about what he calls "illegitimate recombination" is that they "offer insights into how evolution may be able to occur in quantum leaps" as well as by slow accretion. Moreover, "transposable elements may explain how whole blocks of genes can be turned on or off during the development of an organism."

Earlier that day, one of Cohen's Stanford colleagues, biochemist David Hogness, had explained how this can occur in fruit flies, which have a much higher number of transposons than most species. The transposons can cause very striking spontaneous mutations. Because a single mutation may control a significant block of genes, entire segments of the body can be drastically affected: part of an eye will turn into genitalia; an abdominal segment that normally has no appendages could be transformed into a thoracic segment and develop legs.

Genes and gene fragments, then, "jump" in totally unexpected ways, moving about from one part of the genome to another. In fact, chromosomes are known to exchange large packets of DNA. Genes have even been observed to undergo complete flip-flops, executing 180-degree reversals on the DNA chain so that their genetic messages read backward. A group of scientists at Cal Tech has been especially intrigued by this jumping-gene phenomenon. The head of this aggregation is Leroy Hood. Hood and his colleagues have been able to explain how jumping genes serve a vital developmental purpose by making it possible for the body's immune system to turn out an extraordinary diversity of antibodies—a capacity that has always been a source of mystery.

Another surprise that recent genetic research has turned up is the fact that genes are not necessarily, as formerly believed, continuous stretches of DNA that spell out precise instructions for making, say, proteins. Genes have in many cases (some geneticists believe in most cases) turned out to be split up in the most bizarre fashion. Such genes have their intelligible stretches of nucleotides constantly interrupted by what appear to be nonsense sequences—at least they are not part of the gene's instructions for making the given protein. These "intervening sequences" (labeled introns by noted Harvard gene splicer Walter Gilbert, as distinct from exons, the sequences that express the needed instructions) are as long as and frequently longer and more numerous than the "real" genetic sequences. It would be as if we were to take a word such as *inspiration* and spell it *i-n-s-x-q-l-x-x-p-i-s-q-w-a-m-g-r-a-t-i-z-z-q-p-t-x-o-n* and expect the reader's eye to pick out the meaningful letters and ignore the rest.

Apart from introns, there are other perplexing reiterative sequences of apparently unused DNA, including many nonoperating "pseudogenes." Are they really meaningless? If not, what could they possibly mean? Guesses have ranged all over the lot: the untranslated sequences could have something to do with gene regulation and control, or with species differences, or with differentiation during fetal development, or with evolutionary mechanisms. It has even been suggested that a lot of genetic material may be purely self-serving, just taking advantage of the free ride in order to perpetuate itself; as long as it doesn't particularly interfere with other functions, it just replicates along with the rest of the genome during procreation. Clearly, no one has yet begun to solve the puzzle.

Another new wrinkle: DNA, instead of being strung out smoothly, is folded in upon itself in unexpected ways, coiling and "supercoiling" into intricate configurations that still remain undivined. It turns out, too, that mutations and mistakes occur all the time, but DNA fortunately possesses the capacity to repair itself—most of the time, anyway.

With the rise of recombinant-DNA technology, it has become fashionable to say that genetics will dominate the 1980s just as, say, computers dominated the previous decades. But it seems to me that this is understating the case. Rather, we are entering a much more sweeping Age of Genetics, an age that perhaps has no more recent parallel than the revolution that occurred during the Neolithic era, when our ancestors discovered how to domesticate animals and raise crops. We will now be able to create some plants that will photosynthesize more efficiently and others that will fertilize themselves, perhaps via the insertion of nitrogen-fixing genes, thus saving the billions of dollars and the millions of barrels of oil that are now spent every year on synthetic fertilizers. We will create bacterial strains capable of converting wastes into useful products, even into foods and fuels. With our genetic know-how only in the infancy of its development, we can hardly imagine how far-reaching the impact of genetics on every branch of industry and agriculture will be.

But more than any of that, we will become the masters of the molecules we are made of—and therefore of our bodies and psyches and of all living things that creep or swim or run or fly. Hardly any facet of our personal lives will be unaffected, and the political questions raised will often require solutions on a global scale. Every genetic benefit carries with it a concomitant worry. Will attempts to cure genetic diseases, for instance, lead to gene tampering to "improve the race"?

The picture comes back to me of Stephen Lombardi at Genex, holding up his flaskful of pure nucleotides. I had compared him to an alchemist. But I now realize what a pale comparison this is. The alchemists' dreams seemed grandiose and arrogant for their time, but their cravings were really fairly modest. They dealt with gross elements and mixtures of elements, whereas the Stephen Lombardis of contemporary science deal with the tiniest components of the living genetic code—which they know how to read and with which they are learning to write. Whereas the alchemists merely wanted to find the "philosopher's stone" that would transmute base metals into gold, we may soon be able to transmute almost anything into almost anything else. Ethical and moral dilemmas abound at every step of the genetic path along which we are now traveling so rapidly. And our inner voice keeps asking: Are we ready for this?

We have always treasured the advice, "Know thyself," inscribed on the temple of Delphi. To know our genetic selves is to seek an ever more profound self-knowledge—and in no way is that incompatible with the more traditional, spiritual paths to self-knowledge. The control of our genes surely entails risks, but is it really less risky to continue to let our genes control *us*—assuming we have the choice? Do we want to keep arguing about whether or not our genes *are* in control—or would we rather find out? In this time of troubles, it seems we were never more in need of knowledge we do not now possess.

The way we should enter the Age of Genetics is with exhilaration tempered by caution.

YEAST
AT WORK

Along with making dinner rolls and beer,
the venerable fungus turns out
hepatitis B vaccine and cancer drugs.

TERENCE MONMANEY

Terence Monmaney is a staff writer for Science 85.

IT WAS ONLY a matter of time. Biotechnology, the business of getting microbes to produce unusual and valuable substances, is now returning to the one that got it started 5,000 years ago. Right about then, as one version of the story goes, wild yeast cells floated into a vat of fruit juice, and a few days later some unsuspecting Egyptian drank from it.

Today, yeasts are being used in ways that are as surprising as that first drink of wine must have been. Modified yeasts are helping produce fuel for cars, feed for pigs, and cheese for us. Researchers have coaxed human genes into yeast to manufacture copies of human proteins such as growth hormone for healing wounds, interferon for treating cancer, and an enzyme for calming inflammation. And yeast, the lowly, single-celled fungus that puts the bounce in bread and the bubbles in beer, was the organism of choice for producing the first genetically engineered vaccine ever tried on people—a vaccine against hepatitis B virus.

New work on yeast promises not only vaccines and medicines but a more fundamental understanding of disease itself, especially cancer. Prying into yeast with the tools of recombinant DNA, researchers in several laboratories have discovered genes that are nearly identical to some human genes believed to start the growth of certain tumors—a finding that offers a powerful new way to study the molecular workings of cancer.

As some wagers have it, yeasts are racing with bacteria for the lead position in biotechnology, an industry conservatively expected to earn $1 billion a year by 1990, $10 billion by 2000. Yeasts have an edge in that they produce none of the poisons that can foul a batch of bacteria. And unlike bacterial cells that must be burst open to retrieve a protein, yeasts can secrete a ready-made product—an enzyme, say—beyond their own cell walls, essentially delivering it for sale.

Yeasts offer something else that isn't so easy to measure. While people tend to associate bacteria with disease, human civilization long ago warmed up to what yeast can do. "Another advantage of yeast," says Philip Barr, a researcher at the California biotechnology firm Chiron Corporation, "is a psychological one: People eat bread and drink beer."

They do indeed. American brewers sell more than five billion gallons of beer a year, bakers more than 10 billion pounds of bread. That is the labor of roughly 13,000 tons of yeast, on the order of 262 quadrillion cells. Yeasts, one scientist joked, make up the largest nonunionized work force in the world.

Ever since it was discovered 150 years ago that yeasts are living creatures, brewers have taken a certain interest in biology. And now that yeasts are becoming one of biotechnology's leading workhorses, biologists have taken a certain interest in brewers. "One thing that people in the brewing industry understand is how to grow *lots* of yeast," says Graham Stewart, director of research and quality control for the Labatt Brewing Company in Canada, which boasts the largest laboratory in North America devoted to industrial yeast research. Although Labatt scientists have helped genetic engineers grow special yeasts in industrial-sized cauldrons instead of Petri dishes, mostly what they've been thinking about is how to brew the perfect diet beer.

What that amounts to is developing a

Genetic engineers now look for genes in diverse species in the manner of mechanics prowling for spare parts.

strain of yeast that will digest more of the carbohydrates in barley malt so beer drinkers won't have to. During fermentation, a yeast cell breaks down a starch or sugar into carbon dioxide and ethanol—with a dozen or so chemical reactions in between, each catalyzed by a specific enzyme. Of the 500 species of yeast, many can break down more of the complex carbohydrates in barley malt than *Saccharomyces cerevisiae*, or brewer's yeast. So Labatt researchers are mating the common brewer's yeast with more voracious strains. One of these hybrids has already fermented a beer with fewer calories, but it didn't get past Labatt's taste panel, two groups of scientists who meet at 10 and at 12 o'clock each weekday morning to sniff and swallow their experimental results. "It's not that these beers taste bad," Stewart says. "They just don't *match* our established beers."

Crossing cells and waiting for what the hybrids will bring seems quaint, reminiscent of Gregor Mendel crossing peas in his monastery garden more than a century ago. Today, researchers can pluck a gene for a specific trait from one organism and transplant it into another, since the genetic code is nearly universal among living things.

Genetic engineers did not create the molecular tools with which they move genes from one organism to another. They simply borrowed them, taking the snipping and mending enzymes that bacteria naturally produce. And their success with these tools can be traced to a fairly recent event.

In 1973, Herbert Boyer of the University of California, San Francisco, and Stanley Cohen of Stanford did the first practical recombinant DNA experiment. They worked with bacteria, but the same methods are used to affix genes into the more complex cells of mice and yeast.

Boyer, Cohen, and coworkers used a DNA-cleaving enzyme to snip from a bacterium, *Escherichia coli*, the gene that carries resistance to the antibiotic kanamycin. With a different enzyme, they joined, or recombined, that gene to a small loop of DNA, known as a plasmid, that confers resistance to tetracycline and readily passes into bacteria. They then put this recombinant DNA into a

dish containing millions of *E. coli* cells that were not resistant to either antibiotic. The bacterial cell that happened to take up the foreign DNA and process it became resistant to the drugs—a trait that was simple to detect. When the cells were streaked across plates containing the antibiotics, only the recombinant cell survived, multiplying into a colony of identical cells, or clones.

Using the same approach—snip, join, shuttle, select—Boyer and Cohen next transferred into *E. coli* a gene from a different species of bacteria, and then a toad gene. These feats, Cohen once noted, "represented a breach in the barriers that normally separate biological species."

Armed with a few enzymes, genetic engineers now look for genes in diverse species in the manner of mechanics prowling for spare parts in a junkyard. That is the difference between the new biotechnology—Wall Street's biotechnology—and the old, the vintner's and the brewer's and the baker's. As one Labatt researcher explains, there is even a shorthand for making this distinction. "Biotechnology used to be BBC," he says, "*before* Boyer-Cohen. Now it's ABC—*after* Boyer-Cohen."

Only 10 years ABC, recombinant DNA technology delivered its first pharmaceutical product approved for people— human insulin made by *E. coli* growing in 10,000-gallon vats. Meanwhile, it was proving tougher to move foreign genes into yeast.

The obstacles were many. Yeast cells have tough sticky coats that keep stray things like plasmids of DNA out. And unlike bacteria, yeast's DNA is separated from the rest of the cell by a nuclear envelope—yet another layer between a researcher and a gene. Working at Cornell University, Albert Hinnen, James Hicks, and Gerald Fink crossed the yeast's genetic barrier in 1978.

They stripped off the yeast's cell coat with an enzyme and suspended the naked cells—now vulnerable to foreign DNA—in a solution. With a plasmid shuttle, they moved the gene for a particular enzyme from a normal yeast cell into one that lacked the gene—with telltale results. The gene helps make the nutrient

leucine; yeast without it can't survive without being fed leucine. After adding the foreign DNA to yeast cells, the researchers spread the cells across a plate that lacked leucine. Only the cell that integrated the leucine gene flourished. What's more, a remnant of *E. coli* DNA stayed with the genetically engineered yeast and its clones, tagging them with a kind of molecular badge that the scientists could identify with a chemical probe. "That was *the* important moment," says Elizabeth Jones, a molecular biologist specializing in yeast at Carnegie-Mellon University in Pittsburgh. "It provided absolutely irrefutable proof that foreign DNA can become part of the yeast genome."

Although many laboratories have been quick to take up these methods, one is lagging—on purpose. In an industry where success depends as much on image as substance, Labatt's Stewart doesn't want even remnants of *E. coli* in yeast that he might engineer for brewing diet beer. "DNA is DNA is DNA, wherever it comes from," Stewart says. "But we don't want customers thinking they've got bloody bits of coliform bacteria in their beer." To head off this public relations problem, Labatt researchers are trying to build effective plasmids for shuttling foreign genes into brewer's yeast that are made only of yeast DNA. It'll be at least five years, Stewart predicts, before any brewing company bottles a diet beer made solely from genetically engineered yeast.

This means that brewers, the first biotechnologists, may be the last to take advantage of genetically engineered yeast. Recombinant DNA's first yeast product on the market will be a vaccine against hepatitis B virus, which the Food and Drug Administration is expected to approve later this year. About 200 million people, mostly in Asia and Africa, carry hepatitis B virus. Two out of three develop disease symptoms; one out of four eventually dies. The vaccine now used, which is made of a virus protein purified from the blood of hepatitis carriers, is expensive, and supplies are uncertain.

Several university and commercial laboratories are working on alternative sources for the virus protein. But Merck

"DNA is DNA is DNA, wherever it comes from."

Sharp & Dohme, using yeast engineered by Chiron Corporation to make copies of the virus protein, was the first to run clinical trials with a vaccine. In the first trial, with 37 people, yeast-derived hepatitis B vaccine spurred the immune system as effectively as the current vaccine. Maurice Hilleman, a leader of the Merck group, says their success with yeast came only after their failure to extract a pure protein from *E. coli*. "The first go-round to make the vaccine in bacteria was no good. *E. coli* produced too little protein." Besides, Hilleman says, "growing yeast is simple. It's fermentation chemistry. Goes back to Pasteur."

Louis Pasteur didn't discover yeast's role in fermentation, but he did finally convince scientists and brewers alike of its importance. The term yeast referred originally to the stuff that hisses and foams up a barrel of beer or wine; yeast was thought a lifeless thing, like wheat gluten. But by 1837, three men of science had announced that yeast was— make that were—alive, a mass of individual cells. Improved microscopes let them see cells in the stuff added to make beer. The cells appeared to reproduce. And boiling them stopped fermentation dead. One of the three scientists, Theodor Schwann, named the yeast he saw *Zuckerpilz*, or "sugar fungus," later translated to Latin as *Saccharomyces*.

But the notion that the alcohol and bubbles in beer were waste products didn't suit Germany's most powerful chemists, Justus von Liebig and Friedrich Wöhler. They believed that fermentation was adequately explained by simple chemical equations, like those describing a reaction of vinegar and baking soda. In an article that the English biologist Thomas Henry Huxley later called "the most surprising paper that ever made its appearance in a grave scientific journal," Liebig and Wöhler mocked the idea that a microscope revealed yeast's true nature: "Without going into this hypothesis any further, one sees [alcohol] arising unceasingly from the anuses of the animals, and from their enormously large genitals squirt at very short intervals a stream of [carbon dioxide]."

In the early 1870s, after having settled the questions of why vinegar was sour and how to preserve wine, Pasteur went to work on beer. Published in 1876, his *Études sur la bière* helped settle the controversy over yeast's true identity. The Carlsberg brewery in Denmark put a bust of Pasteur in its laboratory.

It was Pasteur's celebrated style to work on mundane problems that led to sublime new ideas. So while brewers began to use microscopes to keep their vats free of the "ferments of disease"—those wild yeasts and bacteria that sour a brew—biologists learned new ways to identify and grow pure cultures of microbes. These tools, in turn, helped drive the last century's great revolution: the theory that infectious diseases were caused by microorganisms with lives of their own.

As in Pasteur's day, any fresh insights into yeast that biologists find usually get put right to work by industrialists. Within a year—to take a recent case in point— genetic engineers at several commercial firms were busy making use of what Ira Herskowitz and Janet Kurjan had learned in 1982 about a yeast's sex life.

S. cerevisiae reproduce in two ways. The yeast cell can simply divide, giving rise to two identical cells, each with a full set of chromosomes. These daughter cells keep dividing—provided there is ample food around. Or in leaner circumstances, a yeast will sporulate, giving rise to two pairs of cells, or spores. Like a human egg or sperm, each cell, designated either *a* or *alpha*, has half a set of chromosomes and mates only with its opposite type. The courtship has fascinated biologists for decades. When freckle-sized colonies of *a* and *alpha* cells grow near each other in a dish, *a* cells suddenly stop dividing—turned off, as it happens, by a chemical signal, a pheromone secreted by the *alphas*. Arrested, they mate.

Herskowitz, a molecular geneticist at the University of California in San Francisco, and Kurjan, now at Columbia University, fished out and cloned the gene responsible for the mating pheromone, known as *alpha* factor, and also explained how the cell creates it. An *alpha* cell first makes a large protein, or precursor. Enzymes slice this precursor into four active pheromones that are then secreted beyond the yeast's cell wall and into the environment—a vat of nutrient broth, perhaps. In manufacturing terms, the *alpha* factor system is assembly line, packaging plant, and delivery route in one. For its simplicity, says Herskowitz, "It's the best protein-processing system known."

This did not escape the notice of genetic engineers at biotechnology firms, whose business is manufacturing proteins. Their idea was to hitch a foreign gene to the gene for the precursor, so that the desired protein would be secreted out of the cell and into the vat. There the protein would be easy to collect and purify, unaltered by any of the cell's natural protein-digesting enzymes. So shortly after Herskowitz and Kurjan explained exactly how an *alpha* cell makes its pheromone, biotechnologists were exploiting the system to make human interferon and interleukin, natural components of the immune system that may be used to treat cancer and AIDS; and beta-endorphin, a human brain peptide and a possible painkiller. And yeast can be engineered to secrete rennin, an enzyme from the fourth stomach of an unweaned calf, which is used by cheesemakers to coagulate milk.

Bacteria, on the other hand, are not such realistic surrogates for mammalian cells. *E. coli*, for example, can't yet make human insulin, normally secreted by the pancreas, in one step. Instead, genetically engineered *E. coli* make two pieces of the human insulin molecule that chemists must then join together.

What really impresses genetic engineers in the food industry, though, is that yeasts don't require much fuss. "We don't even have to purify the rennin all the way because yeast are on the FDA's generally-regarded-as-safe list," says David Botstein, a molecular biologist at MIT and a consultant on the rennin project. "There was never any question about using yeast for this."

But yeasts can do far more sophisticated jobs than standing in for a calf's stomach lining. They are now revealing how human disease may arise when individual genes misbehave. Dauntingly complex, the role of genes in the origins

Yeasts are now revealing
how human disease
may arise when
individual genes misbehave.

of cancer is being simplified by putting suspected human cancer genes in yeast and watching how the cells grow.

Of the many genes implicated in cancer, one family of genes, known as *ras*, has attracted special attention. Researchers have cloned three human *ras* genes and identified the proteins they manufacture. While researchers aren't sure how the *ras* genes are involved in cancer, they have found a mutation in a *ras* gene in 10 to 20 percent of malignant human tumors. In 1983, Edward Scolnick of the Merck Sharp & Dohme Research Laboratories in Pennsylvania and Michael Wigler of Cold Spring Harbor Laboratory discovered genes nearly identical to human *ras* genes in yeast.

In a recent study done in Wigler's lab, researchers snipped out a *ras* gene from yeast cells and inserted in its place a human *ras* gene. The cells germinated as usual, indicating that *ras* genes function alike in humans and yeast. Next they chemically altered a yeast *ras* gene in a way that corresponded to a mutated, and perhaps cancer-causing, human *ras* gene. This time the cells didn't germinate, and like human cancer cells, those with a mutated *ras* gene failed to respond to chemical cues that regulate their growth. "In human cells, cancer seems to happen when the cells start ignoring signals that say, 'Don't go haywire,' " says Scott Powers, a researcher in Wigler's lab. "Normally yeast sense their environment very well and respond to it. But these yeast with mutated *ras* genes—it's like they became stupid. We're now left with the job of seeing how analogous this is to human cancer."

The similarity between yeast and human genes, however, doesn't stop with the *ras* gene. The *alpha* sex pheromone made by yeast closely resembles a protein hormone involved in ovulation. And *S. cerevisiae*, besides making an almost identical version of the muscle protein actin, also makes an estrogenlike steroid. Researchers don't know how all these substances function in yeast. But their discoveries help explain why genetically engineered yeasts are a leading industrial workhorse. "Yeast have all the molecular equipment for making human proteins," says James Hicks. "They just have to be taught how to use it."

Despite all the talk of molecular machinery and equipment, precision tools, and DNA processing, genetic engineers work with far less certainty than their counterparts who build bridges and engines. A few years ago, for example, molecular biologists believed that genetically engineered yeast would routinely be able to attach to mammalian proteins the sugars needed for some proteins to work. (Bacteria can't do it at all, researchers believe.) Proteins that aid blood clotting, for instance, require sugar additions. But yeasts, more often than not, fail to add such sugars, or fasten them in the wrong place. So many genetic engineers are banking more on cultured mammalian cells as the source for complex therapeutic proteins. Then again, sugar additions aren't always necessary. Surprisingly, the yeast-derived hepatitis B vaccine made by Merck is fully effective even though its active ingredient lacks the sugar found on the native virus protein.

Gene splicers have also been disappointed to find that yeast can be stingy, producing a mammalian protein that amounts to no more than eight percent of its protein output. The simpler *E. coli* can crank out a mammalian protein that's 40 percent of its total output. "What I see is a shift toward cultured mammalian cells for producing pharmaceuticals—for things that might need sugar linkages, like blood factors," says Donald Moir, director of molecular genetics for Collaborative Research in Massachusetts. "Otherwise it comes down to a race between yeast and *E. coli*. Is it going to be better to make a product in *E. coli*—at 20 to 40 percent production level—but then have to process the protein in ways that are expensive? Or will it be more economical to make a five percent protein in yeast—but have it fully active already and easy to purify?"

Faced with these trade-offs, biotechnologists are going their own ways. California Biotechnology is said to be cutting back or shutting down its yeast projects, while Chiron is building its up. And ZymoGenetics, a Seattle-based firm, devotes the majority of its research to yeast. As Botstein says, "Each product is a separate problem that will find a separate solution."

University scientists, meanwhile, are having an unabashed love affair with yeast. "Yeast is now the premiere organism to work on," says Leland Hartwell, a geneticist at the University of Washington. The scientific literature on yeast is blossoming. Organizers of symposia about yeast are turning people away for lack of lecture hall seats. Hicks sees biotechnology as a great boon for basic science: "The more the industrial types use yeast, the more interest and money for research." Herskowitz, like a cardsharp holding a hot hand, slyly understates the case. "Yeast," he says, "are going to be *very informative*."

Science Debates Using Tools to Redesign Life

Keith Schneider

Special to The New York Times

WASHINGTON, June 7—Genetic engineering, the most powerful and precise biological tool for manipulating life ever devised, has reached a milestone.

Fourteen years after scientists first spliced genetic material from one microbe into another to create a bit of life that never before existed, genetic alterations once confined to science fiction are becoming ever more common.

Now the United States Patent and Trademark Office has ruled that genetic engineers may patent higher life forms—even mammals. The decision promises to widen vastly the commercial and agricultural applications of novel methods of producing new kinds of life.

Industrial leaders say they must be able to patent new life forms and processes if they are to protect their investments and move forward in a field full of innovation and risk. But the patent office ruling has also revived anxiety about the safety and morality of tampering with life forms.

Of Things to Come

That concern prompted a Congressional committee to schedule hearings this week on ethics and regulations in the field of genetic engineering. Last month, the Senate approved a measure that would prevent the Patent Office from spending money on reviewing patents for animals, but it still faces a conference committee vote.

In the near future biotechnology may see these developments:

• In laboratories across the country, the genes of viruses and bacteria will be placed in plants to enable them to produce their own insecticides or fertilizers. These so-called transgenic plants will be field-tested and farmers will begin using them in place of conventional crop varieties.

• Researchers will manipulate the primordial cells that produce sperm and eggs to enable breeders to select the characteristics of animals, including gender.

• Scientists will routinely transplant genes from one species to another.

A Rust-Colored Pig

As the debate unfolds, many eyes will turn to a rust-colored pig in Beltsville, Md., with the growth hormone gene of a cow. That pig represents success to the genetic engineers and, because of its pathetic infirmities, new reason for concern to those who fear that mankind now has too many tools for meddling in the complex matter of life.

In recent months most of the concern about genetic engineering centered on the release into the environment of newly devised organisms in the form of bacteria designed to help plants resist pests, diseases and bad weather. With the new patent ruling, however, the concern has begun to shift to more complicated genetic manipulation of higher life forms—mammals—resulting in transgenic creatures like the pig with a cow gene.

In the long run, opponents and proponents of genetic engineering see a vast array of potential applications, including plants and microbes designed to produce fuel; cows that produce medicines instead of milk, or even babies destined to have a particular height, hair color or other traits.

The genetic traits of plants and animals have been manipulated for centuries. But until now animal breeding and the hybridization of crop plants have been slow, cumbersome and difficult. Furthermore, until now, breeders were never able to introduce genes from one species into another or to make such extensive changes.

Because many breeding techniques, such as artificial insemination, in vitro fertilization and embryo transfer, have already made their way into medicine, there are some who fear it may not be long until the manipulation of animal traits will extend to human traits as well.

"Important legal, constitutional, and policy issues were raised by this decision," said Representative Robert W. Kastenmeier, a Wisconsin Democrat who heads the House Judiciary Subcommittee on Courts, Civil Liberties, and the Administration of Justice, which will hold the hearing Thursday on the Patent Office ruling.

The commercial applications of genetic engineering are already apparent. Sales of genetically engineered products, most of them new pharmaceuticals, have almost doubled annually in recent years and topped $350 million last year, according to industry analysts. The Congressional Office of Technology Assessment has identified almost 400 companies seeking to develop products with genetic engineering and other modern biological technologies. More than $3 billion, two-thirds of it provided by the Government, will be invested this year in biotechnology research, according to the General Accounting Office and industry analysts.

Yet as the ambitions and accomplishments of genetic engineering increase, awareness of its power and potential is generating a mixture of fascination and hope, aversion and misunderstanding.

A survey of 1,273 American adults published in May by the Congressional technology office found that while a majority of those interviewed believed that the potential benefits of genetic engineering outweighed its risks, they were disturbed by some applications, particularly the release of manufactured life forms into the environment and manipulations in human embryos.

'SOMETHING WONDROUS, AND PERHAPS PERILOUS'

"People understand at a gut level that there is something wondrous, and perhaps perilous, about a technology that changes the blueprint of life and will force us to make choices that are likely to be more profound than anything we, as a society, have ever faced," said Senator Albert Gore Jr., a Tennessee Democrat who has studied the biotechnology industry.

Though scientists generally agree the field offers great promise, there is sharp disagreement over its potential perils.

"We are bringing a completely human-centered utilitarian attitude toward life," said Dr. Michael Fox, a veterinarian and scientific director of the Humane Society of the United States. "All of earth's living things will simply become items to exploit."

Other scientists and many biotechnology industry executives insist that genetic manipulation will hasten the development of cures for diseases like AIDS, lead to solutions for toxic chemical pollution, produce a new agricultural cornucopia and open an industrial era based not on fossil fuels and chemicals, but on new, non-polluting substances produced by genetically engineered plants or microbes.

Genetic engineering was recognized as a momentous development in 1973, when Stanley N. Cohen of Stanford University and Herbert W. Boyer of the University of California at San Francisco snipped a piece of the genetic code out of one bacterium and inserted it into another.

But that experiment was followed almost immediately by a host of safety and ethical questions, many of which remain unresolved. Are living, gene-altered microbes safe to release outdoors? What is the best way to assess the risk from such uses? Is it ethical to alter the genetic codes of animals? What about people? How can a society know whether a new technology should be pursued or ignored?

"The issues range from ethics within universities, to the environment, to eugenics, to definitions of nature, to religious thought, to what it is to be human," said Dorothy Nelkin, a professor in Cornell University's Program on Science, Technology and Society. "Other disputes over technology have been much simpler and mostly focus on health concerns."

Rearranging Gene Chemicals

The source of the excitement and the conflicts is a technique, conceptually simple but in practice quite complex, for rearranging basic hereditary material: the deoxyribonucleic acid, or DNA, that makes up genes.

DNA molecules are long, twisted ladders of chemicals called nucleotide bases: adenine, thymine, guanine and cytosine. More than 30 years ago, scientists determined that adenine always pairs with thymine, and cytosine with guanine. These chemical connections are called base pairs; a single gene, a section of DNA, is typically made up to 10,000 to 20,000 base pairs. Human beings, it is estimated, have between 100,000 and 200,000 genes, or up to 4 billion base pairs, organized on 46 chromosomes.

Though the numbers of genes in mammals, plants, and microbes differ, their ladder-like molecular structure does not. Scientists are now able to identify and isolate specific genes and remove them with proteins, called restriction enzymes, that slice DNA in specific places. The enzymes cause the pairs on either end of the gene to split, leaving nucleotide bases without corresponding mates. Scientists call the unpaired bases "sticky ends" because, seeking the correct chemical fit, they easily merge with another organism's genetic structure.

CURRENT LIMITATIONS AND POSSIBILITIES

Yet simply isolating a gene from one animal and plugging it into another does not mean that the gene will produce the desired result. A gene's functions are determined by its location on a chromosome, the workings of neighboring genes and other factors that are still mysteries.

So far, genetic engineers are largely limited to transferring single genes into microbes, plants and animals, or taking single genes out of bacteria and viruses. Alterations involving more than one gene, such as creating crops that produce their own insecticides and fertilizer, or cows that produce medications in their udders instead of milk, are still years away.

Assertions that genetic engineering will produce unrecognizable plants or monstrous animals are considered by many researchers to be scientifically absurd.

"There are severe limits to the extent of the modifications we can make," said Dr. Bernard D. Davis, a microbiologist at Harvard Medical School. "If you mix genes from genetically distant organisms that don't fit each other well, you will not have an organism that can live."

"We're not going to make weeds out of non-weedy species," said Dr. Winston Brill, vice president of research and development at Agracetus, a plant biotechnology company in Middleton, Wis. "We're not going to have Frankensteins crawling around."

Nevertheless, transfers involving a single gene can yield striking physiological changes.

A Pig Unlike Any Other

For example, the transgenic pig, a rust-colored boar born last November at the Department of Agriculture's experiment station in Beltsville, now weighs as much as its natural cousins; unlike them little of its bulk is fat. But it has trouble walking on short legs swollen by arthritis. Its eyes, peering from a broad and wrinkled face, are slightly crossed. If it is like its father, who was one of the world's first transgenic farm animals, it will not live to be two years old.

Nothing about producing transgenic animals is easy. Genes are injected into fertilized animal eggs. Piercing cell walls kills between half and three quarters of the eggs, said Dr. Vernon G. Pursel, the research physiologist conducting the swine experiments. In four years, scientists injected more than 8,000 fertilized eggs to produce just 43 transgenic pigs.

It is little wonder, then, that researchers at Beltsville consider the birth of the rust-colored pig to be a scientific success. The young boar inherited the gene that scientists inserted into its father, and the gene expressed itself. Scientists are now working to control the gene so that it produces animals that grow fast, eat less, and produce more lean meat, without the complex of crippling diseases afflicting the boar.

The Foundation on Economic Trends, a small public policy group that opposes genetic engineering, and the Humane Society of the United States unsuccessfully filed suit in Federal District Court three years ago to halt the research that produced the rust-colored boar's father. They said the research was cruel, violated the innate dignity of animals and would have significant social and economic effects by producing bigger, more expensive animals that would cause dislocations in the farm economy.

"That kind of scientific reductionism undermines the respect for life and future generations will come to regret it," James Rifkin, president of the Foundation, said recently.

The two groups are also protesting the new Patent Office policy. In this battle they are joined by farm organizations, consumer groups, environmental groups and most major animal welfare groups.

"Farmers believe they will be facing fewer choices in terms of breeds available to them and will be paying far more for animals," said R. Keith Stroup, legal counsel for the League of Rural Voters,

43 Years of Advances in Altered Life Forms

Keith Schneider

Special to The New York Times

WASHINGTON, June 7—The advances that make genetic engineering possible began in 1944, when Oswald T. Avery, Colin MacLeod and Maclyn McCarthy, researchers at Rockefeller University in New York, determined that deoxyribonucleic acid, or DNA, carried the hereditary blueprint of all living things. The same year, Congress passed the Public Health Service Act, which provided grants to universities for medical and biological research.

The work at Rockefeller University and the new funds for exploring its implications ignited a field where the pace of discoveries, slow at first, has accelerated to the point at which important findings are being announced almost weekly.

A major riddle was solved in 1953, when James D. Watson and Francis H. Crick found that DNA was chemically organized in two strands, a double helix. Their findings, which earned them a Nobel prize in 1962, enabled researchers to understand how DNA worked. By the mid-1960's, biochemists and molecular biologists were able to purify fragments of DNA, tag the molecules with radioactive isotopes and analyze the fragments.

1973: Genetic Engineering

In 1970, scientists developed proteins, or restriction enzymes, that cut DNA strands in precise locations. Three years later, Stanley N. Cohen of Stanford University and Herbert W. Boyer of the University of California at San Francisco used the restriction enzymes to isolate fragments of DNA in one bacterium and insert it into another. Genetic engineering was born.

Scientists reacted quickly. In September 1973, and again in July 1974, several of the country's leading biologists warned in eminent journals that the new gene-splicing techniques might present novel hazards. They asked scientists to delay research until the dangers of genetic engineering could be more carefully evaluated.

Three months later the National Institutes of Health established a committee to monitor biotechnology research and to consider new research rules; the following February, 139 scientists from 19 nations met in California to draw up new research guidelines to minimize the potential that gene-altered microbes could escape from laboratories. The guidelines were adopted by the N.I.H. in 1976.

By then, investors and industrialists had seen the commercial possibilities of the new field. In 1976, Genentech in South San Francisco, Calif., became the first company established to commercialize the new technology.

In the early 1980's, pharmaceutical makers, chemical companies and other concerns signed university researchers to long-term contracts aimed at quickening the pace of research with commercial potential. Some scientists and scholars cautioned that such relationships could alter the direction of research and stifle the free exchange of scientific information. But the lucrative contracts proved worthwhile for many universities, companies and scientists.

Oil-Eating Microbe

The first important court decision on genetic engineering came in 1980, when the United States Supreme Court, overturning a policy of the Patent and Trademark Office, ruled 5 to 4 that an oil-eating microbe developed by a General Electric researcher could be patented. Bolstered by this ruling, the pace of discovery quickened.

In 1981, the Food and Drug Administration approved an application by Eli Lilly & Company to market the first drug made from genetically altered bacteria, a form of insulin.

On Jan. 16, 1986, the Department of Agriculture granted the Biologics Corporation of Omaha, Neb., the world's first license to market a living, genetically engineered microorganism, a virus used as a vaccine to prevent a herpes disease in swine. It was tested in Lometa, Tex., in June 1984.

On May 29, 1986, Agracetus, a biotechnology company in Wisconsin, was the first to field test a genetically engineered crop plant, a genetically altered form of tobacco it planted at an undisclosed site 20 miles from its headquarters.

1986 Biotechnology Law

On June 19, 1986, President Reagan signed into law a coordinated set of rules and regulations governing the testing, use and sale of the products of biotechnology. Though some scientists have said some sections of the law are based on incorrect theories, the program is serving as a model for biotechnology regulation in other countries.

On April 16, the United States Patent and Trademark Office decided to extend patent protection from microbes to higher life forms—even mammals.

And on April 24, 1987, after four years of delays because of legal challenges by Jeremy Rifkin, a public policy activist involved in biotechnology issues, the first free release of genetically engineered bacteria was conducted outside Brentwood, Calif.

a family farm advocacy group that opposes the policy. "There's another issue here, too. Most farmers who deal with animals on a day-to-day basis want to be very thoughtful and careful about tampering with life. Clearly we have not explored fully the repercussions, morally and ethically, of what these patent applications seek to do."

Fifteen applications have been filed, but the office does not release descriptions of patent proposals until they are approved.

Dr. Pursel said he was sensitive to the protests but unsure how to respond. "We are not doing anything out here that is cruel," he said, adding, "The research could have a tremendous practical value."

Less Invasive Applications

Other applications of genetic engineering technology are less invasive and also less divisive. Scientists have discovered several methods for moving bacterial and viral genes into plants to make them more resistant to insects and disease. Field tests were conducted last year in Wisconsin, North Carolina and Mississippi and this year in Missouri without protest.

Researchers at the University of California at Davis are working to develop plants that produce their own fertilizers. Around the country, scientists are studying genetic manipulations to increase the rate of photosynthesis to hasten plant growth.

In the pharmaceutical industry, four drugs produced by genetically engineered bacteria have been approved by the Food and Drug Administration. The agency has been asked to approve another new drug made by genetically altered bacteria, tissue plasminogen activator, to be used in treating heart attacks. Its manufacturer, Genentech of South San Francisco, predicts that sales of the drug could reach $800 million a year. Genentech has also inserted human genes into bacteria to produce insulin, growth hormone, and alpha-interferon, which is used to treat hairy cell leukemia.

Until recent weeks, the most raucous battle in biotechnology focused on releasing living, gene-altered bacteria outdoors. Two field tests in California in April had been delayed three years by protests and challenges and then were marred by vandalism as a result of initial scientific concerns that the microbes would spread beyond the test site. They did not.

IMPLICATIONS FOR CHANGE IN HUMAN GENETIC CODE

The conflict took a new turn in April after the Patent Office decision to patent new forms of animal life, a ruling some opponents of genetic engineering characterized as a rifle shot at the human genome.

It is already possible to detect some of the 3,000 genetic diseases. It will be possible, scientists say, to cure genetic disorders by removing or adding genes.

But what should the proper limits be for intervening in the human genetic code? Is asthma a genetic disease, and should that be cured genetically? What about baldness? And if diseases are curable genetically, that may also mean that traits could be added to the human genetic code, such as hair color or height.

Since 1975, the National Institutes of Health has refused to provide funds for research involving human embryos or fetuses. Senator Gore, vice-chairman of the two-year-old Congressional Biomedical Ethics Board, said the panel was planning hearings on whether such grants should be reinstated.

Like every other twist in genetic engineering, the patenting of higher life forms is certain to heat emotions and generate contention among scientists. The technology is powerful and complex; if allowed to blossom, it could enable people to make direct genetic choices for their offspring and begin to determine human evolution.

The gene revolution

Bernard Dixon

BERNARD DIXON, British science writer and consultant, is European editor of The Scientist magazine and was formerly (1969-1979) editor of the British scientific journal The New Scientist. Notable among his published works are Magnificent Microbes (1976), Invisible Allies (1976) and (with G. Holister) Ideas of Science, Man and Medicine (1986).

GIVEN the mixture of benefits and problems spawned by the first Green Revolution two decades ago, it is not surprising that both optimism and apprehension surround the application of genetic engineering now to agriculture tomorrow. Mixed reactions *are* appropriate, because those developments—focused upon so-called recombinant DNA—are destined to have even more far-reaching effects than the techniques deployed in the first revolution. Today's new wizardry could undoubtedly transform agriculture throughout the world. At the same time, its precision in modifying living cells offers a stern challenge to our prudence and wisdom.

At the centre of the stage is deoxyribonucleic acid (DNA), the material which carries in coded form the hereditary instructions responsible for the behaviour of cells and the plants, animals or microbes of which they are part. The astronomically long DNA molecule can be subdivided into regions—genes—which determine particular characteristics. Recombinant DNA is the name given to the product when a piece of DNA from one organism is combined artificially with that from another.

Genetic manipulation of this sort is the basis for the boom that has occurred during the past decade in biotechnology. Such activities were, of course, possible previously. Some, like the art of fermenting sugar to make alcoholic drinks, are almost as ancient as Man himself. Others, including the first mass production of antibiotics, were developed earlier this century. But all of these processes were based on organisms as they occur in nature—albeit with other, equally natural, methods being used to select high-yielding strains.

The arrival of recombinant DNA, however, has altered the rules profoundly. It has already greatly enhanced our specificity and power in tailoring living organisms for beneficial purposes. In future, it will extend our range of options much further.

The breakthroughs which have led to this historic watershed in the fabrication of novel plants and microbes happened during the early 1970s. The key discoveries were made by molecular biologists who learned how to splice into bacteria genes which they had taken from other bacteria, and even from totally unrelated animal or plant cells. They first found out how to locate the particular gene they wanted among the countless numbers on the DNA of one organism. Then they used natural catalysts called enzymes to cut out that gene and "stitch" it into a *vector*. This is usually a virus or a plasmid (a piece of DNA that replicates independently from the nucleus, the main repository of DNA). The vector became a vehicle for ferrying the selected DNA fragment into the recipient. Once inside its new host, the foreign gene divided as the cell divided—leading to a clone of cells, each containing exact copies of that gene.

Because the enzymes used for genetic engineering are highly specific, genes can be excised from one organism and placed in another with extraordinary precision. Such manipulations contrast sharply with the much less predictable gene transfers that occur in nature. They also make it possible to splice genes that would be unlikely to come together naturally. By mobilizing pieces of DNA in this way, genetic engineers are beginning to create pedigree microbes for a wide range of new purposes in agriculture, medicine and industry.

Although genetic manipulation is taking longer to perfect in plants, several techniques are now emerging. The most useful so far is based on *Agrobacterium tumefaciens,* a bacterium that causes crown galls on many flowering plants. It contains a tumour-inducing (Ti) plasmid which is responsible for triggering the disorderly growth that appears as ugly galls. Genetic engineers have learned how to delete the Ti plasmid's tumour-inducing genes and use it as a vector with which to carry new genes into plants.

A serious drawback so far is that while *A. tumefaciens* infects potatoes, tomatoes, and many forest trees, it does not normally attack the monocotyledons such as cereals, which are prime targets for genetic improvement. Progress is being made, however, and recent research indicates that rice in particular can be manipulated using the Ti plasmid. Alternative vectors and other methods of transferring genes are also being developed. One exciting possibility is to use an electric current to promote the incorporation of foreign DNA. This works with maize cells, though scientists still have to persuade the cells to develop into whole plants.

One gene that has been transferred into tobacco by *A. tumefaciens* comes from bacteria and gives the plants the capacity to produce a toxin that is lethal to insects. The inbuilt insecticide makes the plants resistant to insect attack and does not, of course, have to be applied repeatedly. Some plants can mobilize defences against virus infec-

tion through a process analogous to immunization in animals, and this suggests another route for genetic alteration. Incorporation of one virus gene into tobacco has helped to protect this plant against subsequent inoculation with the entire virus.

Another development concerns weeds — a major limitation on crop husbandry in most countries. Although weeds can be combatted using selective herbicides, these often impair the growth of the crop too. It is now possible, however, to introduce resistance genes into tobacco and petunia. One such manipulation results in the synthesis of enzymes in the plant that are no longer sensitive to the inhibitory action of the herbicide glyphosate. Commercial companies now plan to market a package containing both herbicide and resistance seed.

Some 70 per cent of the world's intake of dietary protein consists of cereal grains and seeds of legumes. On their own, neither cereals nor legumes can provide a balanced diet for human consumption, because the "storage proteins" they each contain are deficient in one or more amino acids. Now, added to analyses of the proteins in both cereals and legumes, we have precise information about the DNA sequences coding

for them. This knowledge may well lead to methods of altering those sequences or introducing new genes that code for a more balanced spectrum of amino acids.

The world's energy and food supplies rest upon the ability of green plants to convert atmospheric carbon dioxide into carbohydrates, fats and proteins, using light from the sun. Unfortunately, the mechanism by which they consume carbon dioxide is inefficient in those plants, such as wheat, barley and potatoes, that are cultivated in temperate climes. Oxygen in the atmosphere interferes with the first enzyme involved in the assimilation of carbon dioxide. Considerable efforts are now being made to alter the DNA sequence of the gene coding for this enzyme, to prevent the deleterious action of oxygen. Other researchers are trying to introduce into temperate zone plants certain genes taken from maize, which has a more efficient mechanism of carbon dioxide uptake. In nature this mechanism appears to operate only at higher temperatures, but there are hopes of "switching it on" in cooler areas.

Another atmospheric gas is the subject of parallel efforts to make plants more efficient. Nitrogen constitutes 80 per cent of the air, yet plants cannot use the gas

How to recombine DNA

Drawing shows how a micro-organism (in this case a bacterium) is manipulated to make it synthesize a desired substance. (1) A bacterium contains a plasmid, which is a circular piece of DNA. This plasmid is isolated (2) and, with the help of a restriction enzyme, opened in a precise spot (3). Meanwhile, with the help of other restriction enzymes, the gene for synthesis of the desired substance is isolated from the DNA of another organism (4). Still using enzymes, this gene is grafted onto the previously opened plasmid (5). The plasmid is re-introduced into a bacterium (6). The manipulated bacteria are put into a culture, where they synthesize the desired substance. (7)

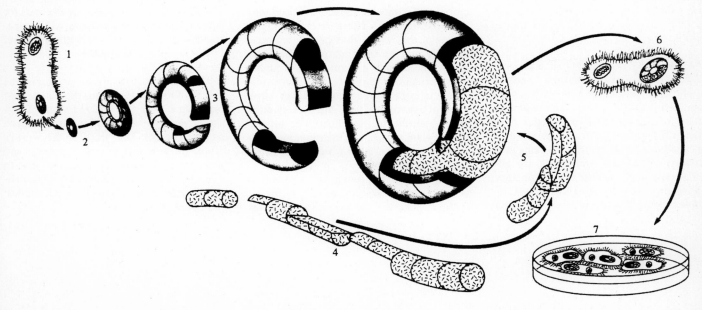

Source FAO-Ceres, Nov-Dec 1984

directly. Hence the heavy dependence of modern intensive agriculture on fertilizers—nitrate, ammonia or urea—synthesized by the chemical industry. Natural nitrogen fixation depends in part on rhizobia, bacteria that live symbiotically with legumes such as peas, beans and clover. The bacteria grow on sugars provided by the plant, and are maintained in characteristic nodules on the plant. There they convert nitrogen directly into ammonia, leading in turn to the synthesis of plant proteins.

Molecular biologists have now isolated and characterized several of the genes required for nitrogen fixation. They have found, however, that many more bacterial *and* plant genes are involved than they first imagined. This makes the manipulation of those genes correspondingly more difficult. So it will be some years before we can enjoy the cost and energy savings that should accrue by providing crops such as wheat and maize with the ability to fix their own nitrogen.

Drought and high temperatures are unwelcome to all plants, despite being better tolerated by varieties that have evolved in such environments. Desiccated soils also often contain high levels of salts and metallic elements, which are toxic to plant growth. Genetic engineers would dearly like to fabricate plants resistant to such stresses, but success is unlikely in the near future. Before being able to identify the relevant DNA sequences for transfer between plants, they require a far better understanding of the many ways in which plants respond to their environment. An additional problem may be the involvement of several different genes, as with nitrogen fixation. Drought resistance which depends on a reduced area of leaf surface, for example, may be caused by the interaction of multiple genes.

Microbes that contribute to healthy plant growth are also on the drawing board for genetic engineering. One possibility being examined is the production and deliberate release of rhizobia that fix nitrogen more efficiently than natural strains. Other bacteria capable of forming nitrogen-fixing partnerships with wheat and maize are also being considered. A third type of prospect follows the discovery by researchers at the University of California, Berkeley, that frost damage to strawberries is triggered by bacteria which act as nucleii for the formation of ice crystals on leaves. The cause is a particular bacterial protein, the gene for which the California biologists have learned to delete. They believe they can prevent the extremely costly frost damage by spraying crops with this "ice minus" strain, which will outgrow the natural flora.

Genetic engineering holds considerable promise too in the improvement of "biological insecticides", microbes that attack pests and have enormous ecological advantages over their chemical counterparts. *Bacillus thuringiensis,* for example, has been used for many years to combat nuisance species, but it and similar bacteria and viruses may well be made more powerful by recombinant DNA. One possibility is illustrated by the pine moth, which damages lodgepole pine trees in northern Britain. In other parts of the country, the moth is controlled naturally by a baculovirus that infects the caterpillars. There are now plans to make the virus more efficient at killing caterpillars and to release it in the pine plantations. The first experiments are being carried out with a virus that has been altered only by having a "marker" introduced into a non-coding region of its DNA. This will allow researchers to follow the virus's distribution and survival after spraying. If all goes well, the virus may be given a gene allowing it to synthesize an insect-killing toxin. The potential for this technique in other countries, against other destructive insects, is clear.

The safety of laboratory and industrial activities using engineered organisms is based on the idea of containment. Facilities are graded according to the degree of risk. New questions arise, however, when microbes and plants are to be introduced into the environment. There is concern, for example, that weeds may be created accidentally and be inordinately difficult to eradicate. If such a plant were drought-resistant, herbicide-resistant, *and* frost-tolerant, it might spread quickly over large areas of agricultural land and be very difficult to eradicate. As illustrated by the Kudzu plant in Asia and the water hyacinth in America, even natural weeds can cause considerable havoc.

The prospect of genetically engineered crops themselves becoming weeds is remote, however, because crop varieties cannot compete well with other plants when left unattended. The inherent difficulties of mobilizing plant genes also make it unlikely that unwelcome varieties will be produced accidentally. And there is always the possibility of destroying by fire or other means an engineered plant, released initially in a defined area, that *did* create problems. Nonetheless, field trials with novel plants, particularly crops able to cross-fertilize with weeds, need to be very carefully monitored.

Greater caution still is required with engineered microbes, which would be broadcast in astronomical numbers and be impossible to trace in their entirety should anything go awry. But it is reassuring that no health, environmental or other dangers have been caused by recombinant organisms since they were first fabricated over a decade ago. Moreover, biologists now agree that there is no significant difference between a microbe that has received a new piece of DNA through artificial manipulation and one that has acquired the same DNA fragment through natural mechanisms of gene transfer. Most experts argue that recombinant DNA manoeuvres are intrinsically safer, because they can be vastly more precise and selective. Certain laboratory manipulations are ruled out anyway by *a priori* predictions that they would generate hazardous recombinants.

Many researchers believe that tests with recombinants should always be restricted to closed environments such as greenhouses. But these "microcosms" can never simulate the richness of the natural biosphere. So they can never provide conclusive evidence about an organism's potential safety or performance in nature. The scientific consensus is now for a gradual approach, *a priori* evidence about a released organism's likely behaviour being used as the basis for successively larger trials during which experience and confidence are gathered about how it actually does behave.

There is one other argument against the much-publicized view in the USA that organisms carrying recombinant DNA should *never* be released for purposes such as pest control. One third of the world's crops are now lost through infection and pestilence. It would be foolhardy not to make use of an ecologically acceptable technique capable of achieving even a modest reduction in that toll.

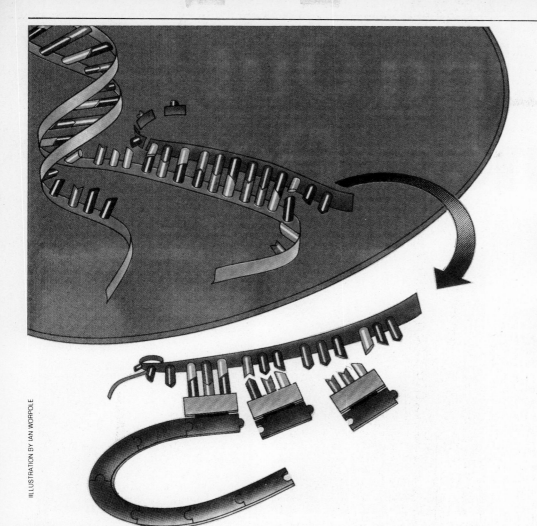

ILLUSTRATION BY IAN WORPOLE

HOW GENES WORK

The wizardry of genetic engineering depends on an understanding of how genes are built and behave. Genes are made of DNA (deoxyribonucleic acid), a molecule that consists of a long chain of subunits called bases. DNA has only four kinds of bases—usually abbreviated A, T, G and C. Their sequence in the chain determines the message they encode. The message of each gene specifies one kind of protein molecule, such as insulin or hemoglobin.

Genes can encode the structure of a protein molecule because both are linear structures. The gene's linear code corresponds directly to the same linear sequence of subunits in a protein molecule.

Each protein molecule is a chain of subunits called amino acids. Once formed, the chain spontaneously folds up into a characteristic shape, ready to do its job. Genes specify a given amino acid by bearing a sequence of three bases. T-T-C, for example, specifies lysine. If the next three bases are A-T-G, the gene is calling for tyrosine to be attached to the lysine.

There are about 20 kinds of amino acids, and the gene's series of three-letter "codons" dictates the sequence in which amino acids are to be chained.

The process starts when DNA's double helix—resembling a twisted rope ladder—splits down the middle, as if each rung had broken in half. The exposed half-rungs, held in sequence along the lengths of rope, each represent one base. After DNA "unzips" in this way inside the cell's nucleus, special molecules in the nucleus align themselves along the exposed sequence of bases, preparing to copy the message so it can be carried out of the nucleus to govern the cell.

These molecules are free-floating subunits, again called bases, of messenger RNA (ribonucleic acid) that bind weakly to the DNA bases, but only in a sequence that corresponds exactly to that of the DNA bases. As the RNA bases line up opposite the DNA, they link to form their own chain, a molecular transcription of the genetic message. This chain then breaks free of the DNA and drifts out of the nucleus to direct protein synthesis in the cell.

Genetic engineers use proteins called restriction enzymes that are known to break lengths of DNA into defined segments. By using the right combination of enzymes, they can snip a gene or group of genes out of the DNA of one species. Using another set of enzymes, they can splice segments of DNA into the existing DNA of an entirely different species. Because the genetic code is essentially the same for all species, genes usually work no matter what cell they find themselves in.

–Boyce Rensberger

Spelling Out a Cancer Gene

JULIE ANN MILLER

The dramatic difference between a normal cell and one that is malignant may be attributable to a very small change in the genetic material. In the gene responsible for malignancy in cells derived from a bladder tumor, the only abnormality is the substitution of one nucleotide (the basic subunit of DNA), according to the work of three independent groups of scientists. This genetic change is expected to alter one amino acid near the beginning of one cellular protein.

The scientists were surprised both that the genetic difference is so small and that it falls within the DNA that encodes a protein. They had expected the change to affect a region involved in turning the gene on and off. "It is hard to understand how such a simple [genetic] change can cause so dramatic a change in a cell's development," says Mariano Barbacid, one of the scientists involved.

This detailed analysis of a gene arises from the intersection of two major aspects of cancer research. One approach focuses on a group of viruses (called retroviruses) that cause animal tumors. The other work centers on the genetics of human tumor cells.

Together these approaches, with recombinant DNA and gene transfer techniques as tools, are providing a new view of cancer at a very detailed level. For certain genes a change in content, location or number of copies present in a cell has been implicated in malignancy. The question remains how to fit these genetic events into a scheme that could explain the multi-step development of human cancer.

Investigation of viruses has gone in and out of the spotlight on cancer research during the last 70 years. The first cancer-causing virus was isolated from chickens in 1911 by Peyton Rous. However, the simple idea that most human cancer can be explained as a viral infection turned out to be unlikely, and much research turned in

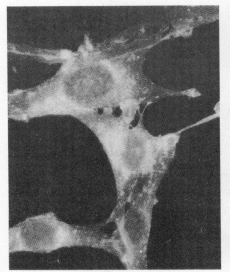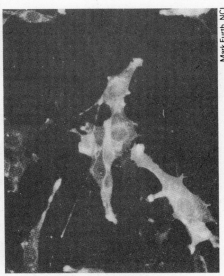

Mark Furth, NCI

A normal rat gene and a normal human gene attached to control sequences from an animal tumor virus can make cells malignant. In these transformed mouse cells growing in laboratory culture, the fluorescent stain binds to the protein product of the rat (left) and human (right) oncogene.

other directions. More recently attention has returned to these viruses because they can provide clues to discovering human genes that appear to be involved in some cancer.

An irony brought viruses back into the mainstream of cancer research. Scientists have discovered that the genes contained in certain viruses — the genes that allow the viruses to trigger rapid malignant growth —had been captured from normal cells (SN: 5/26/79, p. 344). The normal cellular genes, called cellular oncogenes or, more cautiously, proto-oncogenes, are conserved over great evolutionary distances. Very similar proto-oncogenes can appear in human, rodent, bird and even fruit fly cells. The wide distribution of each of these genes (almost 20 have been identified) suggests it plays some basic role in the normal functioning or growth of a cell.

The gene from the normal cell not only resembles the cancer-causing gene, but in

at least three cases it can be made to act in the same way. George Vande Woude of the National Cancer Institute first demonstrated that a cellular oncogene, hooked up to a control region of viral genetic material, can transform normal cells to a cancerous state. So within each animal cell are genes that can cause malignancy if they are put under a different set of controls.

Vande Woude says that this finding may explain how a variety of genetic insults are implicated in cancer. They may all change the chromosomes in such a way that an oncogene becomes active. Viruses, chemical carcinogens and other agents can alter the sequence of nucleotides either in protein-coding or control regions of chromosome. Or they can move pieces of DNA within or between chromosomes. Such changes, he says, may directly activate an oncogene or turn on its control switch. They may also move a gene away

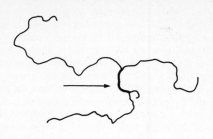

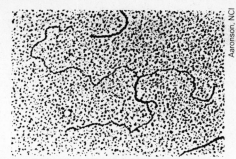

A piece of human DNA (on right in electron micrograph) binds to DNA from a rodent tumor virus. They attach at the region (arrow), 650 paired nucleotides in length, of the oncogene they hold in common.

from its normal control elements, putting it under an inappropriate set of controls. Thus, activated oncogenes may underlie a variety of malignancies.

A distinct line of cancer research begins not with viruses but with human cells that have already turned malignant. In most cases of human cancer, no virus seems to be involved. Robert A. Weinberg at Massachusetts Institute of Technology, Michael Wigler at the Cold Spring Harbor Laboratory and Geoffrey M. Cooper at Harvard Medical School found that DNA taken from certain malignant cells can transform normal cells into cancerous ones, without the influence of any viral genetic material. The DNA taken from the malignant cells thus contains human cancer genes, generally called transforming genes or, again, oncogenes (SN: 9/26/81, p.199).

At least six such transforming genes have been identified. NCI's Barbacid reports that the same oncogene may be present in clinically unrelated tumors, for instance a lung tumor and an embryonic tumor. And tumors with the same clinical diagnosis may contain different oncogenes. Barbacid finds a transforming gene in about 15 percent of human tumors.

The viral and the tumor gene approaches to cancer research were recently brought together by a striking finding. The transforming genes from some human cancers match up with the oncogenes of animal cancer viruses. For example, the transforming gene of the cell line derived from a human bladder cancer is very similar to that of one rodent virus (Harvey murine sarcoma virus) and the transforming gene of several lung and colon cancers is closely related to another rodent virus (Kirsten murine sarcoma virus). These results were reported last spring by researchers at NCI, Harvard Medical School and Cold Spring Harbor.

"The most obvious consequence of these newly discovered homologies is a realization that at least some of the cellular sequence [genetic material] can become activated via one of two routes: either by recombination with a retroviral sequence, or by nonviral, somatic muta-

tional events," Weinberg says in the August 1982 CELL.

Close examination is underway of the 20 or so distinct cancer-causing genes discovered in animal viruses and in malignant cells. Some surprising similarities have already been observed among genes that had been thought to be unrelated. M. Yoshida and colleagues at the Cancer Institute in Tokyo find, for instance, that two avian sarcoma virus genes (called *yes* and *src*) code for proteins that are homologous in 70 percent of their amino acid subunits. A third gene, called *mos,* found in a rodent virus, also seems to fall in this group.

In another group are at least four oncogenes also very similar to each other. This group includes the oncogenes identified in the bladder cancer cell line and in lung and colon cancers. These genes are thought to represent a second family, called *ras,* arising from genes that duplicated before establishment of the modern animal phyla. Esther H. Chang and Douglas R. Lowy of NCI demonstrated that normal human cells contain four different members of this gene family, and other investigators have detected a fifth member.

If the 20 oncogenes fit into a few distinct groups, the problem of discovering their normal cellular functions is simplified. Only a small number of distinct enzymatic functions may be represented by the numerous cancer genes. For one group of oncogenes scientists already know that its product is a tyrosine kinase, an enzyme that chemically alters other proteins by adding a phosphate group to the amino acid tyrosine (SN: 5/26/79, p.344). Another group of oncogenes encodes a protein of 21,000 dalton molecular weight that binds guanosine nucleotides. Weinberg com-

ments, "Maybe the present oncogene jungle is not so impenetrable after all."

An important question is just how the cancer-causing and normal oncogenes differ. Althogh they are almost identical in structure, there is a clear biological difference. In the cases studied, the normal gene, isolated and produced in quantity but unassisted by viral control regions, is far less effective than the gene from a malignant cell in transforming normal cells.

Scientists from MIT and NCI set out to determine the exact difference in the case of the bladder cancer gene studied. They made exchanges of defined segments of the normal gene and the one derived from malignant cells to determine a small region that could confer on a normal gene the ability to trigger malignancy.

In separate laboratories at NCI Ravi Dhar and E. Premkumar Reddy then determined the sequence of nucleotide subunits in this region of the gene. The single essential difference found among the gene's 4,600 nucleotide pairs was a sequence of GGC in the normal gene and GTC in the gene from malignant bladder cells. This change is expected to insert a valine instead of a glycine near one end of the protein encoded by this gene.

In his most recent work, Dhar found the altered gene in cells taken directly from a bladder tumor of a patient. He told SCIENCE NEWS that in this case the altered gene was also detected in blood and other apparently normal tissues of the patient, so the mutation either was inherited or occurred early in embryonic development.

The related gene in three animal cancer viruses also differs from the normal cellular gene at the same location. In each virus the glycine of the protein encoded is replaced with a different amino acid.

This work was performed by three groups. They include Weinberg at MIT, Edward Scolnick now at Merck & Co., Inc., and Lowy, Chang and Dhar at NCI; Barbacid and Reddy at NCI; and Wigler at Cold Spring Harbor.

The scientists were surprised to find the change in such a protein-coding sequence; they had thought it more likely that the change would affect regulation of the quantity of a protein produced. In virally transformed cells production of protein encoded by the viral oncogene is usually greater than production of protein encoded by the homologous gene in nor-

...GTG GGC GCC GGC GGT GTG GGC...
...GTG GGC GCC GTC GGT GTG GGC...

A change of one nucleotide out of thousands is the essential difference between a normal human gene and one that makes cells malignant.

mal cells. This observation had suggested that viral genes transform cells by producing proteins in abnormal quantities. But in the bladder gene case there is a distinct change in the protein, perhaps like the shape change caused by a single nucleotide alteration in sickle cell anemia.

There is recent evidence for changes other than in nucleotide sequence being involved in malignancy. In work on leukemic cells two groups of scientists have found that the number of copies of an oncogene is higher in leukemic blood cells from a patient with promyelocytic leukemia than in cells of normal people. The oncogene, called *myc,* was first identified in a retrovirus infecting chickens.

Steve Collins of the Seattle Veterans Administration Hospital and Mark Groudine of the Fred Hutchinson Cancer Research Center in Seattle report that *myc* is present in multiple copies in a cell line derived from a patient's blood cells. Robert C. Gallo and colleagues at NCI estimate 16 to 32 copies of the gene are present in the same cell line. Gallo also observes extra copies of the gene in cells taken directly from peripheral blood of the patient. However, when they looked at other types of leukemia, the scientists did not find evidence of *myc* amplification.

Chromosomal rearrangement and translocations also seem to have a role in causing cancer. Such gross genetic changes have been linked to many cancers (SN: 9/4/82, p.151; 10/31/81, p.278). An exciting finding reported by two groups last month links a chromosome change with an oncogene. Carlo Croce of the Wistar Institute in Philadelphia and Gallo of NCI have located the *myc* gene on human chromosome 8 in a position that suggests it may be involved in the human immune system cancer called Burkitt's lymphoma.

The hallmark of Burkitt's lymphoma is a chromosome rearrangement in the tumor cells. A part of chromosome 8, apparently including the *myc* gene, is transferred to chromosome 14 (or less frequently to chromosome 2 or 22). At the annual Bristol-Myers Symposium on Cancer Research, held at the University of Chicago Cancer Center, Croce reported that in lymphoma patients the *myc* gene is present on the altered chromosome 14. Similar work was reported there by Rebecca Taub and Philip Leder of Harvard University.

"This finding argues strongly for a role of *myc* in Burkitt's lymphoma," Gallo says. He speculates that in its new position on chromosome 14, the oncogene is expressed at an abnormally high level. The researchers find no correlation between the *myc* gene and the Epstein-Barr virus, which is implicated in Burkitt's lymphoma by other lines of evidence. They suggest that *myc* and the virus may be involved in separate steps of cancer development.

Scientists are now looking for other human cellular genes homologous to oncogenes described in work on animal viruses. Using a radioactively labeled copy of an oncogene to bind to its counterpart on the human chromosomes, Stuart A. Aaronson and NCI colleagues have located the rodent virus oncogene called *mos* on a human chromosome. They identified the chromosome by examining hybrid cells having some human and some Chinese hamster chromosomes. By the pattern of oncogene binding, they report in the September PROCEEDINGS OF THE NATIONAL ACADEMY OF SCIENCES "an unambiguous assignment" of the *mos* oncogene to human chromosome 8.

While feeling satisfaction with the tremendous progress made in the last two years by the application of recombinant DNA and gene transfer techniques to the study of tumor viruses and oncogenes, most scientists feel they are still a long way from explaining human cancers. Barbacid says, "Demonstration that a single genetic alteration is sufficient to confer neoplastic properties on a normal cell is in apparent conflict with the lengthy development of most human cancers." He suggests these genetic events might play either an early, common, potentially reversible role or a late, irreversible role in the multi-step process of human cancer.

THE FATE OF THE EGG

Life begins as a single, highly organized cell. And it unfolds with confounding precision.

STEPHEN S. HALL

THE BIOLOGISTS WHO investigate nature's deepest and longest-running mystery often use the term fate map to describe the startling transformations that lie in store for the fertilized egg. It is one of the more venerable terms in embryology, and one of the most appropriate, too, for destiny and geography indeed intersect within the magnificent speck of DNA and cytoplasm that is an egg on the edge of becoming a organism. In this one cell, the entire genetic bill of lading for an animal, be it fruit fly or human, is stored, waiting to unfold with miraculous precision. It is that process of life unfurling—of cells becoming brain or backbone, of genes selectively flashing on and herding cells toward their certain fates, of tissues aggregating and differentiating toward ever more specific tasks—that both confounds and as surely delights developmental biologists.

"Embryology appears to be unique among modern experimental disciplines in biology in not only still possessing but still celebrating its ancient unsolved

problems," noted Rudolf Raff of Indiana University and William Jeffery of the University of Texas in 1983. "Those are still our problems," Raff says now. "We just have better ways of getting at them." Centuries of uncertainty have made embryologists a cautious lot, yet progress made in recent years has been sufficient to prompt Eric H. Davidson of the California Institute of Technology to say, "One now has a chance, for the first time, of understanding the workings of a process that has been around at least 550 million years."

For more than a century, developmental biologists have charted regions of the egg much as ancient cartographers made their first tentative interpretations of the New World. The amphibian egg, to cite a much-studied example, has been viewed as a large globe with a northern hemisphere and a southern hemisphere, north and south poles, and an equator with latent special cellular destinies. The rough outlines of egg cells like this have been known for some time. But researchers are now in a position to journey, as it were, to the center of the planet, where, it may turn out, geography *is* destiny.

Molecular biology has proven to be the vehicle of choice for this journey, opening up the interior of the egg and the embryo to a degree never before

achieved. But the story begins with the egg cell itself. As a model, John C. Gerhart, a developmental biologist at the University of California at Berkeley, holds a six-inch ball of Styrofoam in his hand. It represents the unfertilized egg of *Xenopus laevis*, the African clawed toad. "There's a wonderful geometry to it," says Gerhart. "It's no mean feat to come up with one coherent fate map in a big egg like this."

The egg arrives for its zygotic rendezvous with sperm in a state of exquisite preparation. (It's sometimes hard, in fact, to view the sperm cell as having anything other than a walk-on role during fertilization.) The northern hemisphere is dappled brown with pigment granules, and the egg's southern hemisphere is pale yellow to reflect its dense inner swirl of yolk platelets. The fates of the regions have been well mapped. The northern, or animal, hemisphere is nearest to the nucleus and destined to become the nervous system and the skin. The southern, or vegetal, hemisphere will develop into the digestive tract, and the equatorial region gives rise to skeletal, muscle, and circulatory tissues. The question, of course, is how.

"It's thought," says Gerhart, "that this hemispheric structure, with pigmented and unpigmented hemispheres, represents a real polarity, so that inside the egg, the cytoplasm may be different in the different areas." Indeed, deep inside the egg lies perhaps the organism's most precious inheritance: molecules passed down from the mother, such as maternal proteins and maternal "messages" in the form of messenger RNAs that are waiting to be translated into protein. It is suspected that these molecules, segregated roughly into the separate hemispheres, wheel into action later on in development by tripping genes into action. Embryonic cells have genes that respond to two major cues, or determinants, as they develop: one is the influence of neighboring cells, known as induction, and the other is the influence of materials in the egg, such as maternal messages. The latter type of activation occurs in certain cells and not in others, possibly because of the localization of these signals in particular areas of the egg.

It so happens in the case of *Xenopus* that the sperm cell can only penetrate the northern hemisphere of the egg. As soon as the sperm enters, however, the fertilized egg undergoes two significant planetary tremors. First, the thin outermost layer, or cortex, of the egg separates from the cytoplasm, creating a structure roughly analogous to the Earth's crust sitting above its core. Then, more critically, the outer sphere slides over the inner one for about 30 degrees. The sloshing motion brings material from the southern hemisphere into contact with northern hemisphere material. This smear of contact, known as the gray crescent, determines no less than the dorsal side of the toad—the side that, in all vertebrates, blossoms into spinal column and brain.

No one yet knows why this exceedingly precise rotation occurs, or why the gray crescent forms 40 to 80 minutes after fertilization. But the egg is doomed to a kind of developmental decapitation if this maneuver doesn't take place. By tampering experimentally with the rotation, researchers found that anything less than the 30-degree movement results in embryonic flops. "If the movement is small, about five degrees, the embryo just has a tail," Gerhart explains. "If it is about 10 degrees, it gets a tail and a trunk. If it is 15 degrees, it gets the tail, trunk, and some hindbrain structures. At 20 degrees, it gets the midbrain structure. At 30 degrees, it gets the forebrain structure. The extent of movement sets the context for later regional gene ex-

pression." In other words, the degree of "slosh" brings elements from the two hemispheres into contact, and this commingling seems to affect later events.

One hundred minutes after fertilization, the *Xenopus* egg divides into two smaller cells called blastomeres. This event, known as cleavage, is a general feature of development that continues with successive cells until the embryo becomes a hollow sphere of cells, or blastula, and reaches a developmental stage known as gastrulation, when groups of cells begin to fork off into one of three major fates: an outer layer, the ectoderm; an inner layer, the endoderm; and an intermediate layer, the mesoderm. The endoderm ultimately gives rise to the gut and digestive tract, while the ectoderm develops into skin, sense organs, and the entire nervous system. Lying between the two, the mesoderm becomes muscle, blood, and sex glands.

Gerhart argues that many of the signals sending cells peeling off to pursue their individual fates—the nervous system here, the gut there—seem to be built into the egg. There are outside cues, but the egg seems to march to the beat of its own drummer a good deal, too. "My impression," Gerhart says, "is that there are lots of scheduled events. This egg, once it's been fertilized, has a whole series of steps that it will proceed through." Indeed, the mere puncturing of a *Xenopus* egg with a needle rather than sperm sets off the same series of events: the cortex around the egg, the rotation, the gray crescent. "The capacity is there from the very beginning. There's an enormous amount the cytoplasm can do before any gene activation actually occurs."

THE DEVELOPMENTAL CUES sequestered in the egg's cytoplasm are what most intrigue Eric Davidson and his group at the California Institute of Technology in Pasadena, where one world view can be inferred from a poster on the wall of Davidson's basement office. "It is not birth, marriage, or death, but gastrulation which is truly the most important time in your life."

Davidson, a leather-vested cell biologist, has chosen to study the developmental cues in sea urchin eggs. One reason is that there's no shortage of raw material, as colleague Frank Calzone demonstrates by injecting potassium chloride into a fertile female *Strongylocentrotus purpuratus*. Moments after the

injection, the urchin, its spines waving, disburses a soft precipitate cascade of yellow-orange eggs into a beaker filled with seawater. Within 10 minutes, there is an eighth of an inch of material at the bottom of the beaker—perhaps as many as five million eggs.

Davidson and his collaborators, notably Roy J. Britten of CalTech, have cast the problem of development in stark molecular terms: Sea urchin cells become committed to a particular fate, they believe, when particular genes switch on, so they want to know why these genes are turned on in certain cells and not in others. To that end, they have focused their attention on three cell lineages—the gut, the skeleton, and the embryonic skin— and are attempting to identify the genes that seem to nudge cells toward the formation of these tissues. They know these cells begin to fulfill their particular destinies within 26 hours of fertilization. But, again, the question is how.

Molecular biology has enabled these researchers literally to see genes turning on in the developing embryo, and consequently the fate of cells can be directly tied to the activation of specific genes, providing a degree of focus almost unimaginable five years ago. James Lee of CalTech and Robert Angerer of the University of Rochester have studied, for example, a gene known as *CyIIIa*, which codes for a protein called actin. About 15 to 20 hours after fertilization, this gene blazes into action but only in one particular cell lineage, the aboral ectoderm, which forms the skin of the sea urchin larva.

Another example concerns the cell lineage that builds the skeletal structure. Within five hours of fertilization, the sea urchin egg has subdivided four times. At this fourth cleavage, a cluster of four cells called micromeres form at the southern end of the embryo. After a subsequent division one hour later, there are four small micromere cells and four large ones. The four large cells are destined to form the entire skeleton of the free-swimming larva.

When the subpopulation numbers 32 cells, these so-called primary mesenchyme cells become uncoupled from ongoing cell divisions elsewhere in the embryo. Like independent creatures, they dissociate from each other and wander around on their own in the hollow interior of the spherical blastula. Finally, as the gastrulation process begins, these cells link back up. Joined together in cablelike structures, they begin to secrete a spicule protein that eventually will develop into the skeleton of the embryo.

These cells clearly run through a complicated, highly synchronized series of events. With the use of radioactive or fluorescent tracers, the appearance of specific gene products such as proteins can be detected in the developing embryo. Like a glimmer of destiny, these glowing proteins intimate the shape of things to come, even before morphological characteristics are completely clear. The appearance of new proteins, of course, means that previously mute genes have somehow cleared their throats and begun to express themselves. It is suspected that molecules inherited from the mother, either proteins or messenger RNA, play a role in activating these genes; the cytoplasmic organization in the egg suggests that some of these molecules might have been distributed to certain neighborhoods of the embryo, so that the changes they exert are local.

The next step, however, is to figure out what it is about these genes—what is intrinsic to the sequence of nucleotides, or base pairs, along a particular stretch of sea urchin DNA—that has them blinking on and off at crucial junctures of embryogenesis. That work is already underway, thanks to an ingenious use of "fused genes," which, for example, combine part of the *CyIIIa* gene for actin with another gene that makes an easily detectable bacterial protein dubbed CAT. Using micropipettes, researchers can inject this recombined DNA into unfertilized eggs. Upon fertilization, the fused gene is incorporated into the nucleus and goes along for the ride during development. When *CyIIIa* genes normally turn on—about 20 hours after fertilization—the CAT enzyme lights up in assays. The same approach will provide an extremely precise timetable for key developmental genes.

"That gives us a spectacular advantage," says Davidson. "We're going to be able to learn just what is the DNA sequence information required for this developmental specification of gene activity. If we can find out, by the introduction of cloned DNA, what regions of genes are required to get them to turn on at the right time in the right place, then we at least have a chance of determining what molecules interact with those regions and cause this event to occur."

AS POWERFUL AS IT IS, molecular biology has not cornered the market on embryo-

logical insight. Since the beginning of this century, geneticists have picked through millions of mutant fruit flies, identifying tiny genetic glitches from white eyes to misshapen wings, in an effort to understand which genes control what. In his third-floor laboratory at Cal-Tech, amid hundreds and hundreds of stoppered flasks filled with fruit flies, working in a cluttered room where photographs of classical geneticists like Thomas Hunt Morgan and William Bateson are propped up against a blackboard, a diminutive, gentle-voiced, silver-haired biologist named E.B. Lewis set the stage for perhaps the most interesting revolution to hit developmental biology in recent years.

For virtually his entire career, the 67-year-old Lewis has studied the bithorax complex of genes, which controls the orderly development of many of the fruit fly's body segments: two thoracic and eight abdominal segments, one after the other. In the patient manner of the classical geneticists, he has created mutant flies, carefully identified the growth disorders they displayed, and began to correlate the insects' errant development with a class of "master switch" genes. Lewis embarked on this immensely complex genetic odyssey in 1946, before the structure of DNA was even known, yet recent advances in recombinant DNA technology have confirmed his predictions and created probably the hottest area of focus in developmental genetics: the homeotic genes.

Homeotic genes appear to have a kind of hierarchical control over other genes in developing fruit flies. Lewis first began by studying mutants of the bithorax complex. "It looked as though what appeared to be one large genetic unit with many controlling properties was actually a cluster of genes, and that's what they turned out to be—40 years later," he says now. "It is a cluster of perhaps 10, possibly 30, lined up in a row in the chromosome. The astonishing feature is that the genes are lined up in the same order that they seem to turn on during development, and that order is the same as the order of segments along the body axis."

Philip Beachy, a graduate student in the Stanford University laboratory of David S. Hogness, has worked out some of the molecular details of the bithorax complex and likens the function of these homeotic genes to a corporate hierarchy. "You could think of them as executive officers in a big corporation," he says. "The vice president in charge of the

third leg may be in charge of a whole slew of other functions." This control appears to be exerted because homeotic genes make regulatory proteins that in turn switch on other genes, initiating the cascade of genetic activity that each distinct fly body segment undergoes, just as an executive is responsible for everything from policy to clerical functions within a particular department.

When homeotic genes fail to flip their switches, major disruption ensues. Legs may grow out of the insect's head where the antennae should be, or a second set of wings may sprout where none should be. "If you delete *all* the genes of the bithorax complex," explains Beachy, "what you see is that all the segments of the fly develop as if they were the second thoracic segment." In other words, development gets stuck on the very first of the 10 body segments under bithorax control, and this segment is reiterated again and again because the embryo doesn't have the genetic know-how to develop the next nine posterior segments. Most of these mutations never see the light of day because mutant *Drosophila* embryos are doomed not to hatch, but since the fly wears its skeleton on the outside, its rigid larval skin, or cuticle, faithfully registers all these odd developments, making them easy to spot.

Lewis predicted the developmental role of these bithorax homeotic genes in 1978. By 1983, a team of molecular biologists led by Welcome Bender of Harvard had isolated and cloned this gene complex. In the meantime, a second homeotic complex in the fruit fly has been identified by Thomas C. Kaufman's group at Indiana University. It is called the *Antp* (or *Antennapedia*) complex, and it seems to control the development of the head.

Once this executive suite of developmental control was breached in fruit flies, researchers obtained a kind of molecular skeleton key that opened up other homeotic sites and provided startling surprises. In 1984, the laboratories of Walter Gehring in Basel and Matthew Scott at the University of Colorado independently reported the existence of several mysterious but much ballyhooed patches of DNA in the neighborhood of the fruit fly's homeotic genes. This strip of DNA has come to be called a homeo box. No one is quite sure what it does, but it is beginning to appear with surprising ubiquity up and down the phylogenetic ladder. Three popped up in the bithorax complex of fruit flies, and two turned up in *Xenopus*. Now, according to recent work by William McGinnis and Frank Ruddle of Yale and Michael Levine of Columbia, virtually the same strip of DNA also occurs in the chromosomes of mammals like mice and humans. The segment is short, measuring only 180 base pairs of DNA, but the odds against an identical segment of that length turning up in more than one species are astronomically high. Needless to say, all the firepower of molecular biology is being trained at the homeo box in an effort to explain its enigmatic role, which many believe has to do with coding for a protein that controls other genes. The idea that so particular a segment of DNA could be preserved and duplicated in such a wide range of species suggests that it derives from an ancestral gene that was so successful in orchestrating development that evolution appropriated it for more widespread use. "Generally, you only have one chance in evolution," notes E.B. Lewis. "You don't have a chance of evolving whole new systems without somehow making use of some common ancestral genes to do it with." The homeo box, he suggests, may be an ancient mechanism for regulating the constellation of genes involved in making diversified tissues in higher organisms.

LURKING IN THE BACKGROUND of all this work is the question of how genes determine the fate of a cell, and implicit in that question is an even more fundamental one: How does the switch get flipped? What are the genetic mechanisms that account for the founding of an entire population of cells, and later, tissues?

Explaining how genes switch on and off, of course, would be of fundamental consequence for all of biology. It would go a long way toward explaining how genes become "hard wired"—permanently activated, like the globin gene in red blood cells, or permanently repressed, like the same globin gene in a brain neuron cell. The theater of activity for such inquiry tends to be the cell, and it hinges on the complex affinities of molecules floating around in the cytoplasm for certain patches of DNA in the nucleus. How do these mechanisms work?

Donald Brown, a researcher at the Carnegie Institution of Washington in Baltimore, Maryland, has balanced these huge questions atop a single, tiny, cellular building block known as 5S ribosomal RNA. Ribosomes, which are partly made of RNA, are the factories in each cell where proteins are assembled. Most genes code for protein products and use messenger RNA to translate their messages at the ribosome. But ribosomal 5S RNA genes code for a structural RNA that is actually part of the ribosome.

It is a complex story but truly at the heart of development's conundrums. In the developing *Xenopus* egg cell, the 5S gene comes in two families. There are nearly 80,000 5S genes active in the oocyte. About 1,600 other 5S genes, called somatic 5S genes, are also active, but, unlike oocyte genes, they remain active throughout the life of the organism. The oocyte and somatic 5S genes make very similar 5S RNA molecules, and they both work flat out, making huge amounts. Then, after the egg cell is fertilized, a dramatic change occurs. The oocyte genes mysteriously shut down. The genes are there but inactive. Only the somatic 5S RNA genes continue to work. "That's the developmental control that we have to explain," says Brown. "Why is the oocyte gene family turned on in oocytes, and why is it off in somatic cells?"

The 5S gene, it appears, turns on when at least three distinct protein molecules, which fit together like a biochemical jigsaw puzzle, sit smack dab in the middle of the 5S gene on a patch of DNA about 50 base pairs long, a patch that Brown calls the internal control region. This superstructure of proteins clamped on the DNA sticks out like a landmark and is recognizable to an enzyme known as RNA polymerase III, which cruises around in the nucleus. This superstructure not only functions as a landmark but somehow manages to guide and align polymerase with the 5S gene so precisely that the enzyme copies—or transcribes—the genetic information for making 5S RNA with great fidelity. Only one of the three protein or factor molecules has been identified. Known as A, it was recently described by Robert Roeder while at Washington University in St. Louis. Factors B and C are being investigated now.

The interaction of these factors with the 5S genes also appears to account for the remarkable shutdown of the oocyte gene family. Brown's group has determined that there is a staggering excess of Factor A in the egg cell: about 10 million molecules for *each* 5S gene. Brown's group has also discovered that the key difference in the control region of the oocyte 5S gene and its somatic cousin is the sequence of a mere three nucleotides of DNA. That infinitesimally small discrepancy, in fact, is sufficient to give so-matic genes a huge competitive edge over the more numerous oocyte genes: Evidence suggests they grip Factor A many times more tightly.

That advantage doesn't make much difference in the egg cell, saturated as it is with millions of Factor A molecules. But after fertilization, the concentration of Factor A molecules falls precipitously until there is only one molecule for every four or five 5S genes. In this situation, the somatic genes have a distinct advantage. Without Factor A, the oocyte 5S genes in a sense lose their landmark status—the enzymes can't "see" them, and the genes remain unread.

In somatic 5S RNA genes, however, this transcriptional superstructure somehow manages to remain visible to the enzyme. And by late development, the smaller family of 5S RNA genes is stably turned on while the larger family is stably turned off. This strange superstructure is still ill-defined, but preliminary descriptions suggest the molecular interactions involved in turning certain genes on and off. Is it possible, asks Brown, that these stable, transcribing complexes provide a kind of memory effect that establishes the destiny of a cell? Gene determination, Brown believes, is "probably the most important next question to be answered about developmental phenomena."

WHAT HAPPENS *after* a cell has become committed to a general fate, such as becoming a blood cell or a liver cell? How does this kind of cell acquire the more specialized traits associated with particular body tissues? This movement toward increasing specialization is called differentiation, and it, too, seems to evolve as a particular ensemble of genes gradually becomes active in the descendants of an early cell.

An incisive attack on this problem has been mounted by Beatrice Mintz, a scientist at the Fox Chase Cancer Center in Philadelphia, who has focused her work on mammals. More specifically, Mintz is trying to find out how a mouse embryo manages to establish its population of various blood cells, known collectively as the hematopoietic system, with only a few founder cells.

All the different cell types of this system, from red blood cells to various specialized white blood cells, derive from a small number of ancestral cells with the intimidating name of totipotent hematopoietic stem cells. These stem cells appear to arise in the liver when it begins forming, on the 10th day of fetal development. In the liver, the cells continue to divide and gradually yield more differentiated cells, which enter the circulation of the fetus. Then, around birth, some stem cells reach the bone marrow, where blood cells are manufactured for the duration of the animal's life. Through a complicated series of maneuvers, including the injection of healthy stem cells into the placentas of mouse fetuses with defective stem cells, Mintz has recently shown that these genetically disabled fetuses could be "rescued" by the transplanted cells, which yield normal blood cells. It has also been determined that the entire blood system of the surviving mice can develop from a single normal stem cell.

"Nobody has actually seen a hematopoietic stem cell or figured out a way to get a potful of them," Mintz cautions. Nonetheless, the work done so far has allowed Mintz to develop a hypothesis that might explain how, from a single set of founder cells, the entire blood cell population can branch out and yet constantly renew itself. It appears that the founder cells and some of their progeny have an extensive capacity to renew themselves, slowly but with great persistence over the lifetime of the organism, so that there are always stem cells in the system. At the same time, this cell has the ability to produce daughter cells that, although still stem cells, have slight differences: they have a shorter self-renewal capacity. Within this hierarchy of stem cells, successive generations of stem cells differ ever so slightly until ultimately progenitors to the specialized cells typical of the blood system population are produced. This model might well apply to other developing tissues as they undergo differentiation, and the Mintz group hopes soon to examine the genes that may be influencing the process.

"I doubt it's a simple one-step process, whether these stem cells are precursors of blood, skin, or whatever," says Mintz. "Going from the most primitive cells that have made the commitment to be a certain class of cell to much more specialized cells surely has to involve a lot of different genes. The idea that all the old ones turn off in concert and all the new ones would turn on in concert is extremely unlikely. Differentiation is not a single-step process but involves step-wise commitments."

A more molecular view of what may be going on in differentiated cells has emerged from the enterprising work of Keith Yamamoto and his group at the

University of California, San Francisco. They have focused on genetic elements called enhancers, which appear to be two or more stretches of nucleotides adjacent to and sometimes within genes. As their name implies, they seem to enhance the efficiency with which a gene churns out its designated protein product.

For 10 years, Yamamoto's group has explored this problem as it pertains to a class of steroid hormones called glucocorticoids, which selectively collaborate with specific genes in, say, liver cells—a classic example of differentiation. What makes these particular genes so receptive to these hormones? When a hormone molecule attaches to a liver cell, for example, it combines with special proteins known as receptors that, in Yamamoto's words, "cruise around areas near the cell membrane." Once they meet, the hormone changes the shape of the receptors and confers on them a unique "go-between" quality that allows them to insinuate their way into the nucleus, seize onto the enhancer patches of DNA, and thereby influence nearby gene activity.

Yamamoto, however, has taken the role of enhancers one step further—at least theoretically. His group and others have suggested that enhancers might have a kind of dual role in cell differentiation. In a trigger role, enhancers would accomplish a one-time, irreversible activation of a gene. Thus, they would hardwire the function of a cell. This could occur when a protein plops down on or near the enhancer. Once it lands, this agent, a transcription factor, doesn't budge, and the gene is permanently on.

In a modulator role, on the other hand, enhancers would control the waxing and waning gene activity, consistent with the periodic appearance of hormones. When hormone-unleashed proteins meet up with a triggered gene, they could form short-lived but powerful interactions that would take a turned-on gene and dramatically boost output. An untriggered gene, by contrast, could never be influenced by these regulatory proteins. To view it another way, Yamamoto's model views enhancers as functioning something like a dimmer switch on a lamp—the trigger enhancers would behave like a basic on-off switch, and the "modulator" enhancers would act like adjustable controls, so that the output of protein, like the output of light, could be made higher or lower or turned off, depending on circumstances.

Molecular biology has shown that genes are the entities that control the

synthesis of proteins, but developmental biology is beginning to reveal a surprising and provocative new function. Changes in genes active during development may drive evolution itself. This is one of the most exciting new domains of the discipline.

Evidence for this dramatic new function is beginning to accumulate, thanks to a tiny nematode named *Caenorhabditis elegans*. These faint, white, wiggling burghers of wormdom graze in the soil of probably every backyard from Bangor to Honolulu. They measure about one millimeter in length, dine on bacteria, and complete their developmental journey, from embryo through four larval stages to adulthood, in about 63 hours. Each adult hermaphrodite contains exactly 959 cells, not counting germ cells, and the origin and fate of each and every one of those cells has been painstakingly mapped, which prompts Massachusetts Institute of Technology biologist H. Robert Horvitz to boast, "We know this animal inside and out."

On the wall in Horvitz's office is the fate map of *C. elegans*, resembling a kind of schematic rendering of a railroad switching yard. The resemblance is apt, for *C. elegans* researchers have been able to identify key developmental genes that, like the homeotic genes in fruit flies, play major switching functions. In the classic manner of genetic research, these genes have been revealed by mutation—the time-honored strategy of breaking or altering a gene and then seeing what changes, if any, pop up in the organism.

In just such a manner, researchers identified so-called lineage or *lin* genes. These genes act like binary switches, meaning that they route a cell's progeny toward one destiny, Fate A, when the gene is on, or toward Fate B if the gene is off. For example, a gene known as *lin-14* exercises powerful control over the type of cuticle, or outer skeletal structure, formed at various times during development. Horvitz's lab has shown that when the *lin-14* gene is on in early development, the so-called seam cells form cuticles typical of the larval stage. When *lin-14* shuts down, the seam cells switch to adult cuticle formation. Many other developmental events are similarly affected by *lin-14*. The most interesting wrinkles occur, however, when mutations disturb this delicate timing. If the *lin-14* gene fails to shut off when it is supposed to, certain cells are unable to escape the larval stage, even if the rest of the organism is maturing around it. If *lin-14* fails to turn on, conversely, certain cells are, in a sense, born adult, completely bypassing the characteristics of the first larval stage and out of sync with the rest of the developing larva. In an organism with so few cells, such mutations can dramatically change the worm's overall structure.

Those disproportionately significant changes are what excite Horvitz and colleague Victor Ambros of Harvard University the most. They believe that genes that act like *lin-14* may play a decisive role in species variation because simple changes in timing can produce profound developmental and morphological changes. "The characteristics of all these genes, the so-called heterochronic genes that affect timing decisions, are *precisely* those characteristics that some evolutionary biologists have been looking for," says Horvitz. Mutations in genes that disrupt the timing of developmental events could be the little engines driving major evolutionary changes.

MOST RESEARCHERS AGREE that it will be a combination of astute genetics, powerful molecular biology, and good old-fashioned cell biology that, in concert, will move us closer to understanding how cells make their crucial decisions and an organism's fate map thus unfolds. "The excitement at the moment," says Robert Horvitz, "is that progress is being made on all fronts. We hope that all of those things can be unified and tied together in a way that will make sense and perhaps even be aesthetically pleasing."

No one promises instant answers. After all, the mystery also puzzled Aristotle. A fitting reminder of the difficulty is suggested by a print on the wall above John Gerhart's desk in Berkeley. It shows an Egyptian mosaic populated by a splendid array of crocodiles, hippopotamuses, storks, and snakes. The goal that developmental scientists seek is nothing less than to explain how a single monochromatic tile called the fertilized egg becomes the vast, extravagantly hued, multiplicitous mosaic that is life.

Life's recipe

This brief is about genes and science's greatest achievement this century: deciphering the code in which are written the instructions for building, running and reproducing bodies. The text is still being read.

In the 1940s and 1950s, several clever physicists turned themselves into biologists. Flushed with their success within the atom, they decided the next problem to be solved was life itself. It was a shrewd move. Between 1944 and 1972, discoveries came thick and fast: first, the nature of hereditary material, called genes; then the ingenious molecular structure of genes; then the code in which their hereditary message was written; and then a way to tinker with that code.

Since then, genetics has become a significant part of the biotechnology industry. It has also invaded the rest of biology as a tool. Medicine, agriculture and the study of evolution have all adopted the techniques and insights of molecular biology, along with its jargon of clones, sequences and mutants.

Since 1972, genetics has made few big breakthroughs. This is not because it has nothing left to discover. Progress is being held up because scientists now know practically all there is to know about bacteria (the simple one-cell creatures on which early genetic studies were done), but little about man and other animals.

In the past few years, however, biotechnology has given scientists the tools to explore the genes of mice and even men. Molecular biologists now hope they can tackle two pressing questions: how does a fertilized egg manage to turn into a human being, and how does a healthy cell turn into a cancer tumor?

Before 1860, ideas about heredity were few and vague. Charles Darwin, for example, thought that a child somehow inherited a blended mixture of vital fluid from its father and mother. Then Gregor Mendel, a Bohemian monk, crossed some pea plants and showed that the offspring were not blended versions of their parents, but that they inherited discrete "factors" from each parent and passed those factors on unchanged to their offspring in predictable ratios.

Mendel's experiments, rescued from obscurity by British and American biologists in the early part of the twentieth century, seemed to prove that heredity relied on particles carried within eggs and sperm. The particles became known as genes.

In 1871, a substance called deoxyribonucleic acid (DNA) was found to be common in the sperm of trout from the river Rhine. It, too, had to be rescued from obscurity by a series of elegant experiments in the 1940s, by a Canadian physician, Oswald Avery, and his colleagues. They proved, to the surprise of most biologists, that genes are made of DNA and not, as expected, of protein—out of whose molecules most of the body's machinery and structure are made.

Yet DNA was known to be just a long, thin molecular string, made of a repeating sequence of sugars, phosphates and chemical bases. How could it possibly carry the instructions for building and running bodies? Clearly, its secret must lie in its structure. In 1953, the race to work out the structure of DNA was won by two young Cambridge scientists, Mr. James Watson and Dr. Francis Crick, working largely in their spare time. The ingenuity of the structure was immediately obvious: it was Mendel's factors made flesh.

Two intertwined helices of DNA are linked by weak bonds between the bases. The bases come in four kinds (A, C, G and T) and links can be made only between given pairs of them: G to C and A to T (see chart on this page). Thus a sequence of bases can spell out a code and that code can be faithfully passed on to offspring by untwining the two helices and copying them.

Code crackers

Next, to crack the code. Were it not for some great leaps of imagination, mainly on the part of Dr. Crick, this problem might have remained intractable for decades. It was already fairly widely agreed that the code must be principally a set of instructions for making proteins. Proteins depend largely on their shapes to do their jobs, so somehow the code must spell out that shape. Proteins, like DNA, are made of long unbranching strings. Dr. Crick realized that the protein shape was determined by the sequence of amino acids in the strings, and he guessed that the sequence was somehow determined directly by the sequence of bases (letters) on the DNA.

He was right. By 1970, the last letter of the code had been cracked: the code is written in three-letter words (called

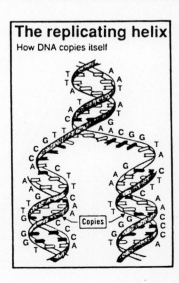

The replicating helix
How DNA copies itself

codons), each of which specifies one of the 20 amino acids in the protein alphabet, except for a few which act as signals to the copying machinery to start and stop. Each gene is faithfully transcribed to make a temporary single-stranded copy from a material called RNA (which is very like DNA, except that it is unable to form double helices).

This transcript, known as the messenger, is then used as the blueprint for building a protein by a neat little machine tool in the form of a cell called a ribosome (see diagram on next page). The ribosome reads the messenger by moving from codon to codon along it. At each stop, a small tug-like RNA molecule called transfer-RNA brings along the appropriate amino acid to attach to a growing protein chain: each transfer molecule has on its head the three-letter word corresponding to the code for the amino acid it carries. The code is the same in man, mouse and all living creatures—with one exception found so far, a tiny slipper-shaped protozoan (called Paramecian) that uses different words for certain amino acids.

One big problem remained—and it has not been fully solved to this day. Every cell in the human body contains the complete recipe for the whole body. Yet cells differ in their sizes, shapes, biochemistry and abundance. Skin cells, blood cells, nerve cells, muscle cells—all must somehow receive instructions to differentiate in the growing embryo. How?

The first part of that question is, how do the cells switch on different sets of genes and so produce different sets of proteins? The hunt for gene switches began in bacteria. A common bacterium like E. coli, from the human gut, is able to switch on the gene for an enzyme called beta-galactosidase when lactose sugar (essentially its food) is present, and switch it off when there is no lactose. The enzyme breaks down the lactose into simpler sugars on which the bacteria can feed.

In the 1960s, two French scientists, Jacques Monod and François Jacob, worked out how the gene is switched on and off. Lactose itself boosts the amount of beta-galactosidase by somehow increasing the activity of another enzyme that transcribes the beta-galactosidase protein which, in turn, digests the lactose. Its abundance controls the means of its own destruction.

However, certain mutant bacteria can produce lots of enzyme whether their lactose food supply is there or not. Jacques Monod came up with an explanation: a specific repressor molecule attaches itself to the DNA, blocking the transcription of the beta-galactosidase gene. Lactose unblocks it by attaching itself to the repressor and putting it out of action. In mutant bacteria, the genes to which the repressor molecule attaches itself do not work, so the main gene is always switched on.

This may sound complicated, but it was in fact the system's simplicity that excited scientists. Similar switches have been found to control most bacterial genes. Do the same mechanisms control human genes? Dr. Monod, in a fit of optimism, declared that "what is true for E. coli is also true for an elephant."

Unfortunately, it is not. Elephant and human genes turn out to be much more complicated—which is not surprising, since bacteria are single cells. Elephants are made up of billions of different sorts of cells.

The first thing scientists found about switches in human cells was that a short distance "upstream" of a gene (i.e, at the end that is transcribed first), there usually lies a short sequence of four bases, TATA. This turned out to be a red herring. The TATA grouping is merely a signal to tell the transcribing mechanism where the gene starts; it is not the switch itself.

To find the switches, scientists had to isolate pure genes in test tubes, add basic ingredients such as the transcribing enzyme and any other proteins necessary to switch on the gene, and then try to work out where the proteins attached to the DNA. In theory, these points could be where the switches are.

What scientists found was a series of short sequences (or "boxes") upstream of the gene and a series of new proteins that attach to these boxes. For example, one gene, to be switched on, requires two proteins called SP1 to attach to boxes that read GGGCGG, and one other protein to attach to a third box between those boxes reading GCCAAT.

Moreover, these boxes (known as promoters) have to be in exactly the right place relative to the gene. Other sequences, known as enhancers, can be anywhere, not even close by, and still control the gene. This gap between the gene and an enhancer took some explaining, until experiments revealed that the protein that attaches to the enhancer can cause the DNA to bend into a loop, bringing the enhancer into contact with the gene.

So some switches have been found. Nobody knows how the proteins that control the switches are themselves controlled, and nobody knows exactly how switches control development. There are tantalizing clues, though. One comes from fruit flies, in which certain genes cause young flies to develop abnormally: to grow a leg instead of an antenna, for example.

In 1984, several scientists discovered that such genes all had a single sequence of 180 DNA letters in them. Even more remarkable, these sequences (called homeo-boxes) were found in the genes that affect development in mice and even human beings. It was as if this sequence were somehow a universal switch for genes that control growth and differentiation.

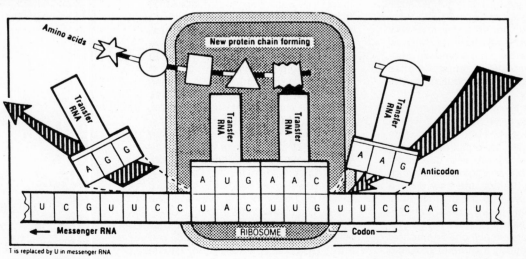

T is replaced by U in messenger RNA

Skeptics point out, however, that the universality of the homeo-box is likely to mean that it is the least interesting part of the machinery. All computers use the same plugs to plug into the electricity supply. Perhaps the homeo-box is the developmental gene's plug.

To understand embryonic development is one of the reasons for studying gene switches. The other is to understand cancer. The discovery that healthy cells harbour special genes called oncogenes that can cause cancer has transformed the way scientists think about this disease.

It now looks as if the proteins made by oncogenes control a series of master switches that are responsible for controlling the cell's growth. Cancer, which is nothing more than a cell gone berserk, growing and dividing continuously, is therefore caused by somebody leaving the growth genes switched on.

Scientists realized this in 1983, when they read the sequence of bases in one human oncogene. It turned out to resemble closely the sequence of a human gene that encoded a protein called a growth factor. In healthy cells, growth factors set off a chain of events via so-called receptors on the cell's surface, so that the cell reproduces itself—for instance, when blood clots, healing cells multiply. Normally, the cells that make a growth factor do not respond to it, and vice versa. If a cell that responds to a growth factor switches on the gene for that factor, it tells itself continuously to grow and multiply. It becomes a cancer.

Two other kinds of oncogenes work in similar ways. One makes a faulty version of the receptor for the growth factor—faulty in the sense that it acts as if it were continuously receiving the factor. Again, it continuously tells the cell to grow. A third class of oncogenes encodes proteins that are themselves switches for other genes.

Gene dreams

These discoveries were quite unexpected. Before 1982, when the first human oncogene was found, nobody expected cancer to be so readily explicable through genetics. Thanks to molecular biology, cancer is now understood, though not conquered. The aim of most molecular biologists is to uncover enough information so that cures can be found to treat hitherto intractable diseases. One possibility is that agents could be developed which block the binding of the growth factors to the receptor. They could turn out to be promising anti-cancer drugs.

These are distant dreams, but they are not fantasies. Already molecular biology has begun to transform medicine in many different ways. It has enabled scientists to track down the genes responsible for inherited diseases like cystic fibrosis. It has enabled them to find the cause of diseases like AIDS and to design drugs and vaccines that fight disease.

Molecular biology has given scientists two especially useful tools: sequencing and genetic engineering. Sequencing is a clumsy term for reading the sequence of letters in a gene and translating it to work out what protein it spells out. As the cancer story illustrates, it is useful mainly because of the coincidences it reveals: a computer first spotted that oncogenes are like the proteins by comparing their sequences.

In 1986, encouraged by the recent invention of a machine to read the sequences of genes automatically, a bold plan was aired by a group of American scientists: to read the sequences of the entire set of 50,000 human genes (called the genome).

The result would be a list of bases 3.5 billion characters long—or enough characters to fill about 5,000 average-sized books. Even with the new machines, it would take about ten years and $1 billion a year to do the job. Moreover, genomes are as idiosyncratic as people: whose would you choose to read?

Genetic engineering means taking a gene from one creature and putting it in another. It was first invented in 1972 and perfected during the late 1970s, for transferring genes from human cells to bacteria. The advantage of doing this is that the bacteria can then be encouraged to churn out large amounts of some human protein that can then be used as a drug.

Insulin for diabetics and growth hormone for young dwarfs are both made this way. Drugs to fight cancer and heart disease have also been mass-produced in bacteria—without any great success, as yet, simply because scientists do not know enough about the body's biochemistry to predict which proteins they want to mass-produce.

Genes can now also be put into plants and even animals. For example, a gene was taken from a bacterium that kills insects and put into a tobacco plant. That plant was then left alone by insects. Putting genes into animals is less easy, but it has been done. Human globin genes, which encode the protein responsible for carrying oxygen in the blood, have been transferred to mice. Scientists hope this will reveal how the switches controlling globin genes work. But they could, conceivably, have more practical ambitions—to make cattle grow wool, for instance.

The most ambitious scheme is to put human genes into other human beings. The point would be to cure certain inherited diseases, such as Huntington's chorea, that are caused by the absence of a working copy of the gene. "Gene therapy" would add a working copy of the gene using a special virus called a retrovirus that inserts its genes into human DNA when it infects a cell.

The virus would first be disarmed by removing the genes that enable it to make more viruses, and the experiment would be done only on the bone marrow, not on the germ cells that carry genes to the next generation. Nonetheless, the idea of playing with human genes and retroviruses—AIDS is a retrovirus—alarms people. The alternative, though, for diseases like Huntington's chorea, is to do nothing.

Behavior

The sum of behavior is to retain a man's own dignity, without intruding upon the liberty of others.

Francis Bacon

Although the study of human behavior is central to the disciplines of psychology and sociology, only in recent years has it become a major focus of biological investigation. Behavior is studied by observing a specific activity of an animal. Biologists, assuming that this activity has a basis in the animal's genetic makeup and past experience, attempt to determine how the behavior helps the organism survive and reproduce. The researchers who have extrapolated this assertion by observing human behavior postulate that if behavior is acquired primarily through learning it is greatly influenced by heredity. Other investigators, however, argue that our acts are uniquely unaffected by our genes and instead are determined by our environment. A number of articles in this section deal with the role heredity plays in human behavior. For example, Daniel Goleman's "Major Personality Study Finds That Traits Are Mostly Inherited" reports that eleven key personality traits are considered to be inherited.

Researchers are now investigating how nutrition affects human behavior. Although the effects are subtle, there is an increasing amount of data suggesting that dietary changes can alter the amount of brain transmitters and, consequently, change animal behavior. New information on recent findings about food and behavior are presented in the article "Food Affects Human Behavior."

Looking Ahead: Challenge Questions

Why is proper nutrition now believed to play an important role in human behavior?

What is the relationship between emotional behavior and health?

Why do plant-derived drugs have medicinal effects on humans?

What criteria are needed for placebos to function on patients?

THINKING WELL

The Chemical Links Between Emotions and Health

NICHOLAS R. HALL AND

ALLAN L. GOLDSTEIN

NICHOLAS R. HALL, with training in psychology, neuroendocrinology, and immunology, is associate professor, and ALLAN L. GOLDSTEIN is professor and chairman, in the department of biochemistry at the George Washington University School of Medicine, in Washington, D.C.

IN 1977, a twenty-eight-year-old Philippine-American woman, feeling weak and complaining of pain in her joints, visited a clinic in Longview, Washington. A physician there ordered blood and urine tests and, based on their results, concluded that the woman was suffering from systemic lupus erythematosus, a disease in which the body's immune system attacks healthy organs with all the ferocity it usually reserves for life-threatening intruders. After various drugs failed to restore her health, the woman sought the advice of another physician, who examined samples of her kidney tissue, confirmed the original diagnosis, and recommended that she follow an aggressive therapeutic regimen using drugs that would dampen the immune system's misguided assault. Instead, the patient decided to return to her native village, in a remote part of the Philippines, where she was treated according to local custom: a witch doctor removed the curse that had been placed on her by a former suitor. Three weeks later, back in the United States, she showed none of the symptoms she had earlier displayed, and two years afterward she gave birth to a healthy girl.

Had this case been reported in one of the nation's tabloids (indeed, it bears all the earmarks of such a sensational story), it might have escaped serious attention. But it was published in *The Journal of the American Medical Association*, in 1981. Significantly, neither the author—Richard Kirkpatrick, of Saint John's Hospital, in Longview—nor, by implication, the journal disputed the basic facts: a serious form of autoimmune disease had been swiftly and, by all indications, permanently reversed. Rather than attempt to explain the transformation, Kirkpatrick concluded his report with a question: By what means did an Asian medicine man cure the woman and, moreover, prevent the medical complications that invariably accompany a precipitous withdrawal from the sort of drug treatment she had been receiving?

The idea that mental states—the target, presumably, of the witch doctor's treatment—influence the body's susceptibility to and recovery from disease has a long and hallowed history. As early as the second century A.D., the Greek physician Galen asserted that cancer struck more frequently in melancholy than in sanguine women. And the belief that disease is a consequence of psychic or spiritual imbalance governed the practice of medicine in both Asia and Europe until the rise, after the seventeenth century, of modern science and its mechanistic view of physiology. But from that point onward, the role of behavior in human disease became a concern proper to none but the superstitious; attempts to empirically investigate the relationship were almost unheard of. Only within the past thirty years, and in particular during the past ten, has the systematic study of this ancient concept been broadly regarded as a legitimate aim of serious medical research.

Scientists have known since the 1950s that the immune systems of laboratory animals can be influenced by behavior. One of the modern pioneers in the field, Robert Ader, of the University of Rochester School of Medicine and Dentistry, has shown that, using the methods of classical conditioning—in this case, by associating the taste of a sweetener with the effects of an immunosuppressive drug—rats can be taught to suppress their own immune responses. And numerous other investigators have demonstrated that when rats and mice are subjected to acute stress—for example, by being confined for short periods of time in crowded living quarters—they become increasingly susceptible to disease.

But in recent years, Marvin Stein and his colleagues, of the Mount Sinai School of Medicine, in New York City, have documented a similar impairment in humans, particularly among men whose wives had recently died. Stein found that during the first few months of bereavement the ability of the widowers' lymphocytes (small white blood cells that spearhead the immune response) to react to intruders sharply diminished. He has also discovered a hampered immune response in people hospitalized for severe depression. In a study done in 1984, Stein documented in such patients not only an ebbing of lymphocyte activity but also a decline in the number of certain kinds of lymphocytes circulating in their bloodstreams.

Less established are the effects on the immune system of positive mental states. Indeed, until quite recently, the data have been largely anecdotal. When, in 1976, Norman Cousins, then editor of *Saturday Review*, described how he

This article reprinted by permission of *The Sciences* and is from the March/April 1986 issue, pp. 34-40. Individual subscriptions are $13.50 per year. Write to The Sciences, 2 East 63rd Street, New York, NY 10021 or call 1-800-THE-NYAS.

overcame ankylosing spondylitis, a chronic, progressive disease of the spine, through the use of vitamin C and laughter (and the buoyant, affirmative frame of mind it produced), there existed no empirical evidence beyond the sheer fact of his restored health to support his claims. As many critics were quick to note, information about the influence of behavior on health was scant and equivocal and, besides, spontaneous remissions of ankylosing spondylitis, inexplicable as they may be, were known to occur.

Since Cousins's recovery, however, a number of rigorously controlled studies have provided evidence of a strong link between behavioral therapy and patient prognosis. In one of the most notable, conducted from 1974 to 1984 at King's College Hospital Medical School, in London, a strong correlation was found between the mental attitudes of women with breast cancer and life expectancy. Of the women who displayed aggressive determination to conquer their disease, seventy percent were still alive ten years after their mastectomies; but among those who had felt fatalistic or hopeless, only twenty-five percent survived. And in our own two-year-long investigation, with Stephen Hersh, Lucy Waletzky, and Barry Gruber, we evaluated the effects of a behavioral approach to disease by measuring certain physiological dimensions of immune responsiveness in cancer patients. Among other things, the therapy consisted of relaxation techniques and guided exercises in which patients imagined that feeble cancer cells were crushed by stalwarts of the immune system. The preliminary results of this investigation show that behavioral therapy amplified the immune system's response to disease; in physiological terms, it accelerated the rate at which lymphocytes mobilized to attack foreign bodies and possibly increased their own numbers. This pilot study, if its results are confirmed by further research, will have provided the first thoroughly empirical confirmation of a correlation between mental states and immunity that as recently as five years ago the medical community viewed as a mixture of pseudoscientific hocus-pocus, self-delusion, and dumb luck.

But for all its novelty and importance, documenting a correspondence between behavioral events and the body's response to disease no more constitutes a satisfactory understanding of immune processes than an observed association between eating and tissue growth constitutes an understanding of digestion and absorption. By what means does stress disturb the functions of lymphocytes? Or, to return to the woman with systemic lupus erythematosus, which physiological systems did the witch doctor engage to relieve her of autoimmune disease? If, as a growing body of evidence suggests, mental states can both retard and enhance the body's ability to fight illness, there must exist a functional pathway that links the organ most closely associated with emotions and ideas—the brain—to the organs and tissues that collectively make up the immune system. In fact, two such pathways—one biochemical, the other anatomical—have been discovered.

THE FIRST LINE OF DEFENSE against viruses and bacteria that invade the body is the phagocyte, or cell-eater. When the skin is breached, by means of a wound or a lesion, swarms of these scavenging white blood cells descend upon the scene and devour the transgressors one by one until the phagocytes literally eat themselves to death.

Though indispensable, phagocytes make up only the most elementary part of the immune response. In humans, as well as in other mammals, a second type of defense also comes into play, one whose chief distinctions are its abilities to recognize particular invaders (phagocytes are indiscriminate in their appetites) and to tailor highly specific chemical counterattacks. The basic functional units of this immunity are the lymphocytes, so named because of the clear fluid in which they are stored and conveyed.

Like all white blood cells, lymphocytes derive from bone marrow; but whereas some remain in the marrow until they reach full maturity, and are therefore designated B cells, others, early in their development, migrate to the thymus—the master gland of the immune system—and for that reason are called T cells. Both types of lymphocytes circulate through the bloodstream before lodging in such lymphoid tissues as the spleen, the tonsils, and the adenoids, where they remain inactive until confronted with any of thousands of antigens, or foreign substances—organic waste, toxins, viruses, and bacteria.

In the presence of a particular antigen, B cells synthesize and release proteins, called antibodies, that are designed specifically to destroy that antigen. T cells also respond to individual antigens, though not by producing antibodies but by performing a variety of special immunological support tasks. Some, called helper T cells, release lymphokines, chemicals that assist B cells in producing antibodies. Others, killer T cells, attack antigens directly, with lethal substances of their own manufacture. Still others, known as suppressor T cells, help protect the body's tissues from being ravaged by its own immune response, by preventing B cells from making antibodies when they are no longer needed (a breakdown of this function is implicated in such autoimmune diseases as systemic lupus erythematosus).

Other chemical mediators of the immune response that are released by T cells and macrophages (a type of phagocyte) include histamine, which dilates blood vessels in preparation for the arrival of legions of lymphocytes; complement proteins, which, by inflaming the afflicted area, create an inhospitable thermal environment for foreign tissues; and prostaglandins and leukotrienes, substances that help start and stop the activities of macrophages and T cells. This complex arrangement lends to the immune system the flexibility it requires to strategically orchestrate its responses to countless new and varied microorganisms that invade the microenvironment of the body.

IN THE 1960s, investigators discovered that in test tubes certain immune functions of lymphocytes transpired spontaneously, suggesting that the immune system, for the most part, functions independently of the rest of the body; that the behavior of lymphocytes is governed solely by the number and kinds of antigens encountered—a system of self-contained biochemical reflexes. But subsequent research, much of it done in the past few years, has disclosed that not only is the immune system heavily influenced by other bodily processes—in particular, by those of the central nervous system, via the endocrine network (the glands that secrete hormones into the bloodstream), and the autonomic nervous system—but it is also impossible to separate the day-to-day functions of the immune system from those processes.

The best-documented influence occurs along a biochemical pathway that links the part of the brain located beneath the cerebral hemispheres—the hypothalamus—with the thymus gland, where the T cells mature, by means of two endocrine organs—the pituitary and adrenal glands. Outlining the cascade of biochemical exchanges that occur along this hypothalamic-pituitary-adrenal axis illuminates well the subtle but infrangible ways in which the brain and the immune system are knit together. In stressful conditions, the hypothalamus produces a chemical called corticotropin-releasing factor, which induces the pituitary to secrete adrenocorticotropic hormone. This hormone, in turn, stimulates the adrenal glands to release steroid hormones (glucocorticoids) into the bloodstream. It happens that T cells are acutely sensitive to glucocorticoids. This is especially true of nascent T cells, which represent about ninety percent of all T cells present in the thymus at any time. Abnormally elevated glucocorticoid levels will either damage or destroy these cells or prematurely induce their migration from the thymus to other immune tissues. The resultant shrinkage of the thymus is so pronounced that the gland has been called a barometer of stress and its weight used as an indirect way to assay the release of adrenal glucocorticoids. Left on their own, the adrenal glands secrete glucocorticoids in a daily tidal rhythm, and the point at which lymphocytes respond most aggressively to antigens has been correlated with the interval during which the level of circulating glucocorticoids falls to its lowest point.

In short, the biochemistry of the hypothalamic-pituitary-adrenal axis is far more complex than anyone ever dreamed. Not only are the participating hormones diverse, including even such agents as those that regulate reproduction and growth, but they produce different effects from one circumstance to another as well. Glucocorticoids, for instance, are biphasic: in high concentrations they mute the immune response, whereas in small amounts they have been shown to activate it. Complicating the issue further is the disparity between the consequences of acute and chronic stress. Whereas acute stress causes immunosuppression, chronic stress sometimes enhances the immune response. Thus, stress-induced modulation of the immune system involves a matrix of factors, any or all of which may be at work in the response of a single lymphocyte.

At the top of the hypothalamic-pituitary-adrenal axis, shuttling electrical impulses across the gaps that separate nerve cells in the brain, are the neurotransmitters. These compounds—in particular, serotonin, acetylcholine, and norepinephrine—regulate the secretion of corticotropin-releasing factor from the hypothalamus, thereby triggering the serpentine chain of events that arouse or pacify the immune response. The same neurotransmitters are now being implicated in the control of the immune system via a second, anatomical pathway—the autonomic nervous system.

The autonomic nervous system is to the central nervous system what an automatic pilot is to a pilot; in some sense the autonomic system can be overridden, but it typically operates on its own, regulating the involuntary actions of the heart, the stomach, the lungs, and other organs through a pervasive network of nerve fibers. Only in the past two years have we learned that branches of this network, chiefly the norepinephrine and acetylcholine circuits, are rooted deep within the body's lymphatic tissues as well. In ground-breaking work at the State University of New York at Stony Brook, Karen Bulloch demonstrated that signature patterns of autonomic nerve fibers radiate into the thymic tissues of reptiles, birds, and mammals, including humans, and that the acetylcholine portions of these circuits appear to originate within the brain stem and the spinal cord. No less striking is the discovery, by David Felten, of the University of Rochester, of a similarly unique mesh of nerves in the spleen, bone marrow, and lymph nodes, as well as in the thymus. These neural fibers follow blood vessels into the glands and radiate into fields profuse with T cells, where they likely manipulate lymphocyte processes. All the biological equipment required for such direct intervention is at hand: adrenaline, norepinephrine, and acetylcholine receptors have been identified on the surfaces of lymphocytes.

The discovery that autonomic nerves interlace lymph tissues has recast our view of the immune system. As a whole, it now seems to resemble an endocrine gland, and, like all endocrine tissue, possesses a direct anatomical link to the brain. Though it remains to be seen whether messages carried by that anatomical link flow in both directions—as they do with the endocrine glands—there now is ample reason to believe that such two-way communication takes place by means of the hypothalamic-pituitary-adrenal axis. Since the substances that exert this influence—adrenocorticotropic hormone and beta-endorphin, an internally manufactured opiate; lymphokines and cytokines, the chemical products of macrophages; and the hormonelike thymosins, which are synthesized in the thymus—originate within the immune system itself, they might be called immunotransmitters.

Unknown until recently, immunotransmitters are now turning up everywhere in the body. Here is a partial list of the disparate phenomena under review: Eric Smith and J. Edwin Blalock, of the University of Texas Medical Branch at Galveston, have reported that adrenocorticotropic hormone and beta-endorphin, once thought to have been released only by the pituitary and the brain, are also secreted by lymphocytes. This suggests that many cells of the immune system act as separate, minuscule glands, dispensing agents that modulate the immune response as they migrate throughout the body. According to a 1984 study by James Krueger and his colleagues, of the University of Health Sciences-Chicago Medical School, the cytokine interleukin-1 (also called T-cell activating factor) induces deep sleep and hyperthermia. Both effects accelerate disease recovery and can thus be regarded as elements of the body's sequence of defensive tactics. And in still another confirmation of the action of immunotransmitters, Hugo Besedovsky, of the Swiss Research Institute, has shown that the firing rates of neurons in the brain can be altered by lymphokines similar to those that assist in the fight against viruses.

But it is the family of substances that originate in the thymus—the thymosins—that best demonstrate how the immune system may influence the central nervous system. When injected into the cavities of the brain, thymosin beta four stimulates the pituitary, via the hypothalamus, to release luteinizing hormone, which helps regulate other endocrine glands. Further, adrenocorticotropic hormone and beta endorphin are released by isolated pituitary cells when cultured in the presence of another group of thymic hormones—thymosin fraction five. But most important, it has been shown that immunotransmitters can exercise a direct influence on a neurotransmitter. Hugo Besedovsky has found that the amount of norepinephrine, the neurotransmitter that suppresses the secretion of corticotropin-releasing factor by the hypothalamus, in the brains of rats drops significantly after administration of a preparation that contains thymosin and lymphokines. If the same holds true for humans, it may be that the thymosins can themselves stimulate the hypothalamus by way of a neurotransmitter and thereby set in motion the chain of chemical reactions associated with the hypothalamic-pituitary-adrenal axis.

The precise function of thymosins and other immunotransmitters in the body's defense against viruses and bacteria has yet to be explained, but they most likely modulate the immune response and therefore constitute the final stage of a feedback loop between the central nervous system and the immune system's lymphatic tissues. The concept of a self-correcting functional circuit that ties the brain to the pituitary, adrenal, and thymus glands has been proposed by many investigators. But missing from this hypothetical circuit has been evidence of chemically defined molecular signals between the thymus and the pituitary, by which the brain adjusts immune responses and the immune system alters nerve cell activity. That link is now known to include such immunotransmitters as thymosin.

CLINICAL PRACTICE cannot long remain unaffected by this research. The discovery of pathways that bind the brain and the immune system rescues the behavioral approach to disease from the shadowy practices of witch doctors and places it squarely within the rational tradition of Western medicine. Aware now of the complex physiological basis for behavioral modification of the immune response, physicians can spend less time fielding criticism and more time exploring which types of therapy are of the greatest benefit. We are witnessing the birth of a new integrative science, psychoneuroimmunology, which *begins* with the premise that neither the brain nor the immune system can be excluded from any scheme that proposes to account for the onset and course of human disease.

Regrettably, much of the present debate is preoccupied with the results of behavioral approaches to the treatment of the most intractable, least understood diseases. Consider the argument that unfolded in the pages of *The New England Journal of Medicine* during 1985. In the journal's June 13 issue, Barrie R. Cassileth and her colleagues, of the University of Pennsylvania Cancer Center, reported that in a study of three hundred fifty-nine cancer patients no correlation could be found between behavioral factors and progression of the illness. The same issue included an editorial in which Marcia Angell, a physician and deputy editor of the journal, cited Cassileth's findings to support the view that the contribution of mental states to the cause and cure of disease is insignificant.

If they accomplished nothing else, the responses, to both the study and the editorial, published in the journal's November 21 issue, called attention to the difficulties facing anyone attempting to unravel the relationship between illness and what one writer called the "dynamic richness and variety of human experience." Most of the correspondents had found (or, at least, acknowledged) evidence for the contribution of mental states to human health, but none could agree as to how such states might affect the myriad immunological factors that come into play during the course of a malignant disease.

Unquestionably, much more research regarding the behavioral approach to illnesses such as cancer has to be done. But in the meantime, the most immediately realizable applications of behavioral medicine lie elsewhere. The first, and, in the long run, the most valuable, clinical spinoffs of psychoneuroimmunology will be in disease prevention—initially, in the development of ways to manage stress. As we study further the relationship between behavior and the biochemistry of immunity, the aim should not be to replace the witch doctor with a Western equivalent so much as to reduce the need for both.

Major Personality Study Finds That Traits Are Mostly Inherited

Data on twins will fuel nature vs. nurture debate.

Daniel Goleman

The genetic makeup of a child is a stronger influence on personality than child rearing, according to the first study to examine identical twins reared in different families. The findings shatter a widespread belief among experts and laymen alike in the primacy of family influence and are sure to engender fierce debate.

The findings are the first major results to emerge from a long-term project at the University of Minnesota. Since 1979, more than 350 pairs of twins in the project have gone through six days of extensive testing that has included analysis of blood, brain waves, intelligence and allergies.

The results on personality are being reviewed for publication by the Journal of Personality and Social Psychology. Although there has been wide press coverage of pairs of twins reared apart who met for the first time in the course of the study, the personality results are the first significant scientific data to be announced.

For most of the traits measured, more than half the variation was found to be due to heredity, leaving less than half determined by the influence of parents, home environment and other experiences in life.

The Minnesota findings stand in sharp contradiction to standard wisdom on nature versus nurture in forming adult personality. Virtually all major theories since Freud have given far more importance to environment, or nurture, than to genes, or nature.

Even though the findings point to the strong influence of heredity, the family still shapes the broad suggestion of personality offered by heredity; for example, a family might tend to make an innately timid child either more timid or less so. But the inference from this study is that the family would be unlikely to make the child brave.

The 350 pairs of twins studied included some who were raised apart. Among these separately reared twins were 44 pairs of identical twins and 21 pairs of fraternal twins. Comparing twins raised separately with those raised in the same home allows researchers to determine the relative importance of heredity and of environment in their development. Although some twins go out of their way to emphasize differences between them, in general identical twins are very much alike in personality.

But what accounts for that similarity? If environment were the major influence in personality, then identical twins raised in the same home would be expected to show more similarity than would the twins reared apart. But the study of 11 personality traits found differences between the kinds of twins were far smaller than had been assumed.

"If in fact twins reared apart are that similar, this study is extremely important for understanding how personality is shaped," commented Jerome Kagan, a developmental psychologist at Harvard University. "It implies that some aspects of personality are under a great degree of genetic control."

The traits were measured using a personality questionnaire developed by Auke Tellegen, a psychologist at the University of Minnesota who was one of the principal researchers. The questionnaire assesses many major aspects of personality, including aggressiveness, striving for achievement, and the need for personal intimacy.

For example, agreement with the statement "When I work with others, I like to take charge" is an indication of the trait called social potency, or leadership, while agreement with the sentence "I often keep working on a problem, even if I am very tired" indicates the need for achievement.

Among traits found most strongly determined by heredity were leadership and, surprisingly, traditionalism or obedience to authority. "One would not expect the tendency to believe in traditional values and the strict enforcement of rules to be more an inherited than learned trait," said David Lykken, a psychologist in the Minnesota project. "But we found that, in some mysterious way, it is one of the traits with the strongest genetic influence."

Other traits that the study concludes were more than 50 percent determined by heredity included a sense of well-being and zest for life; alienation; vulnerability or resistance to stress; and fearfulness or risk seeking.

Another highly inherited trait, though one not commonly thought of as part of personality, was the capacity for becoming rapt in an aesthetic experience, such as a concert.

Vulnerability to stress, as measured on the Tellegen test, reflects what is commonly thought of as "neuroticism," according to Dr. Lykken. "People high in this trait are nervous and jumpy, easily irritated, highly sensitive to stimuli, and generally dissatisfied with themselves, while those low on the trait are resilient and see themselves in a positive light," he said. "Therapy may help vulnerable people to some extent, but they seem to have a built-in susceptibility that may mean, in general, they would be more content with a life low in stress."

The need to achieve, including ambition and an inclination to work hard toward goals, also was found to be genetically influenced, but more than half of this trait seemed determined by life experience. The same lower degree of hereditary influence was found for impulsiveness and its opposite, caution.

The need for personal intimacy appeared the least determined by heredity among the traits tested; about two-thirds of that tendency was found to depend on experience. People high in this trait have a strong desire for emotionally intense relationships; those low in the trait tend to be loners who keep their troubles to themselves.

"This is one trait that can be greatly strengthened by the quality of interactions in a family," Dr. Lykken said. "The more physical and emotional intimacy, the more likely this trait will be developed in children, and those children with the strongest inherited tendency will have the greatest need for social closeness as adults."

No single gene is believed responsible for any one of these traits. Instead, each trait, the Minnesota researchers propose, is determined by a great number of genes in combination, so that the pattern of inheritance is complex and indirect.

No one believes, for instance, that there is a single gene for timidity but rather a host of genetic influences. That may explain, they say, why previous studies have found little connection between the personality traits of parents and their children. Whereas identical twins would share with each other the whole constellation of genes that might be responsible for a particular trait, children might share only some part of that constellation with each parent.

That is why, just as a short parent may have a tall child, an achievement-oriented parent might have a child with little ambition.

The Minnesota findings are sure to stir debate. Though most social scientists accept the careful study of twins, particularly when it includes identical twins reared apart, as the best method of assessing the degree to which a trait is inherited, some object to using these methods for assessing the genetic component of complex behavior patterns or question the conclusions that are drawn from it.

Further, some researchers consider paper-and-pencil tests of personality less reliable than observations of how people act, since people's own reports of their behavior can be biased. "The level of heritability they found is surprisingly high, considering that questionnaires are not the most sensitive index of personality," said Dr. Kagan. "There is often a poor relationship between how people respond on a questionnaire and what they actually do."

"Years ago, when the field was dominated by a psy-

The Roots of Personality

The degree to which eleven key traits of personality are estimated to be inherited, as gauged by tests with twins. Traits were measured by the Multidimensional Personality Questionnaire, developed by Auke Tellegen at the University of Minnesota.

SOCIAL POTENCY   **61%**

A person high in this trait is masterful, a forceful leader who likes to be the center of attention.

TRADITIONALISM   **60%**

Follows rules and authority, endorses high moral standards and strict discipline.

STRESS REACTION   **55%**

Feels vulnerable and sensitive and is given to worries and easily upset.

ABSORPTION   **55%**

Has a vivid imagination readily captured by rich experience, relinquishes sense of reality.

ALIENATION   **55%**

Feels mistreated and used, that "the world is out to get me."

WELL-BEING   **54%**

Has a cheerful disposition, feels confident and optimistic.

HARM AVOIDANCE   **51%**

Shuns the excitement of risk and danger, prefers the safe route even if it is tedious.

AGGRESSION   **48%**

Is physically aggressive and vindictive, has taste for violence and is "out to get the world."

ACHIEVEMENT   **46%**

Works hard, strives for mastery and puts work and accomplishment ahead of other things.

CONTROL   **43%**

Is cautious and plodding, rational and sensible, likes carefully planned events.

SOCIAL CLOSENESS   **33%**

Prefers emotional intimacy and close ties, turns to others for comfort and help.

The New York Times/Dec. 1, 1986

chodynamic view, you could not publish a study like this," Dr. Kagan added. "Now the field is shifting to a greater acceptance of genetic determinants, and there is the danger of being too uncritical of such results."

Seymour Epstein, a personality psychologist at the University of Massachusetts, said he was skeptical of precise estimates of heritability. "The study compared people from a relatively narrow range of cultures and environments," he said. "If the range had been much greater—say Pygmies and Eskimos as well as middle-class Americans—then environment would certainly contribute more to personality. The results might have shown environment to be a far more powerful influence than heredity," he said.

Dr. Tellegen himself said: "Even though the differences between families do not account for much of the unique attributes of their children, a family still exercises important influence. In cases of extreme deprivation or abuse, for instance, the family would have a much larger impact—though a negative one—than any found in the study. Although the twins studied came from widely different environments, there were no extremely deprived families."

Gardner Lindzey, director of the Center for Advanced Studies in the Behavioral Sciences in Palo Alto, Calif., said the Minnesota findings would "no doubt produce empassioned rejoinders."

"They do not in and of themselves say what makes a given character trait emerge," he said, "and they can be disputed and argued about, as have similar studies of intelligence."

For parents, the study points to the importance of treating each child in accord with his innate temperament.

"The message for parents is not that it does not matter how they treat their children, but that it is a big mistake to treat all kids the same," said Dr. Lykken. "To guide and shape a child you have to respect his individuality, adapt to it and cultivate those qualities that will help him in life."

"If there are two brothers in the same family, one fearless and the other timid, a good parent will help the timid one become less so by giving him experiences of doing well at risk-taking, and let the other develop his fearlessness tempered with some intelligent caution. But if the parent shelters the one who is naturally timid, he will likely become more so."

The Minnesota results lend weight and precision to earlier work that pointed to the importance of a child's temperament in development. For instance, the New York Longitudinal Study, conducted by Alexander Thomas and Stella Chess, psychiatrists at New York University Medical Center, identified three basic temperaments in children, each of which could lead to behavioral problems if not handled well.

"Good parenting now must be seen in terms of meeting the special needs of a child's temperament, including dealing with whatever conflicts it creates," said Stanley Grossman, a staff member of the medical center's Psychoanalytic Institute.

THE
CLOCK
WITHIN

Philip Hilts

There is a drawer called "the catastrophe file" in the office of biologist Charles Ehret at the Argonne National Laboratory. It holds reports of various disasters: an air crash that killed hundreds, a ship collision in which crew members drowned, a hospital accident that will bring members of the staff to trial, embarrassing errors in the Pioneer space probes, blunders in Middle East diplomacy.

All the items have one element in common. In each case, human biology was a factor, if not the primary cause of the accident. Or to be more precise, the cause was failure to account for human biology .

Among the catastrophes on Ehret's list is the nuclear accident at Three Mile Island. It was on the night shift, at 4:01 A.M. on a chilly March morning, that three young men sat in the control area at the nuclear power station. The three worked on a shift system called slow rotation—days for a week, evenings for a week, then late nights for a week. If a biologist like Ehret, who is versed in a relatively new discipline called chronobiology, were to design a shift to guarantee the worst possible human performance, slow rotation might well be it.

During the first 100 minutes of the nuclear accident, the control room workers made a surprising series of mistakes. Fourteen seconds into the accident, one controller failed to see two warning lights. A few seconds later, a valve that should have closed did not, but operators did not realize it. As the president's investigating commission later said, "Throughout the first two hours of the accident, the operators ignored or failed to recognize the significance of several things that should have warned them that they had an open valve and a loss-of-coolant accident. . . ." The president's commission concluded that ". . . except for human failures, the major accident at Three Mile Island would have been a minor incident."

One of the relationships discovered by chronobiology is that the many rhythms within the human body normally move in a set synchrony with one another. For example, body temperature, pulse, and sleep-wake cycles follow roughly the same beat, while other processes vary with a quite different beat—and the relations among them may change, slightly but predictably, day after day. Each gland, each organ, each chemical has its own beat, and together they are orchestrated as harmoniously as the players in a symphony.

They are harmonious until some internal change or catastrophe occurs. Illness can throw the body's rhythms out of phase or frequency. Drugs can do it. Jet travel over many time zones can do it. Putting workers suddenly onto the night shift can do it.

Over the past three decades biologists have found that in practically every function of living systems, time is of the essence. The rhythmic frequency of many biological functions operates approximately on a 24-hour cycle, which led Franz Halberg at the University of Minnesota to coin the term "circadian rhythm." In Latin *circa* means "about," *dies* means "a day."

We think of our heartbeat as a constant, but it is not. It will vary by as much as 20 or 30 beats per minute in 24 hours. At one time of day it may beat 60 times per minute, and at the opposite time of the circadian cycle, it may beat 80 or more times per minute. Blood pressure may measure 120 over 80 in the morning, but in the evening it is likely to be higher, possibly as much as 140 over 100, an unusually high reading. Body temperature does not hover around 98.6 degrees, as most of us believe, but varies by one and a half to two degrees over a day—from as low as 97 degrees to more than 99 degrees. The scores of other functions which have been shown to swing up and down widely during the day include more than three dozen separate chemicals in the blood and urine, as well as mood, vigor, eye-hand coordination, counting, time estimation, addition, and memory. Cell division rate in the body has been found to vary by 1,200 percent over a day; one chemical in the rat's pineal gland varies by 900 percent over a day.

After the TMI nuclear accident, several utilities, including the General Public Utilities Corporation which runs the plant, turned to biologists for help in lessening the dangers of erratic human performance.

"We may be able to improve the situation by several orders of magnitude," says Ehret, who has consulted with the nuclear utilities and drawn up alternate night work plans, all according to biologically based rules that have emerged from research in chronobiology.

But there is another, equally critical area where a failure to take into account the circadian fluctuations of the body's natural rhythms will lead to trouble: medicine and medical research.

In medicine, the prevailing biological idea has been that the body seeks equilibrium, a steady state. When it is ill, say with a fever, it tries to return to wellness by sweating and other means of cooling itself. This is the homeostatic view. It states that in each bodily function there is an ideal mean, and that a healthy person's functions will flutter randomly about that middle number. For example, in temperature, 98.6 degrees is the accepted mean. In blood pressure, 120 over 80 is the rough center.

Now it is clear that this approach is faulty, or at least incomplete. Bodily functions do fluctuate, but not randomly. They move up and down quite regularly. They keep in order among themselves, and the varying numbers they produce may result in all sorts of different interpretations by doctors who are unaware of this wide, normal variation.

Since so much of the chemistry and biology of the body changes each day, and changes by as much as tenfold, biologists have gradually come to realize that an animal—whether human or guinea pig—is virtually a different creature, physically and chemically, at different times of day.

This means that we may be more susceptible not only to accidents but also to disease at certain predictable times. It means that a drug taken at one time acts differently than the same drug, in the same dose, taken a few hours later or earlier.

Chronobiology has already begun to transform the way biological research is conducted, and it is expected to have a major effect on all of medicine as well. It can alter results at every stage of practice, from preventing time-linked illnesses, to improving diagnoses, to improving treatment. Some of the conventional medical rules of thumb may have to be abandoned; for example, drugs are now given three times a day or four times a day, completely without regard for the differing effects they produce at different hours.

The evidence from the laboratory is clear: When rats are given a nearly lethal dose of amphetamine, their survival depends chiefly on what time the agent is given. At one time in the animal's circadian cycle, six percent of them die. At another time, 78 percent die.

Certain insects have now been found to be far more vulnerable to some commonly used insecticides in the afternoon than at any other times. Rats given a sleep-causing drug will nap for about 50 minutes when the drug is given at one time, but will sleep twice as long when the drug is given later.

High levels of noise usually cause convulsions in animals. But the probability that an animal will be thus afflicted varies according to its body's clock. At the worst time, the probability is 100 percent higher than the daily mean. At the best time, it is 80 percent lower than the daily mean.

In yet another study, rats were given enough phenobarbital to kill at least half of them. But during the most favorable time, none of the animals died. At the least favorable time, the same dose killed all of them. The list could go on. There are already many drugs and active chemicals whose ability to kill or to cure has been shown to swing widely with the rhythms of an animal's body. There are, in short, "windows" of daily drug resistance and effectiveness.

Because of this, chronobiologists say, the results of some previous drug and cancer research studies are now dubious. Chronobiologists suggest that studies of toxicity, especially of the behavioral effects of toxic agents, must now be completely redone. At the very least, the conduct of scientific research must be changed for all future studies. Time must now be included as a major factor in medical and biological equations.

Colin Pittendrigh, among the most respected of the biological researchers in the field, says that this kind of biological research has begun to have an important impact on medicine. "There are some very important findings, principally from Franz Halberg and his people, on the time-of-day dependence of drug action. That is a first-rate result . . . but I have talked to good pharmacologists, and most of them don't yet know the facts. That is worth reporting: The pharmacological fraternity is not informed on what's been found."

One thing that Pittendrigh suggests should be done immediately is change the federal guidelines that govern research. The time of an animal's cycle must be taken into account in the research that determines the safety and effectiveness of drugs, he says.

Federal agencies provide most of the money for basic research and they also set the guidelines. Thus far, the government has not recognized chronobiology as a major variable in setting guideline policy, and it now appears that a significant factor is being left out of most of the testing.

Despite the evidence that has built up, the common practice in laboratory testing includes such hazardous actions as testing nocturnal animals in the daytime; testing animals that have not had time to adjust to a new laboratory environment; testing animals that are kept in crowded cages, a condition which has been shown to cause altered internal rhythms among animals; testing animals in light and dark cycles that are unregulated; testing animals in conditions subject to frequent disturbances, such as turning on lights during a dark phase; and testing animals whose feeding schedules are not fixed and recorded. All of these can cause altered rhythms in animals, but none are mentioned in federal research guidelines. Meanwhile,

Chronology of chronobiology

The curiosity about rhythmical events in nature is ancient, but probably the first experimental test of the idea occurred in 1729. The French astronomer Jean de Mairan had become curious about a heliotrope plant that opened its leaves to the morning light and closed them at dusk. He discovered that the opening and closing was apparently not controlled by the light in the plant's environment.

Observations like de Mairan's were recorded again and again over the next two centuries by scientists in many fields. At the beginning of this century a major question was whether the cycles in plant behavior were being regulated by a clock within the plant, or by some outside rhythm such as night and day or changing temperature. Experiments that attempted to disrupt the 24-hour rhythms of plants, and later insects and animals, failed. The cycles persisted despite environmental changes.

It was Colin Pittendrigh of Stanford who, in the middle 1950s, finally put into clear terms all the evidence collected up to that time. He and others had proved that clocks exist in life forms as simple as single-celled animals and as complex as man. Internal clocks, he said, are a fundamental property of life.

While clocks are definitely internal, they may follow external rhythms, be reset by external rhythms, or be disrupted by them. It is, he said, as if two oscillators were operating—one inside the body and one outside.

Most recent research in the field has centered on one main question: Where within the body is the clockwork? Is it an organ or one function of an organ? Is it within the cells, and if so, where within the cells?

The sophisticated clocks found operating in single-celled animals settled at least part of the question by proving that one need not look to higher life forms for a fully operating biological timer. Two competing lines of recent research have tried to locate the clock mechanism either in the membrane of the cell, or in the process of protein manufacture.

The first approach holds that the membranes of a cell function by opening and closing channels through which chemicals pass. The process is thought to be electrical: When a certain number of ions build up on one side of a membrane, the flow across it is shut off until the ionic concentration drops again, triggering a feedback mechanism which starts the flow of material through membrane channels again. This theoretical model for an open-and-close rhythm within the cell may be the oscillator, the heart of the ticking clock.

The other approach puts the rate of protein-making at the center of the clockwork. Research has shown again and again that when cells are flooded with substances that foul up the manufacture of proteins, the cells' clocks are reset to a new time. But until recently, it was not certain that the clock was reset only because the protein-making process was disturbed.

Now, in research just completed, Jerry Feldman a former student of Pittendrigh, has demonstrated that one particular chemical that slows the clock does so specifically because it damages protein-making. Cycloheximide injected into the ordinary strain of *Neurospora* fungus normally will foul up the cell's internal clock. But Feldman used a mutant strain of *Neurospora*, one in which cycloheximide does not damage the protein-making machinery. He wanted to see how the cell's clocks would react to doses of cycloheximide.

Since the protein machinery is unaffected by the chemical, the cell clock should run normally despite the addition of cycloheximide. If the clock changed, however, it would mean that something besides the protein machinery functioned as the clock.

The clocks of the fungus ran normally. "We proved that protein synthesis is necessary for the clock to run," he said. "Of course," he added, "since proteins are also important to the function of cell membranes, it is probable that both approaches are correct, differing only in emphasis. The one may depend on the other, and thus the two approaches merge."

---

many far less important factors are mentioned.

Lawrence D. Scheving at the University of Arkansas Medical Center, whose work has established chronobiological effects for a number of drugs, says that time is a variable important enough to affect the validity of some experiments and the accuracy of many.

"We are very careful about controlling the sex, weight, age, and other things. Time is equally important if not more important than the other variables that we rigorously control for," Scheving says.

One kind of study commonly carried out in toxicology laboratories is the "LD50" study. Animals are injected with a poisonous agent to determine what dose is lethal to 50 percent of the animals. Commonly, rats for testing may be flown in from halfway across the country, hurriedly brought to the lab, and given the lethal compound. There is no quarantine, no adjustment to environment, no regular feeding regimen, and no attention to the light-dark cycle. Similar tests for the same drug often provide results that vary as much as 100 per-

cent. Paying no attention to an animal's cycle, says Morris Cranmer, the former director of the National Center for Toxicological Research, results in studies that are consistently inaccurate. "This means the variation in experiments is increased—and as you increase the variation, you decrease the resolving power of the experiments. That is my concern," says Cranmer.

He added another reason to take biological time into account, a reason that may be even more important than inaccurate results. If a strong effect of biological time is ignored by federal regulations, food and drug manufacturers could use to their own advantage an animal's resistance to toxic effects. Toxicity tests could be designed around an animal's highest tolerance period so that resulting effects would be the most negligible.

The attitude of federal officials varies greatly. William D'Aguanno, the Food and Drug Administration's officer on the toxicology of new drugs, says the FDA has no policy on chronobiology. But the attitude of Cranmer's successor at the National Center for Toxicological Research may be seen as a bellwether. Thomas Cairns came into the NCTR when that agency was about to conduct a series of experiments that would have tested the importance of biological time in toxicology. That study got scrapped in the transition from Cranmer's to Cairns' administration. The matter was left there, untouched, for two years. A reporter recently asked Cairns about the im-portance of chronobiology. He was not up on the literature, he said, and felt that it was probably not very important. But he requested a few days to look into the matter.

Cairns has now begun to worry about the effects chronobiology may have on scientific research funded by the government and on industrial research as well. "My problem is . . . that this could easily be used in reverse to a manufacturer's benefit."

A good example of the way in which chronobiology might possibly change medical practice is in the treatment of high blood pressure. Howard Levine, chief of medicine at New Britain General Hospital in Connecticut, and long interested in chronobiology, suffers from high blood pressure. But, he says, "diagnosis of my condition was delayed a couple of years because I used to go to the doctor in the morning. That is when my blood pressure was at its circadian low, and so it seemed normal when actually it wasn't."

Levine also found that late at night the percentage of red blood cells in his blood normally dropped by about four and a half percent, and the total volume of red blood cells in his blood changed by almost ten percent. A five-percent drop in red blood cells often prompts a transfusion in a hospital setting.

Frederic Bartter, an eminent endocrinologist at the University of Texas who has done numerous studies of drugs, blood pressure, and their rhythms, points out that "every drug that has ever been explored for the rhythmicity of its action has been found to have such rhythms." In normal practice, it is quite likely that physicians are prescribing drug doses that are anywhere from half to twice as much as a patient needs.

Once a patient's 24-hour rhythm is established, blood pressure need not be measured repeatedly because the pattern is predictable. "Then you can *know* to what extent a given medication will take away those peaks," says Bartter. "To

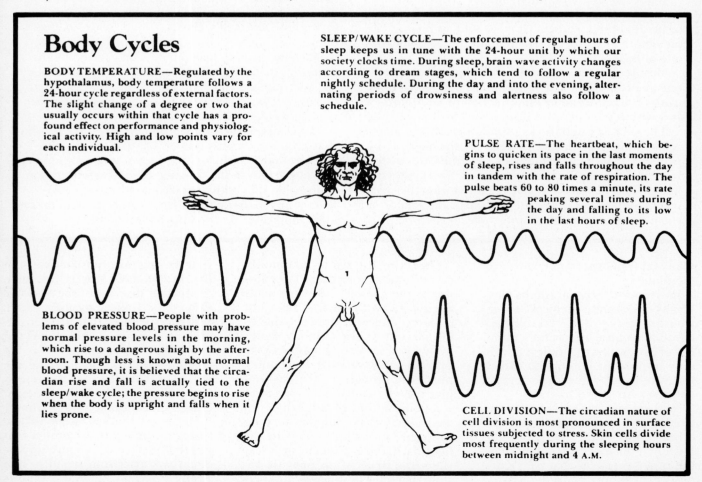

Body Cycles

BODY TEMPERATURE—Regulated by the hypothalamus, body temperature follows a 24-hour cycle regardless of external factors. The slight change of a degree or two that usually occurs within that cycle has a profound effect on performance and physiological activity. High and low points vary for each individual.

SLEEP/WAKE CYCLE—The enforcement of regular hours of sleep keeps us in tune with the 24-hour unit by which our society clocks time. During sleep, brain wave activity changes according to dream stages, which tend to follow a regular nightly schedule. During the day and into the evening, alternating periods of drowsiness and alertness also follow a schedule.

PULSE RATE—The heartbeat, which begins to quicken its pace in the last moments of sleep, rises and falls throughout the day in tandem with the rate of respiration. The pulse beats 60 to 80 times a minute, its rate peaking several times during the day and falling to its low in the last hours of sleep.

BLOOD PRESSURE—People with problems of elevated blood pressure may have normal pressure levels in the morning, which rise to a dangerous high by the afternoon. Though less is known about normal blood pressure, it is believed that the circadian rise and fall is actually tied to the sleep/wake cycle; the pressure begins to rise when the body is upright and falls when it lies prone.

CELL DIVISION—The circadian nature of cell division is most pronounced in surface tissues subjected to stress. Skin cells divide most frequently during the sleeping hours between midnight and 4 A.M.

know how bad it is, you *must* have some estimate of a patient's peaks, and not just his morning pressure."

Probably the most studied of all human rhythms are those in sleep, and one of the interesting findings of researchers Elliot D. Weitzman and Charles Czeisler in New York is that the label "insomnia" covers at least two quite different ailments. Insomnia has been presumed to be a disruption of the normal sleep pattern, but the Montefiore Hospital and Medical Center researchers have found that 10 to 15 percent of "insomniacs" have no trouble at all sleeping a full eight hours. Their trouble is that their bodies insist that they sleep their full night between about 4 A.M. and 11 A.M.

These night owls are unaware of this natural rhythm and spend their lives being analyzed by psychiatrists and drugged by doctors because their natural sleep cycle wreaks havoc with their jobs and social lives. Weitzman calls the malady "delayed sleep-phase syndrome." He corrects the problem by getting patients to go to bed three hours later every day until they have wrapped all the way around and end up awakening early in the morning.

Weitzman's patients, once they have been shifted, have been able to adjust easily and have continued to sleep normally for the several years since the treatment began. After an average of more than a year, none of his patients has relapsed. "This may apply to 10 to 15 percent of all insomniacs. There are 25 million insomniacs, so we are talking about two to three million people . . . ."

One of the most important medical applications of chronobiology may be in cancer treatment. Franz Halberg of the University of Minnesota is perhaps the foremost ad-

vocate of "chronotherapy," and he has led a group in studying the possible application of chronobiology to cancer therapy. The group has proved that an animal's resistance to cancer drugs is high at one time and low at another. It has also proved that if this changing resistance is exploited by changing the drug dosage according to time of resistance, it is possible to improve treatment effects by giving animals much higher total doses of cancer drugs than was possible before. Or, it is possible to give the same dose far more safely.

"We do know that some death occurs in cancer treatment not due to the disease, but due to the drug," says Erhard Haus of Minnesota. The number of deaths caused by the treatment rather than the disease may be as high as ten percent. If drugs were used at different times in different doses, Haus believes, lives would be saved.

It is also hoped that the cycles of cancer cells themselves may be shifted so that their resistance to drugs will be low when the body's resistance to drugs is high. Then the maximum possible dose could be given for maximal therapeutic and minimal toxic effect. These hopes have been realized in practical cancer therapy being conducted by Francis Levi of France, Salvador Sanchez de la Pena of Mexico, and William Hrushesky of the Masonic Cancer Center of the University of Minnesota, in cooperation with Halberg and Haus. "Even now, even with all the money spent by the National Cancer Institute and others to compare treatment effects with different drugs and different combinations of drugs," says Hrushesky, "there has not yet been a single study anywhere in America that has taken time into

account. Not one. We will try to do that next."

Other potential applications of chronobiology include treatments ranging from an improved drug therapy for asthmatics and rheumatics to a cure for jet lag. Chronobiology has shed new light on the cause of manic depression with the work of Thomas Wehr and Frederick Goodwin at the National Institutes of Health. They have achieved temporary cures in bringing depressive patients' wobbling body rhythms back into step.

Though there is resistance to chronobiology from doctors because they will have to make more measurements to diagnose and treat patients, though there is resistance from federal bureaucracy, including such agencies as the FDA, the Federal Aviation Administration, and Environmental Protection Agency, still the advances in the biology of time go on. Chronobiology's advocates maintain that the emerging discipline will gradually take its place alongside others that cut across conventional boundaries of study, such as genetics, evolution, and developmental biology. Each of these is a synthesizing discipline, one that brings together many ideas and fields of work, and knits them into a single pattern of knowledge. Each deals in a different scale of time: Evolution counts in millions of years, developmental biology counts in tens of years, and, says Charles Ehret, chronobiology "deals with the phenomenology of minutes and days, and contains more hardcore rules of general and predictive consequence than any of the others with the exception of genetics."

Slowly the question turns from whether biological time will be accounted for in science and health, to when. It is a matter of time.

CHEMICAL CROSS TALK

Why Human Cells Understand the Molecular Messages of Plants

Jesse Roth and Derek LeRoith

Jesse Roth is director of intramural research at the National Institute of Diabetes and Digestive and Kidney Diseases, in Bethesda. Derek LeRoith is chief of the molecular and cellular physiology section of the institute's diabetes branch.

Six hundred centuries after the death of an anonymous Neanderthal man, his bones were found in a cave in Iraq, surrounded by fossilized grains of pollen. The pollen belonged to eight genera of flowers, and their selection appears to have been far from haphazard: seven of the eight are reputed to have medicinal powers. Indeed, even today Iraqis make use of all seven, including the grape hyacinth, as a diuretic; the roots of the marsh mallow, for intestinal maladies; and the leaves of the shrub *Ephedra*, as a remedy for asthma (its active ingredient, ephedrine, can be found in over-the-counter cold remedies).

The Neanderthals were hardly alone in their exploration of herbal medicine. Cultures the world over have independently discovered that the ingredients of plants have a variety of predictable effects on human health and behavior. The anesthetic properties of the white juice within the young opium poppy were noted two millennia ago by the Greek surgeon Dioscorides, in his pharmacological text, *On Medical Remedies*, and the like properties of its derivatives, including morphine and codeine, have been rediscovered by generations of physicians and drug users since. Peyote cactus buttons, which can induce hallucinations, have been at the center of American Indian rituals for centuries, and various other peoples have discovered the similarly hallucinogenic, and sometimes toxic, effects of belladonna. Even the plant from which licorice comes—*Glycyrrhiza glabra*—has a long pharmacological history. It is one of several herbs listed on stone tablets carved thirty-eight hundred years ago by the civil servants of Babylon, and today we know that eating large amounts of licorice can raise blood pressure.

Most early explanations of these various effects involved spiritual or divine forces (even as recently as the mid-nineteenth century, Oliver Wendell Holmes opined that "the Creator himself seems to prescribe [opium], for we often see the scarlet poppy growing in the cornfields, as if it were foreseen that wherever there is hunger to be fed there must also be pain to be soothed"). And the first "scientific" explanations came little closer to the truth. Thus, Galen, the great Greek physician, has not been vindicated in his belief that herbs do their work by altering the balance of the four bodily humors.

During this century, the mystique surrounding plant-derived drugs has been largely dispelled by a scientific explanation: substances in such plants as *Glycyrrhiza glabra* and opium poppies mimic the body's own chemical messengers. Morphine, one of several alkaloids that account for the effects of opium, attaches itself to receptors on the surfaces of human nerve cells, thus "turning on" a cellular program normally activated by endorphins, chemicals released by the body in response to great pain or fatigue. And glycyrrhizic acid, an alkaloid in the licorice plant, binds to receptors on kidney cells that then respond as they do to aldosterone, which the adrenal gland releases to counter low blood pressure.

These and dozens of other such effects are often attributed to coincidence. Certain plant-derived chemicals, it is said, happen to bear a structural or functional resemblance to certain cellular messengers in human beings. But evidence uncovered over the past ten years suggests that the similarities are due not to chance but to common evolutionary descent. The basic biochemical elements of cellular communication, according to the new view, can be found very near the bottom of the tree of evolution—before the plant and animal kingdoms branched off—in unicellular species ancestral to human beings and plants alike. These primordial messenger molecules, or their corresponding cellular receptors, have in some cases changed little over the course of many millions of years, even as their communicative functions have evolved in various directions. (Though plant alkaloids such as morphine may have developed later in evolution, the new view suggests that natural selection shaped them to fit plant cell receptors that happened to be related to, and therefore structurally similar to, receptors in human cells.)

This view, which can aptly be called the "unifying theory of intercellular communication," aims to explain in a single stroke the many coincidences involving plant molecules and human cells. Indeed, it suggests answers to a variety of perplexing questions about links between the animal and plant kingdoms—why pigs can detect truffles buried three feet underground, for instance, and why spinach plants contain chemicals almost indistinguishable from human insulin. But the theory merits the term *unifying* for a second reason, as well. During this century, the study of cellular communication within vertebrates, especially human beings, has fallen into fragmentation, as scientists have discovered diverse chemical messengers that do not fit easily into traditional categories. The new theory promises to tidy up the conceptual disarray by returning to the roots of the tree of life and revealing the underlying unity of many, if not all, forms of intercellular communication.

DURING THE NINETEENTH CENTURY, the existence of chemical messengers, such as aldosterone, that are secreted by one cell and affect the behavior of distant cells was scarcely suspected. Orthodox opinion held that all coordination of cellular behavior in vertebrates is controlled by the transmission of information along lines of consecutive cells in the nervous system. But in 1902, the English physiologists Ernest Starling and William Bayliss observed behavior that could not be explained in these terms.

It was already known that the entry of undigested food into the intestine causes the pancreas to release digestive juices, and it was believed that nerve cells relay reports of the food's presence to the pancreas. To their surprise, Starling and Bayliss found that the pancreas responded to the presence of food even after the nerve cells had been severed; the pancreas ceased to respond only after blood vessels leading to it had been tied. Thus, they reasoned that the bloodstream, not the nervous system, conveys the information controlling pancreatic activity and that a chemical messenger—secretin, they called it—must be synthesized in the intestine and released into surrounding capillaries.

Suspecting that he had happened on a new class of chemical messengers, Starling gave them a name: hormone, from the Greek *hormon*, meaning "arouse to activity." He defined a hormone as a messenger molecule produced by cells in one organ and carried, via the bloodstream, to target cells in a distant organ. The hormone-producing tissues were termed endocrine glands, to distinguish them from exocrine glands, whose secretions go directly into a hollow internal organ or a body cavity or onto the surface of the skin, rather than into the blood. (Thus, digestive juices and saliva are products of the exocrine system; insulin and adrenaline are endocrine products.)

The conclusion drawn from Starling's work was that there are two basic systems of cellular communication in the higher vertebrates, working separately: the nervous system and the endocrine system. This belief was a natural one, given the data and investigative methods available at the time. In fact, many scientists still subscribe to it. But evidence gathered since has called this simple, binary system of classification into question.

Among the earliest complications was the discovery in the early twentieth century that when nerves send signals to other nerves or to muscles or glands, they typically do so not by strictly electrical means, as had been thought, but with the help of chemicals called neurotransmitters. A neurotransmitter is synthesized inside a neuron, transported along the neuron's axon, and then, upon excitation by an electrical impulse, discharged into the synaptic cleft, the gap between one nerve cell and the next. The neurotransmitter typically travels across the cleft to receptors on the membrane of the adjacent nerve cell, which may then repeat the process, relaying the signal to yet another cell. Thus, a neurotransmitter has some characteristics of a hormone—it is secreted by one cell and carries its message to a target cell—yet does not ultimately qualify as one, since it does not travel long distances through the bloodstream.

The line between neuronal and hormonal communication was further blurred by the discovery of catecholamines, chemicals that can serve either as hormones or as neurotransmitters. For example, noradrenaline, which is secreted by the adrenal gland to help mobilize the body for fight or flight, is also produced by nerve cells at their junctions with involuntary muscles. And, in addition to the catecholamines, there are chemicals that function as hormones yet are synthesized not by endocrine glands but by neurons within the brain, and then discharged into adjacent blood vessels. One such chemical is vasopressin, which causes the kidney to conserve water. It is synthesized within neurons in the hypothalamus and released from their axons in the posterior pituitary.

It is clear in retrospect that these findings posed serious problems for the binary scheme of classification, but their implications did not at first receive the attention they deserved. Only during the past few decades, with further erosion of the traditional scheme of classification, has its inadequacy been widely recognized.

TO BEGIN WITH, there was the discovery of messenger molecules that, although not involved in transmitting nervous impulses, do not quite qualify as hormones, either. Included in this group are the paracrine molecules, which, rather than traveling to distant target cells through the blood, regulate the activities of neighboring cells. Another type of messenger has been designated exocrine, because it is found in exocrine secretions; somatostatin, for example, which inhibits the gut's absorption of certain nutrients, travels via the gastrointestinal tract, as well as the bloodstream. Still other messengers that do not completely satisfy the classic criteria of hormones are insulinlike growth factors, which accelerate cellular reproduction; antiviral substances, such as interferons; and messengers of the immune system, such as interleukins. The great diversity of these messengers made it clear that Starling's two-system classification had outlived its usefulness.

In the 1960s, Anthony G. E. Pearse, of the Royal Postgraduate Medical School, in London, formulated a theory that, as elaborated during the 1970s in collaboration with Julia M. Polak, accounted for some of the difficulties encountered by Starling's scheme. One difficulty was that the neurotransmitter somatostatin, originally thought to be a product of brain tissue exclusively, was found in abundance in the gastrointestinal tract. Conversely, cholecystokinin, a gastrointestinal hormone, had been found in the brain. It turned out, further, that the neural cells that synthesize somatostatin are remarkably similar, in many of their chemical functions, to the cells in the gut that produce cholecystokinin. Pearse suggested that both kinds of cells derive from the neural crest, the embryologic structure that emerges from the inchoate spinal cord. This theory was dubbed APUD, an acronym based on the technical term for the means of neurotransmitter production—amine precursor uptake and decarboxylation—employed in these two kinds of cells.

The APUD theory was at one level a fairly narrow theory about ontogeny; it explained how certain components of the endocrine system—hormonal cells in the gastrointestinal tract, for example—develop in the course of a vertebrate's life cycle. But when expanded and extended, it suggested a broad theory about phylogeny—an explanation of the *evolutionary* development of the *entire* endocrine system. Though the idea that "ontogeny recapitulates phylogeny"—that embryologic development retraces the steps of evolutionary development—is no longer taken to be precisely true, it may have some validity still, and the APUD theory thus prompted scientists to speculate that, in the course of natural selection, portions of the nervous system had evolved into the precursors of the endocrine system. Such a scenario might indeed explain why the boundary between the two systems of cellular communication is not clear-cut; evolution, in carrying cells from one system to the other, could well have left some in intermediate phases along the way. But while the APUD theory is consistent with the crumbling of Starling's two-system classification, it has been found wanting nonetheless.

In the first place, as a strictly ontogenetic hypothesis, the APUD theory was undermined by the discovery that removing the neural crest from rat and chick embryos does not prevent the development of hormone-producing cells in the gastrointestinal tract, as it should if this part of the endocrine system grows out of the crest. More relevant to the unifying theory of intercellular communication is the shaky status of some phylogenetic extrapolations from the APUD theory. For if endocrine cells evolved from nerve cells, hormonelike messengers could not have existed before unicellular life diverged into the plant and animal kingdoms. Yet the evidence suggests they did.

Aɴʏ ɴᴜᴍʙᴇʀ of unicellular organisms contain substances remarkably like the hormones of vertebrates. A chemical quite similar to insulin has been identified in a protozoan, *Tetrahymena pyriformis,* and in a bacterium, *Escherichia coli,* and such neurotransmitters as acetylcholine have also been found in single-celled organisms. While it is possible that these compounds are not involved in communication, this seems unlikely, in the light of evidence that unicellular organisms have receptors similar to those that receive chemical messages in the cells of plants and animals. For instance, the feeding behavior of amoebas is inhibited both by the endorphins produced in the human brain and by morphine, the plant alkaloid that mimics endorphins.

The presence of hormonelike chemicals and receptors in single-celled creatures raises the question of what sorts of communication these organisms engage in. It appears that they exchange signals to regulate two principal activities, feeding and reproduction. The role of communication in *Dictyostelium discoideum,* a cellular slime mold, is illustrative. During their free-living stage, slime mold cells move independently of one another. But when food becomes scarce, some of the cells begin to emit an attractant substance. Other cells move in the direction of the source of this attractant and, in turn, secrete their own. In this manner, all the cells move toward a recruiting center and eventually merge to form a single, sluglike organism, which produces spores that give rise to the next generation of slime mold cells. The attractant substance is cyclic AMP, an intracellular messenger that plays an important role in both the hormonal and the nervous control of vertebrate cells.

The mating of *Saccharomyces cerevisiae,* a yeast commonly used in the preparation of alcoholic beverages, is also coordinated by chemical communication. To initiate sexual union, alpha cells secrete a chemical messenger that binds to receptors on the surfaces of A-cells. The A-cells then respond by secreting a different messenger, which acts on the alpha cells. This exchange induces a mass of cells to cluster together, after which alpha and A-cells pair off and fuse. One of the two chemical messengers that orchestrate this mating, it turns out, is similar in structure to gonadotropin-releasing hormone, a sex-related messenger in mammals. Further, the cell-surface receptors involved in the mating are quite similar to receptors in human tissues.

Note what simple means of communication these organisms have: one cell secretes a substance that travels through a surrounding medium to an adjacent cell or to more distant cells. And note that this process does not precisely parallel either hormonal *or* neuronal communication. Indeed, these evolutionarily primitive systems of communication more closely resemble the paracrine systems of mammals, in which cells release messengers that travel through extracellular fluid to neighboring cells.

To get from this generic sort of communication to either an endocrine or a nervous system required evolutionary adaptations. Cells that specialized in conducting messages to adjacent cells developed electrical means of intracellular communication to supplement the chemical messengers that we now know as neurotransmitters. With the nerve cell's invention of extended axons, this sort of communication could occur even over long distances; there are nerve cells in the toes that, by virtue of the length of their axons, are adjacent to nerve cells at the base of the spine. Other cells specializing in the transmission of long-distance messages took an alternative approach: they coalesced into glands and made use of structures adapted to transporting various fluids long-distance—the blood vessels.

In short, neither the nervous system nor the endocrine system is really prior to the other. Both descended from a common ancestor. And because evolution is a tremendously economical process, one that makes use of whatever materials are at hand, many of the chemicals that served as messengers during the unicellular phase have maintained their essential structure, even as their communicative functions have changed. Thus do single-celled organisms share elements with their multicelled descendants: animals and plants.

WHEN THE NERVOUS and endocrine systems are viewed as specialized descendants of a more general system, the various forms of cellular communication that do not fit neatly into either system become much less mysterious. For example, it is quite possible that some chemical messengers—ancestors, perhaps, of noradrenaline—were adopted for use by both the inchoate nervous system and the inchoate endocrine system, and thus occur in both today. Another class of messengers—the paracrines—may simply have continued functioning in the more primitive fashion, passing from one cell to a nearby cell without benefit of bloodstream or electrical impulse. Still other messengers ended up working in all three of these ways. Thus, somatostatin functions as a hormone in the liver, as a neurotransmitter in the brain, and as a paracrine messenger in the pancreas. In similar fashion, the unifying theory explains why endocrine messengers sometimes turn up in the exocrine system: the gastrointestinal tract, like the bloodstream and

the axon, is just an adaptation that serves as, among other things, a transporter of information; the chemical messengers and receptors are essentially the same, regardless of whether the medium is the bloodstream or an internal cavity.

The theory has the further advantage of explaining the scores of coincidences surrounding cellular communication in plants and animals—not just those involving various medicinal herbs and modern plant-derived drugs but other odd parallels, as well: that the thyrotropin-releasing hormone, common in vertebrates, is found also in alfalfa; that an insulinlike substance is found in spinach; that truffles produce molecules identical to a steroid in the testes of boars—a fact that leads sows to expertly detect and actively seek even deeply buried truffles.

Unfortunately, we cannot say much about the present-day communicative function of such messengers in plants. To be sure, we know that plant cells communicate chemically with one another to regulate growth, the ingestion of nutrients, and responses to changes in weather and daily or seasonal cycles. What's more, some of the chemicals involved in this communication have been isolated. But no one knows exactly what the TRH hormone does in alfalfa, nor what opium and its constituents do for poppies. One of the main reasons for this ignorance is that the unifying theory of intercellular communication is only a few years old. Until recently, scientists, operating under the assumption that the mimicry of vertebrate messengers by plant molecules is mere coincidence, had no reason to suspect that such molecules have any communicative function in plants. We may expect that, in the next few decades, the research project into medicinal herbs begun more than sixty thousand years ago will reach new fruition, as human beings ask not what plant messenger molecules can do for them, but what such molecules do for plants in the first place.

THE INSTINCT TO LEARN

James L. Gould and Carol Grant Gould

James L. Gould, professor of biology at Princeton University, studies the navigation and communication of the honey bee. Carol Grant Gould is a writer and research associate in Princeton's biology department.

When a month-old human infant begins to smile, its world lights up. People reward these particular facial muscle movements with the things a baby prizes—kisses, hugs, eye contact, and more smiles. That first smile appears to be a powerful ingredient in the emotional glue that bonds parent to child. But scientists wonder whether that smile is merely a chance occurrence, which subsequently gets reinforced by tangible rewards, or an inexorable and predetermined process by which infants ingratiate themselves with their parents.

If this sounds like another chapter in the old nature/nurture controversy, it is—but a chapter with a difference. Ethologists, specialists in the mechanisms behind animal behavior, are taking a new look at old—and some new—evidence and are finding that even while skirmishing on a bloody battleground, the two camps of instinctive and learned behavior seem to be heading with stunning rapidity and inevitability toward an honorable truce.

Fortunately for the discord that keeps disciplines alive and fit, animal behavior may be approached from two vantage points. One of these sees instinct as the moving force behind behavior: Animals resemble automatons preordained by their genetic makeup to behave in prescribed ways. The other views animals as basically naive, passive creatures whose behavior is shaped, through the agency of punishment and reinforcement, by chance, experience, and environmental forces.

In the last few years, however, these two views have edged towards reconciliation and, perhaps, eventual union. Case after case has come to light of environmentally influenced learning which is nonetheless rigidly controlled by genetic programming. Many animals, ethologists are convinced, survive through learning—but learning that is an integral part of their programming, learning as immutable and as stereotyped as the most instinctive of behavioral responses. Furthermore, neurobiologists are beginning to discover the nerve circuits responsible for the effects.

Plenty of scientists are still opposed to this new synthesis. The most vociferous are those who view the idea of programmed learning as a threat to humanity's treasured ideas of free will. However, it now appears that much of what we learn is forced upon us by innate drives and that even much of our "culture" is deeply rooted in biology.

As though this were not enough of a shock to our ingrained ideas of man's place in the universe, it looks as though the reverse is true, too: Man is not the sole, lofty proprietor of culture; "lower" animals—notably monkeys and birds—also have evolved various complicated ways of transferring environmentally learned information to others of their own kind.

The honey bee provides entrancing insights into the lengths to which nature goes in its effort to program learning. These little animals must learn a great many things about their world: what flowers yield nectar at what specific times of day, what their home hives look like under the changes of season and circumstance, where water is to be found.

But new work reveals that all this learning, though marvelous in its variety and complexity, is at the same time curiously constrained and machinelike. Certain things that bees learn well and easily, they can learn only at certain specific "critical periods." For example, they must relearn the appearance and location of their hives on their first flight out every morning; at no other time will this information register in the bee's brain. Beekeepers have known for centuries that if they move a hive at night the bees come and go effortlessly the next day. But if they move the hive even a few meters at any time after the foraging bees' first flight of the day, the animals are disoriented and confused. Only at this one time is the home-learning program turned on: Evidently this is nature's way of compensating for changing seasons and circumstances in an animal whose vision is so poor that its only means of locating the hive is by identifying the landmarks around it.

Since bees generally harvest nectar from one species of flower at a time, it seems clear that they must learn to recognize flower species individually. Karl von Frisch, the noted Austrian zoologist, found that bees can distinguish any color, from yellow to green and blue and into the ultraviolet. However, they learn the color of a flower only in the two seconds before they land on it. Von Frisch also discovered that bees can discriminate a single odor out of several hundred. Experimentation reveals that this remarkable ability is similarly constrained: Bees can learn odor only while they are actually standing on the flower. And finally, only as they are flying away can they memorize any notable landmarks there might be around the flower.

Learning then, at least for bees, has thus become specialized to the extent that specific cues can be learned only at specific times, and then only in specific contexts.

The bees' learning programs turn out to be restricted even further. Once the bits of knowledge that make up a behavior have been acquired, such as the location, color, odor, shape, and surrounding landmarks of a food source, together with the time it is likely to yield the most nectar, they form a coherent, holistic set. If a single component of the set is changed, the bee must learn the whole set over again.

In a very real sense, then, honey bees are carefully tuned learning machines. They learn just what they are programmed to learn, exactly when and under exactly the circumstances they are programmed to learn it. Though this seems fundamentally different

from the sort of learning we are used to seeing in higher animals such as birds and mammals—and, of course, ourselves—careful research is uncovering more and more humbling similarities. Programmed memorization in vertebrates, though deceptively subtle, is widespread. The process by which many species of birds learn their often complex and highly species-specific songs is a compelling case in point.

Long before the birds begin to vocalize, their species' song is being learned, meticulously "taped" and stored somewhere in their memory banks. As the bird grows, the lengthening days of spring trigger the release of specific hormones in the males which in turn spur them to reproduce first the individual elements of syllables and later the sequence of the stored song. By a trial and error process the birds slowly learn to manipulate their vocal musculature to produce a match between their output and the recording in their brains. Once learned, the sequence becomes a hardwired motor program, so fixed and independent of feedback that if the bird is deafened his song production remains unaffected.

This prodigious feat of learning, even down to the regional dialects which some species have developed, can be looked at as the gradual unfolding of automatic processes. Peter Marler of the Rockefeller University and his students, for instance, have determined that there are rigorous time constraints on the song learning. They have discovered that in the white-crowned sparrow the "taping" of the parental song can be done only between the chicks' 10th and 50th days. No amount of coaching either before or after this critical period will affect the young birds. If they hear the correct song during this time, they will be able to produce it themselves later (or, if females, to respond to it); if not, they will produce only crude, vaguely patterned vocalizations.

In addition, the white-crowned sparrow, though reared in nature in an auditory environment filled with the songs of other sparrows and songbirds with rich vocal repertoires, learns *only* the white-

crowned sparrow song. Marler has recently been able to confirm that the parental song in another species—the swamp sparrow—contains key sounds that serve as auditory releasers, the cues that order the chicks' internal tape recorders to switch on. Ethologists refer to any simple signal from the outside world that triggers a complex series of actions in an animal as a releaser.

Here again, amazing feats of learning, particularly the sorts of learning that are crucial to the perpetuation of an animal's genes, are rigidly controlled by biology.

The kind of programmed learning that ethologists have studied most is imprinting, which calls to mind a picture of Konrad Lorenz leading a line of adoring goslings across a Bavarian meadow. Newborn animals that must be able to keep up with ever-moving parents —antelope and sheep, for example, as well as chicks and geese— must rapidly learn to recognize those parents if they are to survive. To achieve this noble aim evolution has built into these creatures an elegant learning routine. Young birds are driven to follow the parent out of the nest by an exodus call. Though the key element in the call varies from species to species—a particular repetition rate for one, a specific downward frequency sweep for another—it is always strikingly simple, and it invariably triggers the chicks' characteristic following response.

As the chicks follow the sound they begin memorizing the distinguishing characteristics of the parent, with two curious but powerful constraints. First, the physical act of following is essential: Chicks passively transported behind a calling model do not learn; in fact, barriers in a chick's path that force it to work harder speed and strengthen the imprinting. Second, the cues that the chick memorizes are also species-specific: One species will concentrate on the inflections and tone of the parent's voice but fail to recall physical appearance, while a closely related species memorizes minute details of physical appearance to the exclusion of sounds. In some species of mammals, the learning focuses almost entirely on individual odor. In each case, the critical period for imprinting lasts only a day or two. In this short but crucial period an ineradicable picture of the only individual who will feed and protect them is inscribed in the young animals' memories.

By contrast, when there is no advantage to the animal in learning specific details, the genes don't waste their efforts in programming them in. In that case, blindness to detail is equally curious and constrained. For instance, species of gulls that nest cheek by jowl are programmed to memorize the most minute details of their eggs' size and speckling and to spot at a glance any eggs which a careless neighbor might have added to their nest—eggs which to a human observer look identical in every respect. Herring gulls, on the other hand, nest far enough apart that they are unlikely ever to get their eggs confused with those of other pairs. As a result, they are unconscious of the appearance of their eggs. The parents will complacently continue to incubate even large black eggs that an experimenter substitutes for their small speckled ones. The herring gulls' insouciance, however, ends there: They recognize their chicks as individuals soon after hatching. By that time, their neighbors' youngsters are capable of wandering in. Rather than feed the genes of their neighbors, the parents recognize foreign chicks and often eat *them*.

The kittiwake gull, on the other hand, nests in narrow pockets on cliff faces, and so the possibility that a neighbor's chick will wander down the cliff into its nest is remote. As a result kittiwakes are not programmed to learn the appearance of either eggs or young, and even large black cormorant chicks may be substituted for the small, white, infant kittiwakes.

Simply from observing animals in action, ethologists have learned a great deal about the innate bases of behavior. Now, however, neurobiologists are even tracing the circuitry of many of the mechanisms that control some of these elements. The circuits responsible for simple motor programs, for example, have been located and mapped out on a cell-by-cell basis in some cases and isolated to a single ganglion in others.

A recent and crucial discovery is that the releasers imagined by ethologists are actually the so-called feature detectors that neurobiologists have been turning up in the auditory and visual systems. In recent years, neurobiologists have discovered that there are certain combinations of nerve cells, built into the eyes and brains of all creatures, that respond only to highly specific features: spots of a certain size, horizontal or vertical lines, and movement, for example. In case after case, the basic stimulus required to elicit an innate response in animals corresponds to one or a very simple combination of discrete features systematically sought out by these specialized cells in the visual system.

The parent herring gull, for instance, wears a single red spot near the tip of its lower bill, which it waves back and forth in front of its chicks when it has food for them. The baby gulls for their part peck at the waving spot which, in turn, causes the parent to release the food. First, Niko Tinbergen, the Dutch Nobel Prize winner and cofounder of the science of ethology with Lorenz and von Frisch, and later the American ethologist Jack Hailman have been able to show that the chicks are driven to peck not by the sight of their parent but at that swinging vertical bill with its red spot. The moving vertical line and the spot are the essential features that guide the chicks, which actually prefer a schematic, disembodied stimulus—a knitting needle with a spot, for example.

Though the use of two releasers to direct their pecking must greatly sharpen the specificity of the baby gulls' behavior, chicks do quickly learn to recognize their parents, and the mental pictures thus formed soon replace the crude releasers. Genes apparently build in releasers not only to trigger innate behavior but, more important, to direct the attention of animals to things they must learn flawlessly and immediately to survive.

Even some of what we know as culture has been shown to be partially rooted in programmed learning, or instinct. Many birds, for instance, mob or attack potential nest predators in force, and they do this generation after generation. But how could these birds innately know their enemies? In 1978 the German ethologist Eberhard Curio placed two cages of blackbirds on opposite sides of a hallway, so that they could see and hear each other. Between the two cages he installed a compartmented box, which allowed the occupants of one cage to see an object on their side but not the object on the other. Curio presented a stuffed owl, a familiar predator, on one side, and an innocuous foreign bird, the Australian honey guide, on the other. The birds that saw the owl went berserk with rage and tried to mob it through the bars of the cage. The birds on the other side, seeing only an unfamiliar animal and the enraged birds, began to mob the stuffed honey guide. Astonishingly, these birds then passed on this prejudice against honey guides through a chain of six blackbirds, all of which mobbed honey guides whenever they encountered one. Using the same technique, Curio has raised generations of birds whose great-great-grandparents were tricked into mobbing milk bottles and who consequently teach their young to do the same.

What instigates the birds—even birds raised in total isolation—to pay so much attention to one instance of mobbing that they pass the information on to their offspring as a sort of taboo, something so crucial to their survival that they never question if or why these predators must be attacked? The mobbing call, it turns out, serves as yet another releaser that switches on a learning routine.

Certain sounds in the mobbing calls are so similar among different species that they all profit from each other's experience. This is why we often see crows or other large birds being mobbed by many species of small birds at once. So deeply ingrained in the birds is this call that birds raised alone in the laboratory are able to recognize it, and the calls of one species serve to direct and release enemy-learning in others. Something as critical to an animal's survival as the recognition of enemies, then, even though its finer points must be learned and transmitted culturally, rests on a fail-safe basis of innately guided, programmed learning.

The striking food-avoidance phenomenon is also a good place to look for the kind of innately directed learning that is critical to survival. Many animals, including humans, will refuse to eat a novel substance which has previously made them ill. Once a blue jay has tasted one monarch butterfly, which as a caterpillar fills itself with milkweed's poisonous glycosides, it will sedulously avoid not only monarchs but also viceroys—monarch look-alikes that flaunt the monarchs' colors to cash in on their protective toxicity. This programmed avoidance is based on the sickness which must appear within a species-specific interval after an animal eats, and the subsequent food avoidance is equally strong even if the subject knows from experience that the effect has been artificially induced.

But what is the innate mechanism when one blue tit discovers how to pierce the foil caps of milk bottles left on doorsteps to reach the cream, and shortly afterwards blue tits all over England are doing the same thing? How are theories of genetic programming to be invoked when one young Japanese macaque monkey discovers that sweet potatoes and handfuls of grain gleaned from a sandy shore are tastier when washed off in the ocean, and the whole troop (except for an entrenched party of old dominant males) slowly follows suit? Surely these are examples pure and simple of the cultural transmission of knowledge that has been environmentally gained.

Perhaps not. What the blue tits and the monkeys pass on to their colleagues may have an innate basis as well. The reason for this precocious behavior—and we say this guardedly—may be in a strong instinctive drive on the part of all animals to copy mindlessly certain special aspects of what they see going on around them. Chicks, for instance, peck at seeds their mother has been trained to select, appar-

The cells that bring you the world

There was a time when the visual system was thought of as little more than a pair of cameras (the eyes), cables (the optic nerves), and television screens (the visual cortex of the brain). Nothing could be farther from the truth. We now know that the visual system is no mere passive network of wires but an elaborately organized and highly refined processing system that actively analyzes what we see, systematically exaggerating one aspect of the visual world, ignoring or discarding another.

The processing begins right in the retina. There the information from 130 million rods and cones is sifted, distorted, and combined to fit into the four or so million fibers that go to the brain. The retinas of higher vertebrates employ one layer of cells to sum up the outputs of the rod-and-cone receptors. The next layer of retinal cells compares the outputs of adjacent cells in the preceding tier. The result is what is known as a spot detector: One type of cell in the second layer signals the brain when its compare/contrast strategy discovers a bright field surrounded by darkness (corresponding to a bright spot in the world). Another class of cell in the same layer has the opposite preference and fires off when it encounters dark spots.

The next processing step takes this spot information and, operating on precisely the same comparison strategy, wires cells that are sensitive only to spots moving in particular directions at specific speeds. The output of these spot detector cells also provides the raw material from which an array of more sophisticated feature detectors sort for

lines of each particular orientation. These feature detectors derive their name from their ability to register the presence or absence of one particular sort of stimulus in the environment. Building on these cells, the next layer of processing sorts for the speed and direction of moving lines, each cell with its own special preference. Other layers judge distance by comparing what the two eyes see.

The specific information that cells sort for in other retinal layers and visual areas of the brain is not yet understood. Research will probably reveal that these extremely complex feature detectors provide us with what we know as conscious visual experience. Our awareness of all this subconscious processing, along with the willful distortions and tricks it plays on us, comes from the phenomenon of optical illusions. When we experience an optical illusion, it is the result of a particular (and, in the world to which we evolved, useful) quirk in the visual mechanism.

Feature detectors are by no means restricted to the visual system. In birds and bats, for instance, specialized cells have been found that recognize many nuances in sound—locations, repetition rates, time intervals, and precise changes in pitch—that allow the creatures to form an auditory picture of the world.

There is every reason to suppose that our experience of the world is based on the results of this massive editing. Since neural circuits differ dramatically from species to species according to the needs of each, the world must look and sound different to bees, birds, cats, and people.
—*J.L.G. and C.G.G.*

ently by watching her choices and copying them. In the case of many mammals, this drive is probably combined with an innate urge to experiment. The proclivity of young animals, particularly human children, to play with food, along with their distressing eagerness to

put virtually anything into their mouths, lends support to the experimentation theory. Perhaps it is the young, too naive to know any better, who are destined by nature to be the primary source of cultural innovation. The more mature become the equally indispensable defenders of the faith, the vehicles of cultural transmission.

Patterns, then, however subtle, are beginning to emerge that unify the previously unrelated studies of instinct and learning. Virtually every case of learning in animals that has been analyzed so far depends in at least some rudimentary way on releasers that turn on the learning routine. And that routine is generally crucial to the perpetuation of the animal's genes.

Even the malleable learning we as humans pride ourselves on, then, may have ineradicable roots in genetic programming, although we may have difficulty identifying the programs, blind as we are to our own blindness. For example, you cannot keep a normal, healthy child from learning to talk. Even a child born deaf goes through the same babbling practice phase the hearing child does. Chimpanzees, by contrast, can be inveigled into mastering some sort of linguistic communications skills, but they really could not care less about language: The drive just is not there.

This view of human insight and creativity may be unromantic, minimizing as it does the revered role of self-awareness in our everyday lives, but the pursuit of this line of thinking could yield rich rewards, providing us with invaluable insights into our own intellectual development. The times we are most susceptible to particular sorts of input, for instance, may be more constrained than we like to think. The discovery of the sorts of cues or releasers that might turn on a drive to learn specific things could open up new ways of teaching and better methods for helping those who are culturally deprived. Best of all, analyzing and understanding those cues could greatly enrich our understanding of ourselves and of our place in the natural order.

Food Affects Human Behavior

GINA KOLATA

Research on whether food and nutrients affect human behavior is gaining serious attention these days. On 9 November, the Center for Brain Sciences and Metabolism at the Massachusetts Institute of Technology (MIT) held a meeting to bring together reputable investigators who are studying this subject.* "To my knowledge, this is the first meeting on this subject that was not held by and for the true believers," says Richard Wurtman of MIT, who was the conference organizer.

The problem is that serious researchers have tended to steer away from this field which has, Wurtman notes, "a dubious reputation." Yet there is good biochemical evidence that, in laboratory animals at least, changes in diet can change the amount of various neurotransmitters synthesized in the brain and can thereby alter behavior. As was apparent at the meeting, there is strong preliminary evidence that the same phenomena occur in humans.

The effects of nutrients on human behavior are subtle, however, and are not necessarily the effects so avidly believed by many members of the general public. The folk wisdom, for example, is that refined sugars and carbohydrates cause children to be hyperactive and cause criminals to act aggressively. In fact, the more likely effect of a junk food diet is to make people sleepy.

The studies of the effects of food on brain biochemistry began about 10 years ago when Wurtman and his associates initiated animal experiments. Since then, they and others have firmly established that half a dozen nutrients can alter the synthesis of the neurotransmitters serotonin, dopamine, norepinephrine, acetylcholine, histamine, and glycine.

These neurotransmitters are precursor-dependent. The rate at which brain enzymes synthesize the transmitters is limited by the availability of precursors that derive from food and are transported into the brain by carrier molecules.

*The papers from the conference, "Research Strategies for Assessing the Behavioral Effects of Food and Nutrients," will be published in the *Journal of Psychiatric Research.*

The most often-cited case is that of serotonin and its dependency on tryptophan. Serotonin is made directly from tryptophan, and the body's only source of this amino acid is dietary protein. When protein is digested, tryptophan enters the blood and joins a pool of amino acids available for transport to the brain. But the carrier molecule that transports tryptophan transports eight other neutral amino acids as well. The nine amino acids compete for the carrier. Thus the more tryptophan in the blood relative to the other competing amino acids, the more tryptophan enters the brain and the more serotonin is made.

High protein meals do not increase brain serotonin because they do not increase the relative amount of tryptophan in the blood. High carbohydrate meals, on the other hand, do. After a high carbohydrate meal is eaten, insulin is released and this hormone facilitates the uptake into body tissues of all the amino acids except tryptophan.

Serotonin neurons participate in a wide range of behaviors, including sleep, feeding, locomotor activity, aggression, and pain sensitivity. Dietary manipulations that alter brain tryptophan concentrations can, in animals, affect many of these behaviors.

Similarly, other dietary precursors of neurotransmitters seem to have the predicted effects in animals. For example, tyrosine is a precursor for dopamine and norepinephrine, which play a role in regulating motor activity and mood. (Tyrosine concentrations in the blood and brain increase after a high protein meal.) Tyrosine causes an increase in motor activity in animals.

Based on these animal studies, it seems at least plausible that diet may affect human behavior. But the question of whether it actually does so is especially difficult to answer because it is not at all clear how to elicit behaviors in normal humans nor how to measure them. Thus it is somewhat surprising that investigators at the meeting found even subtle effects. But many of these studies with humans have been incomplete—researchers have given subjects an amino acid or a high carbohydrate or high pro-

tein meal, for example, and then tested their behavior. Unfortunately, however, they generally have not measured changes in plasma amino acids to see if these correlate with the observed changes in behavior.

The best studied behavioral effect of nutrients in humans is sleepiness. Animal studies have clearly established that serotonin is used in the regulation of sleep. When investigators destroy neurons of the midbrain raphe where the cell bodies of many serotoninergic neurons lie, animals' sleep time is significantly decreased. When they give animals a substance that blocks serotonin synthesis, the animals sleep even less.

As early as 1963, Ernest Hartmann and his associates at Tufts University School of Medicine began testing tryptophan as a possible inducer of sleep. More recently, Hartmann gave subjects high carbohydrate or high protein meals to see how this affected sleep.

In his review at the MIT meeting of these and dozens of similar studies by other researchers, Hartmann concluded that tryptophan in doses of at least 1 gram (which is in the range that can be supplied by diet) does make mildly insomniac patients fall asleep more quickly and wake less frequently during the night. Normal good sleepers tend not to be affected by this amino acid and neither do seriously ill insomniacs. Hartmann proposes that since it takes a good sleeper only about 10 minutes to fall asleep, and since these people do not wake in the night, it would be hard to detect much of an effect from tryptophan. Mild insomniacs take about 30 minutes to fall asleep, and tend to wake in the night, so it is easier to observe tryptophan's effects.

When Hartmann tested the effects of a high carbohydrate evening meal as compared to a high protein meal, he found that subjects who eat the carbohydrates are significantly sleepier 2 hours afterwards than those who eat the protein. Perhaps the optimal way to use tryptophan would be to give even lower doses along with a carbohydrate meal, Wurtman advises. High doses of any large, neutral amino acid, including tryptophan, will nonspecifically lower the amounts of all the other large neutral amino acids in the brain. So there is good reason to keep the amino acid doses as low as possible.

But the effects of carbohydrate and protein meals may be more complicated than Hartmann's data indicate. Bonnie Spring of Harvard University and her associates reported at the meeting on a study in which they gave 184 subjects carbohydrate or protein meals for breakfast or lunch. They then measured these subjects' performance on tests of selective attention and asked them to rate their moods. In the test, called dichotic shadowing, subjects heard strings of words on a "main channel" of a stereo headset. These words were transmitted through only one of the earphones. On the other earphone were distracting words or sounds. Subjects were asked to repeat the main channel words, syllable by syllable. The persons aged 40 and older felt more calm after a carbohydrate lunch but also did worse on the performance test. The younger subjects did worse on the performance test after a carbohydrate breakfast. Spring says she was quite surprised to find any significant effects on performance at all after people ate a single meal. "The fact that we could detect an effect after one meal is quite extraordinary," she remarks.

There is some suggestion that the sleep-inducing effects of tryptophan might apply even to newborn babies. Michael Yogman of Harvard Medical School, Steven Zeisel of Boston University Medical School, and Carolyn Roberts at MIT gave tryptophan or valine (which competes with tryptophan for uptake into the brain) to 2- and 3-day-old infants in a double-blind study. Infants given tryptophan in their bottles of formula entered both quiet and active sleep more quickly than infants that had been given valine.

Intrigued by these results, Yogman now wants to follow up with studies of breast-fed infants. He strongly suspects that the carbohydrate content of breast milk varies according to the mother's diet and thus different breast-fed babies may be getting different amounts of carbohydrates and, possibly, brain tryptophan. This may affect their sleep patterns, Yogman suggests.

A commonly held belief is that carbohydrates, and particularly refined sugars, make children hyperactive. But this effect is inconsistent with the known biochemical actions of carbohydrates,

and two studies presented at the meeting failed to confirm it.

Judith Rapoport of the National Institute of Mental Health sought out boys whose parents were convinced that they were immediately made hyperactive by sugar. The children were given, in a double-blind test, drinks with sucrose and saccharin, glucose and saccharin, or saccharin alone. (Rapoport added saccharin to the sugar mixes so as to make the tastes of the three drinks indistinguishable.) The boys wore activity monitors that recorded their physical activity. In addition, trained observers, who did not know whether the children had consumed sugar or just saccharin, assessed their behavior. By both of these measures, the boys were not more active when they drank the sugar mixtures. In addition, when Rapoport analyzed separately the results for the normal and the psychiatrically disturbed boys in the study, she found that each group was significantly slowed down by the sugar, but the slowing effect occurred in 3 hours for the normal group and in 1 hour for the disturbed group.

But even the best of these studies of carbohydrates or tryptophan and sleep or performance do not establish that carbohydrates or tryptophan in the diet exert their effects through brain serotonin. Says Hartmann, "Although I and others have assumed that brain serotonin was responsible at least in part, none of the tryptophan studies examined this directly. There have been no studies attempting to block the effects using a specific serotonin receptor blocker."

In animals, increased brain serotonin causes decreased sensitivity to pain. In humans, according to two groups of investigators at the MIT conference, tryptophan causes decreased pain perception. Harris Lieberman and Suzanne Corkin of MIT and their associates conducted a double-blind study comparing tryptophan, tyrosine, and a placebo for their effects on pain sensitivity in normal male subjects. (The test was a standard one used to evaluate drugs that may alter pain perception.) Tryptophan caused the men to be significantly less sensitive to moderate pain but did not diminish their sensitivity to intense pain.

Dorothy Dewart and her associates at Temple University gave tryptophan or placebos to patients with chronic pain of the head or neck. These investigators reported that tryptophan significantly decreased the subjects' reported pain. Of course, tryptophan is hardly a competi-

tor with drugs such as the opiates. "Tryptophan won't put morphine out of business," Wurtman says, "but it looks like the relationship between tryptophan and pain is holding up."

Although most of the studies of nutrients and behavior have focused on tryptophan, the recent discovery that tyrosine also can alter brain biochemistry is leading to studies of that amino acid. Alan Gelenberg of Harvard Medical School and his associates, for example, have begun studies of whether tyrosine can alleviate depression.

One of the theories of depression is that it results from a deficiency of catecholamines, specifically norepinephrine. If this theory is correct, tyrosine might be useful as a treatment for depression.

In a small pilot study, Gelenberg and his associates tested tyrosine against a placebo in depressed patients. They also measured plasma tyrosine and the urinary excretion of a norepinephrine metabolite to see whether the administered tyrosine had the expected biochemical effect. They found that not only did plasma tyrosine concentrations increase 27 percent in the patients taking tyrosine but the concentration of the norepinephrine metabolite increased 24 percent. The clinical results, although not conclusive, are suggestive that tyrosine really is relieving depression, at least in some patients. These investigators have now begun a more definitive study. In any event, says Gelenberg, "We hope that research with tyrosine will be more systematic, rigorous, and therefore conclusive than corresponding work with its sister amino acid tryptophan."

The serious study of the effects of nutrients on human behavior has barely begun but, says Wurtman, it is encouraging that the studies are being done at all and that the results are as positive as they are. One thing that can be said, according to Wurtman, is that now "There is no longer any real controversy over whether nutrients can affect behavior." But researchers certainly could better plan their studies. For example, when they ask whether eating breakfast affects performance, they should specify whether the breakfast is high in protein or high in carbohydrates—something that, up to now, investigators did not always do. Wurtman would also like to see researchers pay more attention to levels of amino acids and other neurotransmitter precursors in the blood in these studies.

The Placebo Effect: It's Not All in Your Mind

Robin Marantz Henig

Robin Marantz Henig is a science writer based in Washington, D.C.

The placebo effect has for years been considered a medical joke, proof of the amusing gullibility of the masses. In the 19th century, Sir William Gull, a noted physician, published a satirical report of the usefulness of mint in the treatment of rheumatic fever. He had meant to mock the many purported cures for rheumatic fever that were bandied about at the time, but to his dismay mint-water treatment soon became one of the most popular of these cures. To his greater dismay, even though he had thought of mint at random, the treatment worked. Gull's mint water, like so many nostrums and potions of his day, worked because patients and physicians believed it would. It worked because of the placebo effect.

Placebos, we now know, can be serious medicine. In recent years, the benefit of the placebo—a pharmacologically inactive drug, usually in the form of salt water injection or a milk sugar pill dressed up to look like the real thing—has been documented and studied around the world. Contrary to folk wisdom, a response to a placebo does not prove that a patient is either a fool or a hypochondriac. Persons who respond to placebos may be well-read, well-educated, and indeed sick. Inactive "drugs" have been shown to cure certifiable physical ills ranging from allergies to warts, and have initiated physiological changes in such hard-to-fake readings as blood cell counts, pupil size, respiratory rate and blood pressure. Scientists studying it now view the placebo effect as a window on the mind-body relationship, and hope to learn from it just how the mind stimulates the body to mount an internal attack against the things that ail it.

The placebo's clinical clout comes not from chemistry, but from conviction. Belief in the pill—and in the doctor prescribing it—has been shown to lead to some rather astounding cures. In one patient, an impressive-looking sugar pill was able to lessen dramatically the tremors associated with Parkinson's disease—but when the same substance was mixed with milk and given to the patient without his knowledge, the tremors returned. Placebos need not be inert substances; they may be drugs with known pharmacological properties that simply are not appropriate for the condition for which they're prescribed, as when antibiotics are given to treat a cold. Dr. Stewart Wolf of the Cornell Medical Center demonstrated this when he treated a group of patients with nausea with a drug he assured them would make them feel better. All the nauseated patients improved, even though Wolf had given them syrup of ipecac—which is normally used to induce vomiting.

The placebo effect is activated not only by drugs, but by other interventions as well. Certain surgical procedures, such as those used to treat angina and the hearing disorder called Meniere's disease, have been compared with sham surgery (skin incision but no real surgery) to see which is more effective. The result: Both real and fake operations were equally successful, leading researchers to conclude that the fuss and attention of surgery, rather than the procedure itself, may have caused the improvement. The placebo effect also has been shown to play a major role in the benefits of psychotherapy.

Faith is essential in mobilizing a placebo effect: faith in the drug, faith in the doctor, and faith in the doctor's faith in the drug. For example, in a study by Dr. Thomas C. Chalmers of the Mt. Sinai Medical Center in New York, designed to see whether vitamin C helped prevent colds, a belief in the vitamin was found to be the best cold preventative. The group of subjects given vitamin C was told they were the placebo group, and they developed more colds than did the real placebo group, who were told they were being given vitamin C. In another study, two groups of patients with bleeding ulcers were given placebos under different conditions: either by a nurse, who told them they were to receive an experimental drug with uncertain effects, or by a physician, who presented the pill enthusiastically and said it was sure to improve the patients' condition. In the nurse's group, 25 percent of the ulcer patients improved with the placebo; in the enthusiastic doctor's group, the improvement rate was 75 percent.

Despite the incredible power of the placebo, many physicians hesitate to prescribe it. Some think its use violates medical ethics, involving as it does the deception of the patient. Some believe it denigrates the patient, since the placebo (which in Latin means "I shall please") is usually used to placate rather than to search for the cause of the patient's complaints. And some, perhaps, fear that an acknowledgement of the effectiveness of placebos is a tacit admission that much of what doctors do—from psychotherapy to drug treatment to surgery—entails a little bit of voodoo.

How Does It Do It?

However much they may resist it, doctors can't escape the fact that placebos work about 30 to 40 percent of the time. A classic literature review by Dr. Henry K. Beecher, a Harvard anesthesiologist, reported in 1955 that of more than 1000 patients treated in 15 clinical studies, an average of 35 percent had improved on placebos. These patients had suffered from conditions including severe postoperative pain, seasickness, headache, cough, hay fever and anxiety.

A lot of poetry and not much hard science has been invoked to explain the placebo effect. As journalist Norman Cousins phrased it in "Anatomy of an Illness": "The placebo . . . works not because of any magic in the tablet but because the human body is its own best apothecary and because the most successful prescriptions are those filled by the body itself." Doctors are often equally vague in their descriptions. "The placebo mobilizes the doctor within," says Dr. Arthur K. Shapiro, a psychiatrist at Mt. Sinai Medical Center, paraphrasing Albert Schweitzer. "The placebo effect is an instinct; it's a built-in, genetically determined will to live. If you think about it in terms of survival of the species, you can see the adaptive value in developing a placebo effect, a rationale for living. Gods, religion, doctors—they're all placebos."

But this leaves us wondering just *how* the body goes about filling its own prescriptions, just *how* a pill—either active or inert—can help turn the will to live into a physiological healing force. The search for the placebo mechanism has been undertaken by only a few researchers, and has taken them inside the human brain.

"The mind is located in the brain," says Dr. Jon Levine, a neurologist at the University of California at San Francisco. "If you take that assumption, then the changes that occur when no active intervention is involved must be a manifestation of the intrinsic healing systems in the body. The most logical place to look for these systems is in the brain."

Levine and his co-workers have demonstrated that the placebo effect may be related to the brain chemicals known as endorphins. These "natural morphines,"

several times more powerful than the drug itself, are released in response to pain, and have been shown to be stimulated by acupuncture, certain drugs, and, Levine theorizes, by placebos as well. Studying patients experiencing pain following extraction of impacted wisdom teeth, Levine showed that the administration of naloxone, a drug known to inhibit the effectiveness of endorphins, worsened the pain in those subjects who previously had responded to placebos. Naloxone had no effect on the pain levels of patients who had not responded to placebos, presumably for the same reason that the placebos themselves had no effect: because the internal painkilling systems of these patients were not switched on.

Scientists are intrigued by Levine's findings, which have since been replicated by researchers at the National Institute of Dental Research. But some have criticized his methods, especially his categorization of subjects as either placebo responders or placebo nonresponders. This, they say, falsely implies that response or nonresponse is a fixed trait determined by an individual's physiology or personality. But the placebo response seems to depend more on the situation in which a placebo is given than on the quirks of the individual who receives it.

"Dummy Pills" Are Not for Dummies

Placebo scholars now know that there is no such thing as a placebo responder. The 30 to 40 percent of subjects consistently found to respond to placebos are not the same 30 to 40 percent time and again; any individual, given the right context, can be a placebo responder. Traditionally, physicians have held an image of a placebo responder as a gullible, uneducated, hypochondriacal patient, typically female, who invents distress and really needs only a pat on the hand and a harmless prescription. Scientists say otherwise. "It's just not true that only dumb people react to placebos," says Arthur Shapiro. "Over the past 10 years, we've conducted studies on more than 1000 patients. And our studies say that with any class of people, no matter what their IQ, educational level, cultural background, or academic achievement, we can, if the stimulus is appropriate to that class, elicit a placebo effect." The most important factor, he says, is the patient's attitude toward the physician. The more the patient likes the doctor, the more competent, attractive and compassionate the doctor seems, and the more likely the patient is to respond positively to a pill the doctor says will work.

A placebo response tends to indicate a healthy doctor-patient relationship, one in which the patient and doctor like and respect each other. Most placebo responses have been noted among patients who are well-educated, cooperative, and resourceful—just the kind of patient that doctors most appreciate. Studying cancer patients, Dr. Charles G. Moertel and his colleagues at the Mayo Clinic found that those who responded best to placebos for pain relief had one thing in common: They were fiercely independent, either by choice or by circumstance. "To be forced into a condition of severe dependency was something they would not tolerate," Moertel says. "They reached out for any kind of help." The men in Moertel's study were more likely to respond to placebos than were the women. Persons who were widowed, divorced, or separated were twice as likely to respond as were those who were married or never had been married. Women with children were twice as likely to respond as were women without children. And professionals and farmers—who are self-motivated and independent in their work—were more likely to respond than were unskilled laborers or housewives.

But old medical myths die hard, and one of the slowest to die is the myth that placebos help only those whose complaints were imaginary all along. Despite the evidence, doctors continue to use placebos, if they use them at all, as punishment, not therapy. "Placebos are used with people you hate," notes one medical resident in New Mexico, "not to make them suffer, but to prove them wrong."

Ironically, even though a placebo effect depends upon a healthy doctor-patient relationship, it often is built upon one thing that can shatter that relationship: deception. Many bioethicists say placebo therapy almost always infringes on the patient's rights. "A placebo can provide a potent, although unreliable, weapon against suffering, but the very manner in which it can relieve suffering seems to depend on keeping the patient in the dark," Dr. Sissela Bok, a Harvard University philosopher, wrote in a landmark article in *Scientific American* in 1974. "The dilemma is an ethical one, reflecting contrary views about how human beings ought to deal with each other, an apparent conflict between helping patients and informing them about their condition."

Some scientists, though, point out that placebo therapy works even when patients are told they've been given a placebo. Studies have shown that deception need not be a part of the placebo package. At Johns Hopkins University, for instance, 14 subjects reported improvement in their anxiety after receiving pills that they knew were placebos. But some of them, despite their doctors' honesty, had to provide their own deception. Even though eight patients believed they had indeed been given placebos, the researchers said, six patients—also told their pills were inert—believed the pills must have contained some active drugs. Two of the six were quite certain of it.

Shapiro points out that not only does honesty *not* wipe out the placebo effect—it sometimes heightens it. "I tell my patients what I know," he says, "including the results of our studies on psychotherapy [which show that much of psychotherapeutic success comes from the placebo effect]. But because I believe in the value of the placebo effect, I can state this information with integrity and treat my patients intelligently. So I get the placebo effect no matter what."

Another argument against overuse of the placebo, quite apart from ethical concerns, is that placebos, like their pharmacologically active cousins, carry certain risks. Side effects from placebos can be as serious as those due to active drugs, including headache, nausea, dizziness, rash, severe clinical shock and drug dependency. Adverse effects may be so severe that one researcher, Dr. Steven F. Brena of the Emory University Pain Clinic, invented a new label for it: the "nocebo" response (from the Latin for "I shall hurt"), which he says is more noticeable and lasts longer in patients receiving disability payments during the time they are sick.

In addition, the widespread use of placebos to treat the otherwise untreatable can perpetuate our false belief in the magic of medication. Many patients do not feel satisfied until a visit to the doctor is sealed with a prescription, even for conditions for which the only cure is time—or hope. When placebos are prescribed merely to please, the patient's reliance on drugs is reinforced.

Norman Cousins says that to focus on the pills, rather than on the placebo response itself, diverts us from the true wonder of the placebo effect—its mobilization of our incredible will to live. "The placebo," he writes in 'Anatomy of an Illness'," "is only a tangible object made essential in an age that feels uncomfortable with intangibles, an age that prefers to think that every inner effect must have an outer cause. . . . The placebo, then, is an emissary between the will to live and the body. But the emissary is expendable. If we can liberate ourselves from tangibles, we can connect hope and the will to live directly to the ability of the body to meet great threats and challenges." We can, he says, learn that the most remarkable thing about the placebo effect is what it shows us about the powerful doctor within.

Suggested Readings

(1) Cousins, Norman, "Anatomy of an Illness," W.W. Norton, New York, 1979.
(2) Rhein, Reginald W., Jr., "Deception or Potent Therapy?" *Medical World News,* pp. 39-47, Feb. 4, 1980.
(3) Shapiro, Arthur K., "The Placebo Effect," in Clark, W. G. and del Guidice, J., "Principles of Pharmacology," 2nd ed., Academic Press, New York, in press.

The Brain

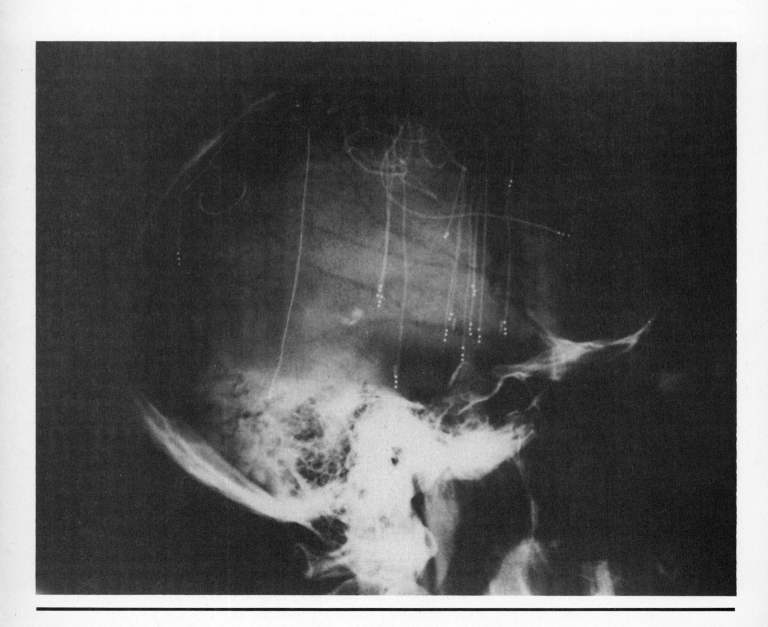

Were I so tall to reach the Pole or grasp the ocean in my span. I must be measured by my soul. The mind's the standard of the man.

Isaac Watts

Historically, humans have always had a difficult time reconciling the physical presence of a "brain" with the range of thoughts, emotions, and memories of the human mind. This "mind-brain problem" has occupied generations of poets, philosophers, and, more recently, biologists. During the present decade, increased interest has developed regarding the study of the brain and central nervous system. This scientific investigation into the nervous system has evolved into its own distinct area of the biological sciences called neuroscience. Neuroscience focuses on all aspects of the nervous system and its relationship to the total behavioral pattern of the organism.

We are oversimplifying when we say that we "see" with our eyes or "smell" with our noses. The brain actually determines what we observe, taste, smell, and feel. While our eyes do send electrochemical signals to the brain in response to various stimuli, what we actually "see" is the product of the organization, analysis, past experience, and interpretation of these signals within the brain. Some neuroscientists have even suggested that those things that make us human are nothing more than interactions be-tween chemicals and electrical impulses within the nervous system. Dysfunctions involving electrochemical impulses have been implicated with brain disorders and abnormal behavior.

This unit considers some of the findings of recent research regarding learning ability, memory, dreaming, and mood and movement disorders. For example, investigations dealing with human memory and the importance of this brain function are reported in "Lab Team Enjoys a Cosmic Cruise Inside the Mind." The author explores the complexities involved in the function of memory, something most people take for granted and many people lose. Another selection proposes a new kind of intelligence test, one which is derived from current neuropsychological research dealing with individual thinking styles.

Looking Ahead: Challenge Questions

Is there any validity in using intelligence tests to measure an individual's ability to learn?

Is it possible to approach memory capabilities from an experimental approach?

Is it rational, as some scientists suggest, to expect to be able to explain brain function solely on the basis of the interactions of chemicals and electrical impulses?

Why is dreaming so essential to our lives?

Intelligence Test: Sizing Up a Newcomer

WRAY HERBERT

In 1904, the Paris school system was severely overcrowded, and the French minister of education made a decision of fundamental social import. He asked psychologist Alfred Binet and psychiatrist Theophile Simon to come up with a way to sort out the children with learning potential from those without — a way, that is, of measuring their intelligence. It was basically an economic decision: By removing from the classrooms the children who were unable to learn, more time could be spent on helping other children to reach their full potential.

Society has never really wavered in its support of mental measurement as a diagnostic tool for educators. Millions of dollars are spent every year on the time-consuming process of assessing individual IQ. Yet in recent years, intelligence testing has come increasingly under fire. Research began to reveal significant racial differences in IQ scores, and it became apparent that as a result of these differences children from racial minorities — especially blacks — were being labeled "retarded" and removed from regular classrooms in disproportionate numbers. In 1972 critics took the testers to court, charging that IQ tests were racially and culturally biased, and in 1979 the U.S. District Court in California ruled that IQ tests could no longer be used for educational diagnosis. Some saw that landmark decision in the *Larry P.* v. *Riles* case as the coup de grace for the mental testing industry and the concept of IQ.

In the midst of the judge's deliberations in the *Larry P.* case, the American Guidance Service, an educational publisher, contacted Alan S. Kaufman and Nadeen L. Kaufman, two psychologists now at the California School of Professional Psychology in San Diego, and asked them to design a new intelligence test. Five years later the so-called Kaufman Assessment Battery for Children (KABC) is almost ready for publication, and it is being hyped by its publishers as a "revolutionary *new* way to define and measure intelligence!" Announced this fall at the meeting of the

American Psychological Association, it is the first new individual intelligence test to be published since the 1930s, and, according to Alan Kaufman, it is the only existing intelligence test to draw on recent advances in neuropsychological research. Doing away with the notion of IQ, the Kaufmans have designed the KABC to measure a purer form of "mental processing ability"; and in doing so, they claim, they have gone a long way toward minimizing the racial and cultural biases that plague existing tests. The testing community is plainly skeptical — about the test's scientific foundation, its validity and its usefulness as an educational tool.

The first intelligence test, the Binet-Simon, used such tasks as naming the days of the week, counting coins, and comprehending written material to arrive at an assessment of learning capacity. Almost without exception, testers today recognize the Binet-Simon (and its American cousin, the Stanford-Binet) as a test of achievement — not a test of intelligence. The Stanford-Binet has fallen into disuse in many schools, and the test that has filled its place, the revised version of the 1939 Wechsler Intelligence Scale for Children, marked an attempt to measure reasoning ability as something separate from language ability — and thus to reduce the biasing effects of culture-bound language. Alan Kaufman played the principal role in the 1974 revision of the WISC.

Today, Kaufman says that the revision of the WISC was not radical enough. "Ten years ago, it was fine, but even then I had the feeling that the publisher was too conservative," he says. "The WISC has too much of a tie-in with tradition. It's time to start looking at a lot of the research and theory in psychology that has happened since the 1930s and take advantage of it in measuring the intelligence of children."

The theory that the Kaufmans have taken advantage of primarily is that of the Russian psychologist Aleksandr R. Luria. Luria theorized that certain areas of the brain were responsible for certain cognitive, or intellectual, processes — specifically, that the frontal and temporal areas were the seat of successive or "sequential" processing, the occipital and parietal re-

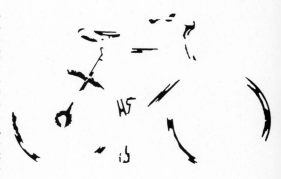

The "Gestalt Closure" task (partial bicycle, above) is intended to tap simultaneous reasoning ability, which emphasizes the synthesis of information presented in space.

gions the seat of "simultaneous" reasoning. Based on the idea that some people tend toward a sequential processing "style" and others toward a simultaneous "style," the KABC is designed specifically to match Luria's cognitive dichotomy.

The test has two cognitive scales: the sequential scale is based on tasks—recalling strings of numbers, for example—that require children to serialize stimuli; the simultaneous scale is based on tasks—interpreting partially completed ink blots, for example—that tap spatial reasoning and the ability to synthesize and integrate information. The tasks involved in both parts of the test are predominantly visual, minimizing the effects of linguistic ability on test performance. "What we're trying to do is be fairer," Alan Kaufman told SCIENCE NEWS. "Our goal is to measure children's ability to process new information and solve new problems, without basing so much of the evaluation of intelligence on what they've already learned."

Whether or not the KABC is fairer than other intelligence tests is open to discussion — and the answer depends on how bias is defined. The Kaufman test does yield a numerical score — a composite of the sequential and simultaneous reasoning scores — and it is this measure of "intellectual potential" that the Kaufmans offer in place of the traditional IQ. The

In the "Word Order" task, the tester speaks a series of words out loud, and the child is instructed to repeat the words in order — either by speaking the words or by pointing to the corresponding images. The task is designed to measure the ability to think sequentially — that is, to process information that is presented through time.

Illustrations: American Guidance Service, Inc.

Kaufmans have conducted some 40 validity tests on the KABC and have found among other things that the test reduces the "mean difference" between blacks and whites by half — from about 15 points to seven; similarly the KABC produces a negligible difference between Hispanic and white children, suggesting that the test has overcome the language barrier that handicaps Hispanics.

But according to psychologist Harold W. Stevenson of the University of Michigan, the fairness may be illusory. "It's absolutely impossible to eliminate cultural bias," he says. "You build a test in a culture, and how to transfer it across cultures has eluded everyone who has ever thought about it." He notes that bias is not carried exclusively in language. Working in Peru, he recalls, he designed a "horrible test," which required the child to imitate an adult's movements. "The task had no verbal part to it, but the child was so upset at calling attention to himself that the test was not very good," Stevenson says. The KABC includes a movement imitation task very much like the one Stevenson describes.

Still others dispute the significance of any reduction in group scoring differences. Arthur R. Jensen, psychologist at the University of California at Berkeley and author of *Bias in Mental Testing,* says that designing a test that eliminates group differences is easy but meaningless; the real proof of a mental test, he says, is if it can predict performance. Jensen argues in his book that the existing intelligence tests are not racially biased, that they do predict academic performance quite well. He points out that the National Academy of Sciences came to the same conclusion in its review of ability testing last year; it follows, he says, that if a test is designed to mask a real difference between groups, then it will predict poorly. "I'm not excited about the Kaufmans' test," Jensen concludes. "I'm most impressed with how non-innovative it is."

Jensen, who is best known for his controversial view that intelligence is genetically determined, also says that he is not surprised that the kind of test the Kaufmans have designed minimizes racial differences. The KABC, he says, seems very much like other tests (such as the Queensland Test, designed to test Australian aborigines) that deliberately circumvent language and rely heavily on short-term memory. "I pointed out years ago that this is one ability on which blacks and whites don't seem to differ much at all," Jensen told SCIENCE NEWS. "The purer the memory test is, the less difference you get between blacks and whites. But they're not a very good measure of intelligence, using the core definition of the term as some kind of reasoning or problem solving ability." Jensen says that Orientals, in contrast, do relatively poorly on memory tasks; as a result, he says, the KABC may systematically depress their scores. It remains to be seen, he says, whether or not the Kaufman test will be a reliable predictor of classroom performance.

Kaufman rejects Jensen's criticism, in part. While conceding that the KABC does contain many short-term memory tasks, he says that it is not memory per se that is being tapped; instead memory tasks are being used to tap the sequential and simultaneous abilities that define intelligence. But more to the point, Kaufman says, Jensen and others are overvaluing prediction. "There has been this feeling that one of the main justifications of intelligence tests is that they predict school achievement. Our feeling has been that if you want to predict school achievement, why not measure school achievement. If your sole purpose is prediction, why beat around the bush?"

While conceding that the KABC does not predict as well as existing tests, Kaufman argues that it should be more useful in designing educational plans for individual students. "It's our basic feeling that if the intelligence scales are to be used properly, you have to be thinking about the present," Kaufman says. "Ours is a more active intelligence scale. It can be used to learn more about the child's approach to solving problems and his approach to learning. It will help you intervene and do something about the future rather than sit back, predict, and let the future happen."

Others, while agreeing with Kaufman's philosophy of testing, question whether or not the KABC is grounded well in theory or applicable in practice. There is disagreement, for example, about Luria's work and what it offers for the testing of normal intelligence. Johns Hopkins University neuropsychologist Alphonso Caramazza, while an admirer of Luria, says that his work is too "theoretically impoverished" to be used as a foundation for an intelligence test. "There is information indicating that various kinds of sequential patterns are controlled by various parts of the brain," he says, "but there is no evidence of a separate independent neurological center responsible for non-sequential organization." Luria's claims about the breakdown of intellect are shallow and invalid, Caramazza says. "They were made during the 1950s; neuropsychologists today don't hold to them any longer."

Charles Golden, a University of Nebraska psychologist who has also incorporated Luria's theory into his Luria-Nebraska Neuropsychological Battery, counters that Luria's work offers a "sound theoretical approach to normal intelligence." He agrees that the notion of simultaneous and sequential "styles" of cognition remains theoretical, and he suggests that the breakdown is probably more complicated than that. "But in normal individuals, the idea of general simultaneous and sequential processes is more than likely true. The Kaufman [test] is an advance over what we've got to date. Whether it's the ultimate test is another issue."

Kaufman, in reply, says that the KABC is not founded entirely on the work of Luria; it has also drawn on the work of cognitive psychologists and split brain researchers. But he also argues that Luria's dichotomy is meaningful—and especially meaningful when it comes to testing for educational intervention. "For an intelligence test to be practical, it needs to be based on a fairly simple model. We're not after a research tool, but a practical clinical tool that is simple and straightforward," he says. "The more complex the theory, the more complex the measurement and, therefore, the more impractical for the real world."

Others challenge the Kaufmans' strongest claim — that the KABC scores enable teachers to tailor curricula to individual students. University of Georgia psychologist Asa Hilliard says that the KABC is essentially based on the "same old model of testing, in which the sole function is prediction. Prediction has no educational value. The students are still going to show up on Monday and the classroom teacher will say, 'So he's got this or that particular cognitive style ... what should I do now?' The Kaufman test is an attempt to save a practice that has made no positive contribution to education."

Kaufman seems more surprised by this line of criticism than any of the others. He says that Hilliard is simply not aware of the research that he and his wife have conducted — research that, he says, provides strong evidence that teaching strategies can be designed differently for sequential thinkers and simultaneous thinkers. Contrary to the traditional model of trying to correct children's specific deficits, the KABC has been designed to provide information about a child's strengths so that teachers can capitalize on these strengths in teaching. Kaufman says he has demonstrated the superiority of this so-called "strength model" in teaching reading to children.

In the end, reactions to the KABC seem to reflect a fundamental philosophical rift in the testing community. Those who think mental measurement should above all tell how a child will perform won't like the KABC; for those who think a test should be designed so that children of different backgrounds do equally well, the KABC (more than any existing test) accomplishes that goal. Kaufman concedes that eliminating group differences in test scores was one of his major goals. He says, however, that the equalizing of scores is not artificial, but rather the result of creating a fairer test, one that circumvents environmental advantage. On that point, Hilliard agrees. The test does seem to measure what children are able to learn rather than what they know, he says, and the equalizing of scores across racial groups should help to prevent the invidious racial comparisons that have been made in the past. But the main advantage of the KABC, Hilliard and others conclude, is political. Whatever its weaknesses, it may be the only test around that will suit the courts and, as Hilliard says, "It'll keep the bias people off his back."

What the Brain Builders Have in Mind

Two years ago, we drew attention to a new mood of excitement in the science of machine intelligence. It centred on researchers calling themselves connectionists, who believe that intelligence will emerge without being programmed in computers that imitate the densely linked networks of cells in real brains. Since then, connectionism has become a crowded band-wagon. Will it roll?

It ought to be possible to make a computer intelligent in the same way that you make a child intelligent: by educating it. "Expert systems", briefed this way, become quite good at a few tasks like medical diagnosis, but they are by no stretch of the imagination intelligent. For those aspects of human intelligence that people take for granted—recognising a face in the crowd, interpreting the words of a telephone conversation—it is impossible to teach a computer the rules humans use, because humans do not know what rules they use.

The old solution to this dilemma is to seek rules that work. From time to time, though, a more seductive approach is aired: simply to build a computer that looks like a brain and make it learn by itself. This approach is called connectionism, neural networks or neurocomputing. Not everybody is seduced. Dr Tommy Poggio, who divides his time between the Massachusetts Institute of Technology and a company called Thinking Machines, jokes about a virus that infects brain scientists, starting a new epidemic every 20 years. The epidemic takes the form of uncritical enthusiasm for a new idea. In the 1920s, the idea was *gestalt* psychology; in the 1940s, cybernetics; in the 1960s, perceptrons. In the 1980s it is connectionism.

Connectionists have four articles of faith. First, they believe in parallel computers: machines that do lots of things at once. Second, they believe that intelligent computers should resemble the networks of

cells of which brains are made. In other words, the processors themselves should be like neurons (brain cells) and the connections between them like synapses (cell junctions). Third, connectionists believe that such a computer's programs should consist of varying the strengths of connections between processors, just as brains seem to work by varying the efficiency of synapses. Fourth, and most boldly, connectionists believe that, given certain conditions, the "programs" (rules, solutions, whatever) will somehow emerge within the machine. It will organise itself.

The first claim is uncontroversial. Almost everybody agrees that a brain is a parallel computer whose processing power is dispersed throughout its network of cells. The second claim, that computers should be built of neuron-like objects, is in some ways peculiar. Compared with the transistors on chips, neurons are big and slow. Transistors switch perhaps 1m times as fast. But whereas a transistor is an on-off switch, a neuron is subtler. Neurons add up the signals they receive, and, depending on whether the sum reaches a threshold, decide whether to send on a signal themselves—like sluice gates that open when all the streams leading to the dam reach a certain height.

AT&T's Bell laboratories in New Jersey have built "neurons" from silicon. Their most complicated circuit has 512 "neurons" on a single chip. Each one is an amplifier and it is connected to others by

resistors. The resistance in these varies with use, which is a direct emulation of the synapses between neurons. Studies of learning in rats and slugs have suggested that learning consists principally of improving the efficiency of synapses as they are used more. In other words, a slug that learns to avoid something nasty is improving the efficiency of the connection between the cell that detects the nastiness and the cell that initiates evasive action.

That leaves the connectionists with their most startling claim: that such neural networks will not have to be programmed. Set up in the right way, they will learn by themselves. The claim is not quite as mystical as it sounds. After all, brains compute and they were not programmed. They evolved the means to learn.

Evolution is not a bad metaphor for what goes on inside a learning "neurocomputer", of the kind connectionists would build. It would try different connection strengths, reject those that are bad and select those that are good. "Good" can mean one of several things, depending on whose machine it is. Dr John Hopfield and Dr David Tank of AT&T's Bell Laboratories use the criterion of energy—when the condition of the whole computer is at lowest energy, the connection strengths are closest to solving the problem. Others, such as Dr David Rumelhart of the University of California at San Diego and Dr Geoffrey Hinton at Carnegie-Mellon University in Pittsburgh, use the "back-propagation" of errors as their criterion for success. This method compares each solution with the expected answer, corrects the errors, sends those corrections back to "hidden" processors in the machine and tries again.

Some of the people who might be thought most likely to be connectionist are not. Two examples are Dr Danny Hillis and Dr Tommy Poggio, who both work at Thinking Machines. Dr Hillis invented the company's product, the Connection Machine—which, with its 64,000 interlinked processors, is a good machine to do connectionist things on. Yet the the coincidence of the names is an embarrassment to him. Dr Poggio works on the most parallel of the brain's functions—vision.

He and Dr Hillis take issue with two things. First, that neurons are anything special. Intelligence, they believe, can be independent of the machinery in which it resides. They, too, are interested in enabling machines to learn. But they do not treat the computer as a black box. Where a connectionist would tell the computer to find out some way of combining 2 and 2 to end up with 4 ("and don't bother me with the details"), a "computationist" like Dr Hillis

or Dr Poggio would also want to work out what method the computer had hit upon and then see where else this new trick called addition might prove useful.

The connectionists, by evading the step of understanding how the computer solves the problem, avoid two difficulties: that you may not know the mathematical formula that works, and that you can solve problems that are intractable to ordinary mathematics. One such is the travelling-salesman problem. A salesman wishes to visit 50 cities. He does not mind in what order he does so, but he would like to spend the least time travelling. To work out his shortest route takes conventional computers forever. A connectionist device has solved the problem fairly quickly. Dr Hopfield and Dr Tank programmed a neural network with the basic rules—forbidding the salesman to visit a city twice and forbidding more than one city from being, say, the tenth city visited. It starts with a random route and is told that whenever it finds a shorter route, it is to strengthen the connections that gave it that route and weaken the others. Eventually, it ends up with a good solution.

A good solution, but not the right one.

The computationists are not impressed. Dr Poggio points out that conventional computers can also arrive at second-best solutions to the problem fairly easily: it is the right answer that is hard to get. Anyway, is the travelling salesman an interesting problem? It is clearly one suited to parallel computers, since many different legs of the journey have to be juggled at once. It is also of possible practical use. Airlines, for instance, may find it helpful to solve such a problem. But it is in danger of becoming an obsession with computer scientists.

Shedding light on the brain
To the computationists, understanding how the machine works is the interesting part: after all, real brains are black boxes and brain scientists devote their careers to making them less black. Dr Andras Pellionisz of New York University straddles the divide between connectionism and computationism. He is interested in the control of movement. It teaches him the importance of parallel devices linked up by programmable connections, but it also teaches him how to program those connections by working out what the brain is computing.

The missing links

Connectionism is one way to let computers teach themselves intelligence. There may be another: to let them evolve their programs.

Dr Danny Hillis of Thinking Machines has begun in the past few months to try "genetic programming" on the Connection Machine, with startling results. His eventual aim is to take a problem for which it is virtually impossible to write the program—say, an enormously long sequence of instructions, none of them contradictory, to a starwars battle station—and simply let the machine evolve the best program by natural selection. It is worth following a simple example in detail. He gives the machine 64,000 randomly selected "creatures". Each one consists of a pair of "chromosomes"; each chromosome has, say, 100 "genes"; each gene consists in this example of a number. His ideally "fit" creature is one in which the genes are numbers 1-100 in ascending order.

He then sets the creatures breeding with each other. Their offspring inherit mixtures of their parents' genes and their chances of surviving to breed depend on how many steady sequences of increasing numbers they have in their chromosomes: the more they have, the more likely they are to survive to breed. Mating can be random, or local, or assortative (meaning, like seeks out like). After enough generations, the perfect creature,

with genes reading from 1 to 100, may have taken over the whole machine.

The startling thing about Dr Hillis's program is that many of the hard-won wisdoms of evolutionary theory pop out of it ready-made. For example, with local mating he sees species evolving and competing with each other. He also sees reproductive isolation: sharp boundaries between species whose hybrid offspring are not viable (the fact that mules cannot breed is what keep horses and donkeys distinct). Dr Hillis has even shown that he might be able to settle a long-running dispute in evolutionary biology: why sexual reproduction evolved at all. After all, many creatures reproduce simply by copying themselves or splitting in half. Most biologists suspect that sex—mixing the genes of two creatures—only pays when conditions are unstable, for it enables creatures to throw up new gene combinations that might suit the changing circumstances. In stable conditions, sex would die out. This happens in Dr Hillis's machine.

So why does sex persist? The answer to this may well hold the key not only to genetic programming but to connectionism as well. Evolutionists are much concerned with a problem called the "local minimum". This basically reiterates the old Irish nostrum "if you want to get there, I wouldn't start from here". An animal

evolves a feature that is not ideal for its job, but any small change to it is a deterioration. To get to the ideal, it must temporarily deteriorate. It is like trying to get to the top of a mountain by always moving upwards: you will soon get stuck on some subsidiary crag from which the route to the peak goes first down and then up.

Sex is probably a subsidiary crag. In an ideally adapted world, animals would abandon it for asexual reproduction until the going got tough. A few creatures, like aphids, do exactly that, breeding asexually in summer and sexually in winter. But people cannot suddenly start breeding asexually—there is simply too massive a change of body organisation required. Dr Hillis's program runs into the same problem. In our example, it might reach a subsidiary crag in which the numbers read 50-100, then 1-49: all except one are in ascending order, but any change makes more than one wrong.

Dr Hillis admits that if genetic programming cannot escape local minima, then it may not be much good at writing programs. Mathematicians have ways of escaping it. One, called simulated annealing, involves random juggling that, if you like, stands a chance of throwing the mountaineer off his subsidiary crag and on to the bottom of the longer slope. And there is some evidence that evolution can escape local minima itself because of similarly random effects. It can keep travelling hopefully.

Move your head from side to side. Did your eyes move, too, or did they remain fixed on the paper? The instinct that holds a gaze steady despite movement in the head and the rest of the body is a remarkably strong reflex, and one whose circuitry is well known. In the ear is an organ of balance, consisting of three semi-circular canals at right angles to one another. Around the eyes are six sets of muscles that control its angle. Between them is a network of nervous connections that automatically ensures that every change registered by the balance organs causes a compensating movement in the eye. Some of the connections are stronger than others, so that, for example, an up-and-down movement of the head causes the muscles controlling the vertical angle of the eye to change more than the other muscles.

That much has been known for decades. What Dr Pellionisz and his colleague, Dr Rudolfo Llinas, realised is that that network is doing a simple mathematical trick that could apply to all movement control within the body. The trick is called tensor analysis.

The position of an object in space can be identified in several different ways. The standard one is Cartesian co-ordinates: measuring from, say, the corner of the room, the object is x feet away, y feet up and z feet to the left. There are all sorts of others. The object's position can be described as the six instructions given to the six eye muscles to make the eye look at the object, or the many instructions given to the muscles of the arm to make the arm reach out and pick up the object. Tensors convert one set of coordinates into another.

Dr Pellionisz thinks that all the connections between the balance organs and the eye muscles are doing is to execute a tensor transformation. He believes that this is also the function of the cerebellum, the separate organ at the back of the brain. The cerebellum has only four kinds of cells and throughout the organ they are repeated endlessly in the same pattern. It is, says Dr Pellionisz, like a crystal. Or like a connectionist device. His approach, though, has been to understand its computations, not to treat it as a black box.

Several companies are already trying to market products based on connectionist ideas. Among the most ambitious is Synaptics, a company founded on the man who tracked down memory in the rat's brain, Dr Gary Lynch of the University of California at Irvine, and a guru of chip design, Dr Carver Mead of the California Institute of Technology. Synaptics has a device that imitates the retina of the eye. Its aim is to make neural chips.

Less ambitious but further forward is a small company in Providence, Rhode Island, called Nestor. Founded in 1975 by two physics professors from Brown University, Dr Leon Cooper and Dr Charles Elbaum, Nestor has invented, among other things, a device that reads handwriting directly into a computer.

The NestorWriter is presently being tested with insurance companies in France and America, and the company expects to be selling the product for $1,500 apiece by the end of May. But do not throw away your keyboard yet. NestorWriter's abilities are still fairly limited. It can learn to read an individual's way of writing single letters either as they are written or after the event. Joined-up writing is still beyond it, and the price does not include the special pen and pad required (though the company has also learnt how to incorporate the pad into the screen of a portable computer).

That limits the product's usefulness, but Nestor hopes to succeed in several markets, where scribbled but brief notes are a big part of the job. One is insurance, where agents fill in forms "in the field"; another is Japanese word-processing; a third is stock trading. The trader (always on the telephone, so unable to dictate) scribbles down on a slip each transaction; it is then transcribed and typed into a computer. Both these steps could be cut out by a machine.

Reading hand-writing is a game that Nestor has largely to itself. Recognising speech is one where it has some competition. Yet connectionists admit that this is the task in which neurocomputers must win their spurs: it requires parallel processing of lots of different kinds of data (pitch, volume, harmonics, context) and it is something at which brains are patently brilliant. Yet no human knows how he does it, so a computer cannot be programmed in the same way. It is the ideal task for neurocomputers, which would learn by simply trying lots of examples of spoken words.

The trouble is that the computational approach is already doing well. Several small companies are on the point of launching devices that enable computers to take dictation. IBM is even further ahead technically, though not ready to sell anything. They all use the computational way: figuring out what rules might work for interpreting speech—grammatical and acoustic—and then programming them into the computer. Can the neurocomputers learn better rules? Japan's Asahi Chemicals may soon know the answer. The company plans to sell a speech recogniser that is based on connectionist principles. It was invented by Dr Teuvo Kohonen of the Helsinki University of Technology, an early connectionist.

The purest form of connectionism is a dream: that you can put together a million-neuron network and set it free to organise itself into intelligence. It will get nowhere. Some kind of programming of the structure is necessary. But most connectionists are more realistic. They believe that there will be parts of parallel computers whose workings will somehow be hidden from scrutiny because they learn from examples using rules that humans cannot decipher. In that, they are likely to be right.

Lab team enjoys a cosmic cruise inside the mind

Kathleen Fisher
Staff Writer

The building squats like a warehouse under the majestic trees of the Bethesda campus and seems to have only back doors. Inside, the corridor folds back on itself in a square. Boxes and tables and refrigerators clog the hall, and a repairman's drill shatters concentration.

But this is the right place: the air is pungent with primate. Most of the white doors in the corridor bear hand-lettered signs: "Thank you for not entering or knocking when the red light is on."

Behind those doors, a half dozen psychologists and one former pediatrician use worn toys, empty cookie boxes and tubes of Tang to unravel just a few of the mind's mysteries.

"I don't think there is anything more fascinating," says neuropsychology lab chief Mortimer Mishkin, "except maybe the cosmos." He pauses. "But then you need to know how you're able to go about contemplating the cosmos."

Consider that it takes millions of steps for today's most sophisticated computers to do what the brain can probably accomplish in 100. True, the computer performs its computations faster, but its speed only partially compensates for its inefficiency.

It takes 110 milliseconds for an image to travel to the brain's limbic system after it has registered on the retina. It takes about 300 milliseconds for us to respond to that image. In that time, millions of neurons have probably fired to bring us a message as simple — and as complicated — as, "That's my favorite coffee cup."

The puzzle Mishkin has chosen to tackle is enormous, insolvable in his lifetime. Yet each additional piece laid in place is as exciting to him as the discovery of a new planet. The picture beginning to emerge from his lab, Mishkin believes, may provide "a way to resolve the most fundamental debate that has gone on in psychology since its inception as a science." If he is right, behaviorists and cognitive scientists may descend from their theoretical Tower of Babel and communicate productively.

"You can hold simultaneously both a cognitivist, or mental, and a behaviorist, or non-mental, view by attributing these different ways in which the brain works to different systems within the brain," Mishkin said.

The memory system he has investigated most, by observing how monkeys behave without it, is the cortico-limbic system. "Its kind of information can be stored in the absence of any response," Mishkin said. "You can just sit and watch the world go by and learn a lot, by storing information about the relations between things, events, and so on."

Without the limbic system, organisms can still learn. But in this case, "there is no way to learn without responding, which is what the behaviorists have always claimed. Because what is learned is the response to a stimulus situation."

"We think the system responsible probably involves interaction between the cortex and the other big subcortical system in the telencephalon, which is the basal ganglia," Mishkin said. "We still have very limited evidence, but damage to the striatum has been found to interfere with pattern discrimination habit formation. That is, in the part of the neostriatum — that's the tail of the caudate nucleus and the caudal part of the putamen — to which the visual cortex projects."

Ticking off these brain structures is second nature for him now. But when Mishkin, 58, began his career the brain was still a big black box and many behavioral scientists believed that its components, if they existed at all, were irrelevant.

Mishkin came to the National Institute of Mental Health 30 years ago, soon after brain research experienced a philosophical revolution. Much as the Big Bang theory turned cosmology on its ear, Mishkin believes Donald Hebb's *Organization of Behavior* threw open the doors of neuropsychology. Mishkin was one of a dozen graduate students privileged to read Hebb's book while it was still a manuscript.

Mishkin credits Hebb with overturning Karl Lashley's view, which predominated during the 1930s and 1940s, that the brain was a blob of homogenous tissue.

"The notion that the brain was jelly — or porridge, you can pick the image you like best — turned away any serious consideration of the brain as a subject for research on behavior," Mishkin said.

Besides Hebb, Mishkin had a second mentor at McGill, H.E. Rosvold. In 1954 Rosvold joined the NIMH psychology lab in Bethesda, and one year later persuaded Mishkin to follow. What began as the animal research section became the neuropsychology section and then a separate neuropsychology lab. Rosvold served as its chief until Mishkin took the helm in 1980.

Today, Mishkin and his staff have mapped a sequence of brain "stations" that relay increasingly complex messages from the eyes. In the first station — the striate (or visual) cortex near the back and base of the brain — a single neuron takes in about one-half of one degree of the visual field. As the message passes successive stations and moves toward the temporal lobes in the front of the brain, the receptive field become larger and processes more information. "It may be that somewhere, at the end, one neuron sees the whole world," Mishkin muses.

The NIMH investigators have also found that the triggers to cue the message from one station to the next build in complexity. It is as though the neuron is asking, "Is this really worth talking about anymore?"

All in all, a far cry from porridge.

Working for peanuts

The neuropsychology lab is just one of 19 intramural NIMH labs, one of six basic research laboratories. All compete for two precious commodities: positions and space.

Room is a treasured resource for neuropsychology. A great deal is needed to house monkeys comfortably, which sometimes leaves little for researchers.

"We really have more people than we have room for," said Mishkin. "Throughout NIMH, and NIH generally, people are crowded in like sardines."

Mishkin said the size of this lab has been stable for a decade. In addition to five staff fellows, the lab encompasses eight technicians and visiting scientists, who may work anywhere from a few months to a few years. They also may become permanent, parttime collaborators with fulltime researchers.

Mishkin can hire a new researcher only if he's allotted a new position. Given current fiscal constraints, that can happen only if another lab loses a slot. "This is not a good time for taking off in new directions," he said. "It's hard enough just to keep from being cut."

Nor does turnover create many new opportunities. Staff fellows come from three to seven years; after seven years, they may become tenured. "Once people arrive, they tend to stay if they can," Mishkin

said. "This is a life sentence," he laughed.

By mid-afternoon four or five of the red lights in the north corridor are glowing. Staff fellow Betsy Murray is behind one of the doors, working the classic Wisconsin apparatus with a rhesus monkey whose amygdala has been removed.

The Wisconsin general testing apparatus, first developed by Harry Harlow at the University of Wisconsin, is a big wooden box with space for the caged monkey on one side. The researcher, seated on the other, covers with various objects two little wells on a tray in front of her. They can be almost anything — toys, kitchen tools, old tin cans — as long as the monkey finds them intriguing. A sliding panel hides the tray from the monkey's view between trials while one of the objects is replaced. If the monkey's memory is intact, it should recall that a peanut waits under the new object.

Habit formation

Murray is exploring the limbic system, that little group of structures near the brain's base that appears to hold a crucial key to learning and memory. Current theory holds that the limbic system processes and sorts everything that comes in, holds it up against past experience and decides if it's worth hanging onto.

There is a possibility that the limbic system also serves to integrate space perception and object recognition. One of its components, the hippocampus, had been assigned the leading role, but Murray and Mishkin recently showed that another, the amygdala, may co-star in memory.

Monkeys taught to recognize an object by touch in the dark, they found, could later recognize it by sight. Animals without a hippocampus were still able to identify the object, but those without an amygdala were not able to make the mental connection.

The limbic system appears necessary to the type of memory that cognitive scientists study, but not to the type that interests behaviorists. Miskin's team has found that monkeys without limbic systems rapidly forget objects presented to them only once. However, if two objects are presented several times, one with and one without reward, the monkeys are able to learn to choose the right one even if the trials are separated by as much as 24 hours. The researchers call this nonlimbic learning "habit formation."

The memory loss pattern they see parallels what clinicians see in humans with global amnesia. Global amnesia wipes out memories gleaned through all the senses, yet there remain many tasks the patients can learn and retain. "Patients can learn, but the learning is characterized by a strange, bizarre aspect, namely that the patient doesn't know anything about when, where, what, how or who . . ." Mishkin said.

Some scientists have been trying to tap this residual memory in an effort to improve the quality of life for amnesia victims. Psychologist Endel Tulving and his colleagues at the University of Toronto have used a computer to teach amnesics words or phrases. The patients have made considerable progress in learning the phrases, but at each lesson they have no recollection of having seen the teacher or the computer.

"The profoundly amnesic patient can't transfer information, because it's not information per se that has been acquired. It's a skill," Mishkin said. What behaviorists call stimulus response bonds, or habit, appear to proceed beautifully "in the absence of the very system that is responsible for informational memory."

To Murray it is miraculous that monkeys are still able to perform so well after losing so much of their limbic systems. "The challenge is to find out how these things work," she understated.

Murray came to NIMH five years ago from the University of Texas Medical Branch at Galveston, where she was a neuroscience student. It was the intellectual pull to explore the structure undergirding memory that drew Murray to the NIMH reservation, as its denizens call it.

While behavior had always fascinated her, psychology seemed to her to contain too much theory and not enough substance. "To me, neuropsychology is really a branch of biology," she said.

Staff fellows agree that the NIMH lab is one of the few places in the country where they can indulge their curiosity about behavior and retain the feel of hard science. They certainly didn't come here for comfort: Murray's office is about the size of a closet.

But while the body may be cramped here, the mind is not.

"This is the best place in the world to do research," Mishkin said. "We have no other responsibilities, such as teaching, and we're not a commercial enterprise. So all of our work is investigator-initiated research of a fundamental type. Each researcher can ask, 'What are the most important questions? What can I do that will provide the best scientific pay-off?' "

Murray agreed that the "Eurekas!" are few and far between. Her behavior studies can take about a year from surgery on the monkeys to journal-ready data. Pure neurology or clinical work would yield results faster, she said.

"But our results are nice and solid. We have control in ablation studies. We *know*, for example, that it is a medial temporal lesion that leads to impairment in visual or tactile memory."

Seeking an impact

In contrast to Murray's bent toward basic research, staff fellow Tom Aigner draws much of his enthusiasm from the hope that his work could have eventual application to drug abuse and to Alzheimer's disease.

The most recent addition to the team, Aigner came to NIMH three years ago from the University of Chicago. As a postdoctoral student he studied drug abuse; as a research associate he became interested in the effects of narcotics, opioids and pain killers on cerebral blood flow.

His current work revolves around the central cholinergic system and acetylcholine receptors. He's attempted to mimic Alzheimer's disease with lesions and by injecting monkeys with a drug, scopolamine, that acts on the cholinergic system and impairs recognition memory.

He's found evidence that another drug, physostigmine, sometimes used to treat Alzheimer's, may not be effective. Although it stimulates acetylcholine action in normal animals, it seems to have no effect on recognition loss in monkeys with basal forebrain lesions — surgery that simulates Alzheimer's cell loss.

Aigner emphasizes that his work with acetylcholine involves only the memory-loss component of Alzheimer's. "Alzheimer's is a much more complex defect, and we're going to need to look at other neurotransmitters."

Because autopsies of Alzheimer's patients do show actual destruction of neurons, Aigner said that drugs alone will probably not be able to turn this memory loss around. "There is really nothing left there to synthesize anything. I suspect that what happens is that there are a few cells

left that we crank up. But they don't grow back."

In addition to his Alzheimer's work, Aigner is cooperating with the National Institute on Drug Abuse in a study to see if there is scientific backing for anecdotal accounts of memory loss in marijuana users. In an initial 21-day study, he gave monkeys moderate oral doses, comparable to what a human might ingest. He failed to find any chronic effects on recognition memory.

Aigner said that choosing the government as one's boss, rather than opting for academia, leads to some trade-offs. "There's no grant writing," he observed, "but there's little technical assistance. I'm lucky because I have an assistant for my work with NIDA."

Mapping the mind

Leslie Ungerleider has been at NIMH the longest of the current five fellows. She came on board 10 years ago from Stanford, where she had done a postdoctorate with Karl Pribram.

Mishkin also studied with Pribram. After obtaining his Ph.D. from McGill in 1951, Mishkin went to Yale, where he, Rosvold and Pribram studied the effects of lobotomies on I.Q. and personality, as well as how frontal lesions affect the behavior of monkeys.

Ungerleider's task is to map what Mishkin has called the stations of the visual system.

Neuroscientists have attempted to visualize the workings of the brain in two ways.

The 2-deoxyglucose method, developed by NIMH's Louis Sokoloff in 1977, allows them to see which areas of the brain are most active in the course of a particular behavior. It involves adding a radioactive glucose isotope to the blood of an experimental animal, then exposing sections of the animal's brain to medical X-ray film. The busiest brain areas will be highlighted in the resulting photo.

The discovery that paved the way for brain mapping, Mishkin said, was the silver-stain technique developed by Walle Nauta about 25 years ago. The substance stains neural fibers that degenerate after damage to the cell body, so that neural transmission can be traced anatomically.

Before that, researchers employed a strychnine preparation that increased electrial discharges from neurons by eliminating neuroinhibi-

tors. "But it was really only guess-work," Mishkin said.

Today, the silver-stain technique has been replaced by axonal transport techniques. One such technique employs radioactively labeled amino acids that actually become part of a neurotransmitter. This allows researchers to zero in on a single bundle of neurons. Some substances, after injection into the area being studied, are taken up by the cell body and captured on photographic film as they streak down an axon. Others are absorbed at terminals and transported back to the cell body. These fluoresce or appear opaque, and can be seen under a microscope.

Ungerleider holds up a color photo of what appears to be Halley's Comet. In fact, it is the illuminated projection of a single neuron bundle, probably about 10,000 cells.

Neuron-by-neuron tracking began after behavioral studies, such as Murray's, indicated that damage to certain areas resulted in different types of impairment.

Ungerleider's current map of visual response shows two diverging paths, one for object recognition and one for spatial perception. Both start in the striate cortex. But the path of messages spelling out "what?" takes off toward the base of the brain and the temporal lobes, while the line of neurons telling the animal "where?" takes a higher road to the parietal lobes.

Ungerleider has spent most of this day hammering out a manuscript on a word processor. A more typical day finds her hunched over a microscope, peering at fluorescent tracings, or at her desk, piecing data from the three-dimensional brain onto a two-dimensional facsimile.

By early this fall, she hopes to have access to a computer program that will stack these two-dimensional maps back into a three-dimensional, rotatable model: in essence, a brain globe.

Search for links

One baffling problem in Ungerleider's current mapping efforts has been to discover how the object-recognition system of the temporal lobe communicates with the spatial perception system of the parietal lobe: How is it that when we look at our favorite coffee mug, we see that it's blue and curved and, at the same time, that it's resting on a table in front of us near our right hand?

Mishkin's work points to the hippocampus as a potential coordinator for all this information. But to get there, the messages from the temporal and parietal lobes would have to pass through several more brain areas. "It's one of the biggest questions for us. We just don't know," Ungerleider said.

As though it weren't difficult enough to puzzle out the hook-ups within the adult primate brain, evidence indicates that there may be connections in the infant brain that fade with development, and other pathways that develop gradually over the first months or years of life.

Jocelyne Bachevalier has wrestled with such developmental puzzles since coming to NIMH four years ago from the University of Montreal, where she was studying the effect of vitamin B deficiency on learning and memory in rats.

Bachevalier has found what appears to be a model for autism in infant monkeys that have had their limbic systems removed. As in adult monkeys with this lesion, the infants' memories are impaired. But in addition, the animals show bizarre motor behavior, such as rocking and hopping, and have severe socialization problems.

It seems that something is damaged which is needed during development to lay down social skills. It may be the same mechanism which holds or shapes our long-term cognitive memories, those that remain even in amnesia, Mishkin suggests.

Bachevalier explained that there are visual memory tasks monkeys cannot master at all when they are three months old. That performance improves gradually through the first year of life.

Apparently, this is because the infant monkey's limbic system, so crucial to visual recognition and associative memory, isn't fully developed. It would seem, therefore, that the anatomical foundation for information memory — the province of cognition — is laid down slowly, while the structures that allow habit formation — the realm of behaviorism — are open for business from birth. Changes in the infant monkey's glucose utilization indicates that brain metabolism also seems to develop gradually over the first four months.

"It may explain why we can't remember the first years of life — any of us," she said. "But we're still not sure if the memories are not there, or we simply can't retrieve those memories."

Development of this habit system is sexually dimorphic — it is faster in females than males. "It's transitory, but it's there and it seems significant." Bachevalier is probing it by manipulating hormonal levels in infant monkeys.

The right haystack

At this point in his exploration of the visual system, Miskin said, "we know the gross circuitry and we are looking for ways to go about studying it molecularly. But first we have to know what haystack to look in."

Two of the laboratory's staff fellows are looking at the individual straws, or neurons.

Barry Richmond began his scientific life as a pediatrician, he said, "because I thought at one time I was interested in development." He now describes himself as a neurologist who is trying to discover how neuronal structures are related to global properties. "How does a two-dimensional image on the retina translate into useful behavior?"

A fundamental and intriguing question for him is how a single neuron transmits information. If one neuron is telling the next neuron, "The cup is blue," how do humans recognize that message if they see it?

Scientists who measure electrical activity in neurons generally assume that, when their recordings show more spikes, the cells are busier relaying messages. But Richmond has found evidence to challenge this purely qualitative measure. "I'm either going to be famous or everyone's going to write me off as crazy," he said.

Richmond, in collaboration with Lance Optican of the National Eye Institute, has applied information theory to compare more precisely what goes into the brain and what comes out. The input, or visual stimulus, is a set of 64 black and white patterns that are based on what mathematicians call a two-dimensional orthogonal basis set. It is such function sets, called Walsh descriptors in this case, that enable computers to generate graphics. Richmond speculated that neurons might use a similar system to assess shape.

When Richmond compared firing patterns generated by different descriptors, he found in some cases that the number of spikes in a given time period was exactly the same. They differed only in the gaps between them.

"That gap had to be essential, because the stimuli were different, and otherwise there was no difference in the pattern," he said. "That meant

that the peak firing rate had no more importance than the conditions under which the neuron didn't fire. The firing rate and the amount of information are not correlated."

It really makes sense, he suggested, that spike quantity isn't necesarily the most informative thing being generated. He offered an example based on recordings from oculomotor neurons, which cause the movement of eye muscles.

"If you know that when the eye is at five degrees you have a firing rate of 50, and at 10 degrees the firing rate is 80, which is the more important number?" Obviously, more is not better, only different. "People working in auditory systems have known this for years," he added.

Richmond said that, if this theory holds up to scrutiny, it will take vision out of the realm of multivariate statistics and provide a way of seeing what neuronal codes actually look like.

Discarding stimuli

Staff fellow Robert Desimone also measures the output of individual neurons. He says of Richmond's theory, "Let's say we have some very intense discussions."

But Desimone's own work is showing something comparable: Researchers should not ignore what the brain is doing when it would seem to be at rest. In this case, he is finding that neurons are active in "throwing out" an irrelevant stimulus. They don't simply turn off in disinterest.

Before Desimone came to NIMH five years ago, he was assisting Princeton psychologist Charles Gross in similar research. He and Jeffrey Moran, a graduate student from the University of Oregon, spend much of their day in front of a TV screen upon which pulses little red and green bars and a black dot.

Behind a curtain on the other side of the room is a monkey in a Plexiglas chair facing the same dot and bars through a large frame apparatus called an eye-coil. A tiny wire from the eye-coil will tell Desimone and Moran whether the monkey is fixating on a black dot on the center of the screen.

On each side of the black dot, red and green bars are flashing. The monkey is expected to attend to the bars on one side and signal when they change shape, while ignoring the bars on the other. At the same time, it's crucial that it continue to fixate on the dot.

Through electrodes to the temporal cortex, Desimone and Moran are recording responses to the relevant versus the irrelevant stimuli.

"We can see from the neurons' responses that, even though the stimulus is the same on the retina, it's different in the monkey's head." Looking at station TE, in the monkey's temporal lobe, Desimone and Moran have seen that the neuron's firing is greatly reduced when the stimulus becomes irrelevant. The response, Desimone explains, is being actively suppressed.

Their findings add to evidence of the brain's increasing sophistication at each "station." "You don't get differentiation like this in the primary visual cortex," Desimone noted. "Activity there doesn't reflect any cognitive process. It's a satellite of the retina; it's just hard-wired."

Desimone said that the monkeys are rewarded with water because the study involves as many as 2,000 trials a day. The scientists would have barely begun before the animals were sated on peanuts. There is an exception, Desimone notes. "We have one monkey who has shaped *our* behavior so that we add Tang to his water."

Neuroscience today requires the input of many minds with differing expertise, Mishkin said. There is little time for meetings to that purpose, however. Researchers said a 40-hour week is rare for them, but that Mishkin's energy level is hard to match. The quality of that effort is widely recognized: this year he was elected president of the Society for Neuroscience, and this summer he received one of three awards from the American Psychological Association for distinguished scientific contribution.

Finding an anchor

Despite the other demands on his time, Mishkin can always find time to talk about the brain. Easing into what would stretch into a two-hour interview, Mishkin noted that Lashley's principles, mass action and equipotentiality, held that an organism's performance depended only on its total mass of brain tissue. If you removed some of that tissue, performance might be diminished, but it wouldn't matter where you cut.

So little was known about the physiology of the brain in the first half of this century that a neurological examination involved only the primary

sensory projection areas and the motor cortex. The other cortical areas were considered "silent cortex" because researchers were unable to evoke sensations or movements by stimulating them. Then in the 1940s, Wilder Penfield showed that their stimulation could trigger vivid memories and thoughts.

The area, it seemed, had some "higher" function, and behavioral scientists were the only ones with the tools to study it, Mishkin said. When he and Rosvold came to NIMH, they continued the research they had begun at Yale by looking at the prefrontal cortex. "But there was no anchor to the work," Mishkin recalled.

He eventually discovered this anchor in the temporal lobes. He observed that animals with lesions to the inferior temporal cortex became impaired on tasks involving vision: they could no longer tell stimuli apart or associate them differentially with rewards.

The impairment affected only vision, yet the animals were not blind and suffered no loss of visual acuity. They simply couldn't remember what they had seen. Here was an incredibly specific point at which to begin exploring the neural maze.

"This was a part of the brain sitting in no-man's land," according to then current dogma, Mishkin said. But these findings were pointing to a functional connection between a sensory input and "brain tissue that was far away." At the same time, Roger Sperry was gaining attention for his work on forebrain commissures and hemispheric specialization. The porridge doctrine was cooling fast.

Back to the cortex

Current evidence may point the way back to the prefrontal cortex. A lot of information, he said, is probably stored in the cortex by the action of the limbic system back upon the cortex. That is, the cortex and limbic system have reciprocal connections.

"The cortex and the basal ganglia, on the whole, do not have reciprocal connections. It's one way. The cortex activates the basal ganglia, and the basal ganglia can activate the cortex in part through an indirect route, by which it gains control of the motor system.

"The anatomy, therefore, fits the ideas that we have proposed, that information may be stored in the cortex

as a result of action of the memory system. What is stored in the habit system is not information, but simply a changing probability of evoking a response to a stimulus, which sort of grows incrementally by trial and error. And that storage is probably not in the cortex but the basal ganglia. So the site, as well as the product of what is stored, is very different in the two systems."

The theory explains why we at times take the route to work when we wanted to go to the grocery store, and why we utter Freudian slips.

"Most often, you may be responding in a learned fashion on the basis of stored products in both systems simultaneously," Mishkin said. "But occasionally they may not cooperate and may come in conflict."

These two systems, Mishkin said, may some day be able to be manipulated, probably through pharmacological agents, to treat memory loss due to age, brain damage or "any kind of pathology that brings about loss of the ability to establish new memories, or the ability to retrieve old information or habits. After all, that is what the nervous system is really good for, why it has evolved in the way it has. For adaptation."

But before such clinical steps are taken, many theoretical miles must be traveled. And an answer must be found to the question of what all this work with monkeys, intellectually fascinating as it is, tells us about the mind of humans.

"The monkey brain and the human brain are very much alike," Mishkin answered. "They're also very different. There is probably as much difference between the human brain and the monkey brain as there is between the monkey brain and *aplysia*."

"Unlike *aplysia*, the monkey is conscious of the world in much the same way we are," Mishkin continued. "But the monkey has no language, no defense mechanisms or ego, no self-awareness."

To use Mishkin's imagery, if the brain is comparable to the cosmos, then perhaps we have gotten as far as comprehending that the earth revolves around the sun. We know that the brain is neither porridge nor a little black box. But how much do we really know?

Mishkin answers with his own variation on an old limerick: "Every box has a little box inside itself to light 'em, and the little box has a littler box and so ad infinitum."

Such Stuff as Dreams Are Made On

"And yet could this millennia-old search for meaning in dreams be merely an exercise in human conceit? Could it be that dreaming itself is devoid of meaning?"

Robert Kanigel

Robert Kanigel is a free-lance writer living in Baltimore.

It is a small, unlovely creature, native to Australia, with a long conical snout, spiny coat, and sticky tongue well-suited to lapping up ants and termites. It is called an echidna and it belongs to a rare species of egg-laying mammals known as the monotremes.

The echidna intrigues neuroscientists. For one thing, the part of its brain called the prefrontal cortex is larger, compared to the rest of its body, than that of any other mammal including man. The prefrontal cortex is the place where such higher functions as planning, interpreting and strategy forming are thought to reside. Yet in these traits the echidna is hardly blessed. Why, then, its outsized brain? The question has mystified science at least since 1902, when anatomist Elliott Smith labeled the phenomenon "quite incomprehensible."

The echidna, from which the main evolutionary line of mammals pulled away about 140 million years ago, can boast one additional distinction: *Alone among mammals, it does not dream.* Scientists base that conclusion on the same kind of evidence that convinces them that other mammals do dream: They attach electrodes to the brain and search for the distinctive brain waves that are the telltale sign of dreaming.

In 1953, Eugene Aserinsky and Nathaniel Kleitman, then at the University of Chicago, observed that sleeping human subjects showed, periodically through the night, a characteristic pattern of rapid eye movements, or REMs; if immediately awakened, they would report vivid dreams.

The sure sign of this dream-laden REM sleep was an unmistakable, low-amplitude, high-frequency brain wave.

REM sleep is part of the general body of knowledge among educated people today, along with *The Iliad,* DNA and $E = mc^2$. But back in the 1950s its discovery caused considerable stir. Looking back on the period a decade later, Gay Gaer Luce and Julius Segal noted that "a furor was created that dreams could be spotted from outside and captured. Hundreds of laboratory nights were recorded and truckloads of dream tapes analyzed. The unbelievable regularity of the dream periods, so startling when Aserinsky first noticed it, gave an even more startling picture *en masse. . . .* There was worldwide exhilaration at the discovery of a handle for extracting dreams."

By now much more is known of REM sleep. Sensory input to the brain shuts off, yet the brain is in a state of high arousal. Blood pressure rises. Muscles lose all tension, yet penises become erect. Human adults engage in REM dreaming for perhaps 100 minutes a night, spread across four or five increasingly lengthy periods; the first might be five or 10 minutes, the last 40. Dreams from early segments typically bear the stamp of recent waking experiences, while later ones show complex integration of the recent past with older memories.

The three-quarters of the night spent in non-REM sleep, meanwhile, also reveals brain activity, but of quite a different sort. Wakened subjects report fewer "dreamlike" episodes, but those they recall make more "sense" and are less vivid and hallucinatory.

Both REM and non-REM sleep are necessary. Fail to get one or the other and the next night you tend to compensate with more of that kind of sleep.

Mammals other than man also dream, or at least engage in REM sleep. Mice, dogs, monkeys, chimpanzees and elephants all show the low-amplitude brain wave pattern that, in humans, corresponds to dreaming.

All except the echidna.

Among dream researchers, that slim fact makes the echidna special—a mystery to be solved by any who would explain just why all other mammals do dream.

"Every dream," said the group leader with perfect assurance, "has a message." The four sat on pillows in a room in an old Baltimore mansion doing service as a holistic health collective. A massage table stood in one corner. Acupuncture and reflexology charts hung from the walls. The four were there for a workshop, "Invitation to Dreamwork," offered through one of the many informal learning networks that have sprung up since the 1960s. All participants, the catalog promised, would get a chance to tell their dreams and learn to get at their meanings.

"But how do you know which interpretation is correct?" a participant wondered at one point. "Is it just emotions?"

"No," replied the slim, goateed group

leader, a veteran of 15 years as a holistic health practitioner, "when you're onto something there's a sense of something clicking into place. It's as if the light dawned, as if the parts suddenly fitted, where they didn't before."

In the workshop was being reenacted an ancient, almost archetypal scene. *What does it mean?* has been asked of dreams since humans first evolved. Now, though, scientists are asking questions that go beyond the meaning of any single dream: *Why do we dream at all? What functions—biological, psychological, or any other—do dreams serve?*

Once, answers seemed more certain.

Four thousand years ago in Egypt the gods made known their will through dreams; that was known fact. Those judged able to interpret dreams were in much demand—which is how the biblical Joseph rose to high estate in the Pharaoh's court. Dreams were thought to warn of danger, urge repentance of sin, supply solutions to life's problems. If, in dream, a woman "gives birth to a cat," a papyrus now in the Cairo Museum advised, "she will have many children."

Ancient Hindu and Chinese medicine looked to dreams for help in the diagnosis of illness. Thus, dreams of music, song and festivities spelled malfunction of the spleen. Drowning meant kidney problems.

In ancient Greece, gods came to men in dreams. At one point, the Greek world reputedly boasted 400 shrines to which sick pilgrims could come, make offerings, follow set rituals and sleep in the temple—there to be visited by a god and, through dream, be cured.

Aristotle saw dreams as sensory impressions left behind from wakefulness which, without the material world and conscious thought to get in the way, left an even deeper mark. He discounted the prophetic dreams then in vogue as "mere coincidences." And he judged dream interpretation a tricky business. "Anyone may interpret dreams which are vivid and plain," he wrote. But dreams are like "forms reflected in water. . . . If the motion in the water is great, the reflection has no resemblance to the original."

For most of this century, through the vision, brilliance and strength of personality of a Viennese genius, it came to seem that Aristostle's "water" had at last stilled, that dreams served a purpose satisfactory to science, that their interpretation might itself be reduced to a science. The year 1900 saw publication of Sigmund Freud's *Die Traumdeutung*, or *The Interpretation of Dreams*.

Dreams are neither messages nor fantasies, said Freud, but they're profoundly meaningful. Beneath their bizarre surface they express hidden, repressed wishes, protecting sleep from the often painful knowledge of the unconscious

mind. Though superficially built up from events of the day, they've actually been twisted and reshaped by a hidden censor into safer form. The psychoanalyst, encouraging techniques of free association, helps his patient root out the real meaning from behind the dream's distracting facade.

In its first six years, *The Interpretation of Dreams* sold only 351 copies. But by the time of its fourth edition in 1914, Freud's theory had captured the intellectual imagination of the world. Today, Freud's special vocabulary is part of common parlance, making it hard to imagine "wish fulfillment," "repression" and "the unconscious" unraveled from the collective psychic fabric.

Among Freud's proteges was Carl Jung, son of a Swiss clergyman; though Freud groomed him to succeed himself as leader of the world psychoanalytical movement, Jung later broke with his mentor. Jung shared Freud's sense of the primacy of the unconscious mind, and he, too, saw dreams as a window into it. But dreams, he maintained, did not invidiously hide their true meaning through a variety of psychic devices, as Freud said. They were a product of nature, rooted in fantasy, legend and myth, that spoke to the dreamer through symbols.

Not long ago, a journal given over to the study of myth, *Parabola,* devoted a special issue to dreams. In an introductory essay, Paul Jordan-Smith noted that "from today's bookstands one could amass a considerable collection of works purporting to unravel the apparent messages of dreams." While biblical dreams had but one source of authority, today "our heads are filled with notions of meaning derived from civil education, Sunday School, Freud and the Freudians, Jung and the Jungians, whimsical books of popular psychology, religion and occultism, fiction and fantasy, and politics. . . . In short, there is not one source for the images of our dreams, but a multiplicity of sources, each vying with the others for authority."

But while only stubbornly yielding their secrets, Jordan-Smith went on, dreams continue to exert a hold on us—"because there is in them an unmistakable call to something beyond the ordinary, something, as Rudolf Otto called it, 'wholly Other. . . .' "

And yet could this millennia-old search for meaning in dreams be merely an exercise in human conceit? Could it be that dreaming itself is devoid of meaning, no more than a side effect of certain physiological processes, perhaps the product of random discharges of electrical energy?

J. Allan Hobson and Robert M. McCarley, two Harvard neurophysiologists, don't say quite that. What they did say, in a theory first proposed in 1975, is that "the primary motivating force for dreaming is

not psychological but physiological"—an idea that is quite revolutionary enough. Extending research by the French investigator Michel Jouvet, Hobson and McCarley described what they call a "dream state generator" in a primitive part of the brain called the pons which, every 90 minutes, issues signals to the higher brain centers to launch the process of dreaming.

Say what you will about dream content, argues Hobson today, dreaming must be seen for what it is, the natural physiological response of the forebrain to stimulation from the pons. To say, as Freudians do, that we dream *in order* to repress emotions too painful for the conscious mind to bear is nonsense, says Hobson. "It would be like saying that you breathe in order to avoid your death wish."

Why do we see things in dreams? Because the signal from the pons excites the brain's visual system. The rapid eye movements that give REM sleep its acronym correspond to signals from the pontine brain stem. We see in dreams, says Hobson, "not because we are invaded by spirits, but because the visual system has been activated. Period."

Then why is what we do see so often bizarre? Simple; because brain systems which are rarely active simultaneously during wakefulness are stimulated together during REM sleep. And scene shifts, time compression, symbol formation—how account for them? "The forebrain," Hobson and McCarley write, "may be making the best of a bad job in producing even partially coherent dream imagery from the relatively noisy signals sent up to it from the brain stem."

But what explains just *what* we see in dreams, the specific content? Now that's another question, says Hobson, one he doesn't much address. Theories of dream content need not conflict with his at all. It's just that most other dream research is "looking where the light is"—at dreams themselves—while sidestepping more fundamental questions of basic biology. The Hobson-McCarley theory merely redresses that imbalance. "I don't want people thinking," says Hobson, "that I don't think dreams are meaningful. I do."

The hedge to the contrary, the Hobson-McCarley theory has generated "skepticism and downright hostility," according to David Foulkes, professor of psychiatry at the Emory University School of Medicine and a distinguished dream researcher himself. Many in the field, Foulkes says, "feel it's presumptuous to reduce dreaming to brain stem physiology." Moreover, he believes the Harvard researchers ignore a large body of dream research. For example any model that sees dreams as chaotic and bizarre, with little narrative continuity, "better fits the stereotype of dreaming than it fits empirical observation."

Whatever the neurophysiological foundation of dreams, Hobson and McCarley took care to note in their original paper, it doesn't mean "that dreams are not also psychological events; nor does it imply that they are without psychological meaning or function. . . ." Saying otherwise would have run roughshod across human experience.

To men and women, almost universally, dreams *mean*. In one sense or another, they are seen as messages—from the gods, from earlier lives, from the unconscious, from the future, from a higher power or a lower force—in any case something beyond normal waking consciousness. Something in dreaming pulls on the imagination and the intellect, leaving us sure that something important, meaningful, central to our lives is going on during sleep.

It was one day in 1963 while vacationing in Wales that the late psychologist and dream theorist Christopher Evans came to a similar conclusion. He was walking along a stretch of deserted beach when he spied a large black bird—a cormorant, he later decided—perched on a single leg, head tucked beneath a wing, standing on a rock. Approaching it, he realized it was sleeping. So oblivious was it that Evans could get near enough to hear its breath and examine its feathers; had he wished, he thought, he could have killed it. After a while he tapped the bird on the shoulder and said "hello"—whereupon the bird awoke in fright and wildly flapped away.

"In biological terms," Evans would conclude later in his posthumously published *Landscapes of the Night,* completed by Peter Evans, "sleep was an immensely dangerous exercise—perhaps the most dangerous single thing an animal could do—and yet all animals indulged in it. What fundamental process could it possibly serve?" What about sleep is so necessary that it could compensate for the risk of sudden death?

Sleep is "possibly the most important single function of an animal's life," Evans concluded. Dreams, likewise, are much more than "gratuitous, purposeless, meaningless exercises performed by a brain with so much spare capacity that it chooses to wallow in episodic acts of self-indulgence."

"If sleep does not serve an absolutely vital function, then it is the biggest mistake the evolutionary process has ever made," veteran dream researcher Allen Rechtschaffen, of the University of Chicago, has concluded. "Sleep precludes hunting for and consuming food. It is incompatible with procreation. It produces vulnerability to attack from enemies. Sleep interferes with every voluntary, adaptive motor act in the repertoire of coping mechanisms. How could natural selection with its irrevocable logic have 'permitted' the animal kingdom to pay the price of sleep for no good reason?"

What is the "good reason?" Why sleep? Why dream?

There's no want of theories; Rechtschaffen, in fact, sees dream research as afflicted with a kind of "theory of the year" syndrome. But no model has won universal acceptance or even provided a common frame of reference. Scientists understand dreams the way one might a grapefruit sliced open to see what it looks like inside: Depending on where you cut, you get a neat, geometric array of segments or a mess of pits, pulp and rind.

One recent Ph. D. candidate who studied the physiology of dreaming for his doctoral research at the University of California at Santa Cruz, tells of being so frustrated with the field that he left it altogether. "Every-

---

> "The forebrain," Hobson and McCarley write, "may be making the most of a bad job in producing even partially coherent dream imagery from the relatively noisy signals sent up to it from the brain stem."

---

body was just sounding off with their theories," complains Joseph Palka, now a television producer on health- and science-related topics in Washington; it was all "just so much armwaving."

As they have been all through history, dreams are still a mystery. About the only consensus among researchers, says Joe Palka, "is that dreaming is not what Francis Crick says it is."

It was Francis Crick who, with James Watson, unraveled the structure of DNA in 1953 and ushered in a biological revolution. Later, having turned to the study of mind and brain, he and a young British scientist, Graeme Mitchison, propounded a theory of dreams: We dream not to consolidate our memories, as some would have it, but to rid our brains of useless ones; we dream not to remember but to forget.

According to one substantial line of evidence, the brain stores information not in individual brain cells but within close-woven nets of cells—just as an array of fine dots may together form a newspaper photograph. The memory of a golden sunset, for example, might reside in one particular constellation of neurons. And those same neurons might also participate in the storage of, say, "dusk" and "glow" and "horizon." In information-processing parlance, such a network is "superimposed." It is also "distributed," a memory being lodged in many neurons, not just one. And it is "robust," in that the loss of a single brain cell doesn't mean loss of the whole packet of information.

Both theory and computer modeling predict that such networks can overload when too many patterns are stored, as it were, on top of one another. The result is "pathological" behavior that in humans might correspond to fantasies, obsessions or hallucinations. Crick and Mitchison call them "parasitic modes." In a properly functioning system, needless to say, they must be regularly purged.

And that, Crick and his colleague suggest, is what dreams do. "We propose," they wrote in *Nature*, "that such modes are detected and suppressed by a special mechanism which operates during REM sleep and has the character of an active process which is, loosely speaking, the opposite of learning. We call this 'reverse learning' or 'unlearning.' "

Among the evidence to which Crick and Mitchison turn in support of their theory is the echidna.

Why does the Australian anteater need so large a prefrontal cortex? So its neural nets don't overload, degenerating into parasitic modes; simulated on computers, it turns out, such nets prove more resistant to overload the bigger they get. A large cortex, then, was the cumbersome solution with which the echidna, more than 140 million years ago, got stuck. Evolutionarily more advanced mammals hit upon a more elegant solution: dreams. Able to "tune" their brains every night through dreaming, thus ridding them of parasitic modes, they could get by with a smaller cortex.

The attention paid the Crick-Mitchison model owes as much to the luster of the Crick name as to any enthusiasm for it among dream researchers. Graeme Mitchison, Crick's colleague, says response to their theory has been highly negative. "We delight that someone of his stature has become interested in REM sleep," says longtime sleep researcher Rosalind Cartwright of Crick. "But he's got a lot of homework to do." Adds Emory University psychiatry professor David Foulkes: "I don't think anybody takes it seriously."

One criticism, as Mitchison reports it, is "that our theory fails to account for the rich, subjective quality of dreams, that it seems impoverished and mechanistic." For example, the theory seems to hold that dreams should fairly ooze with meaningless, trivial material. Yet dreams more often pack a powerful emotional punch. Say you attend a lecture: Does your later dream about it hinge on the baldness of the lecturer, the brownness of the auditorium's paint job, or other such seemingly perfect candidates for Crick and Mitchison-style "unlearning"? No, says Ernest Hartmann, director of Sleep Research at Lemuel Shattuck Hospital in Boston. More likely the dream evokes memories of your father, say, or other emotionally charged material. Declares Hartmann: It's the actual content of dreams that most persuasively argues against the theory.

But while faulting it, Hartmann points to important common ground between the Crick-Mitchison model and others. For its "unlearning" actually serves what he calls *"the interests of learning."* And in one guise or another it's learning—processing information, consolidating memory, assimilating experience, sifting through strategies—that is common to most theories of dreaming. Dreams help "tie up loose ends" in the brain, says Hartmann; they integrate recent experience with past knowledge. To the extent there's consensus about dreams, that's it.

David Cohen, of the University of Texas at Austin, notes that "a mammal requires a lot of learning to be a mammal. Everything isn't there on the genes." We have to learn language and how to interact with others. To do that, "you need experiences, exposures. It's all got to be integrated. Some of this learning," the evidence convinces him, "goes on in sleep.

"What you are is not just why you perceive during wakefulness," he says, but what you make it into through dreams. And that nighttime brain activity, he suspects, is not just "like playing it back on a tape recorder." Rather, information is funneled into an active memory process, where it's organized in novel ways, reworked and restructured.

To Cohen, then, dreaming is a creative act. Awake, we accommodate to the world around us. Then sleep cuts off the outside world and, through dreams, we *assimilate* the day's experiences in ways unique to each of us. "It's our own personal way of integrating experience."

Rosalind Cartwright also sees dreaming as an active process, performing important emotional "work"—as her research into the dream life of women going through divorce tends to confirm. The chairman of psychology and social sciences at Rush-Presbyterian-St. Luke's Medical Center in Chicago places herself "among those folks who say dreams have memory storage function, that they take the day's affective material and process it into long-term storage by finding a match for it in earlier-stored memory." Dreams, in her view, help work through the emotionally charged material of the dreamer's life.

In a study of divorced women aged 30 to 55, the happily wed women she used as a control proved to have "extremely dull

dreams": taking the kids to the swimming pool, selecting vegetables at the supermarket, and such workaday matters. These women had little work for their dreams to do. The divorcees, on the other hand, had "long, complicated dreams, loaded in terms of emotional content." They were undergoing a "critical life event that forced them to reorganize who they were. They weren't Mrs. So-and-So anymore."

Moreover, those divorcees whom standard tests marked as most "traditional"— most centered on marriage and children— dreamed the most. That was hardly surprising, says Cartwright, since such women had the most need for emotional reorganization. Their dreams helped them take the raw, painful reality of divorce and assimilate it.

"Something in dreaming pulls on the imagination and intellect, leaving us sure that something important, central to our lives, is going on in sleep."

"Dreaming," summarizes Cartwright, "works to renew and rehearse, reaching backward and forward in our lives, on emotionally important issues that we have no time to cope with in wakefulness." By providing "an interval of off-line work, it gives us the opportunity to update the file on who we are and how to cope in the world."

"Off-line"?

"File"?

Ubiquitous among dream researchers these days are concepts and terms borrowed from the world of information processing and computers. "It seems to be the language of cognitive psychology today," notes Cartwright. Says Graeme Mitchison of the computer metaphor: "It's rather like the Victorian age, when the metaphor of The Machine" was inescapable. Emory's David Foulkes also sees promise in the information-processing models: "They have a certain kind of plausibility. Maybe there'll turn out to be one among them that will yield the functions dreaming serves."

It was Edmond Dewan who in 1969 first likened dreams to "off-line processing," when a computer closes itself off to new data and manipulates instead what's already in storage. Simple organisms, Dewan said, are preprogrammed, as it were, with instinctive, stereotyped responses to their

environment. Mammals, on the other hand, must first learn about their environment, then decide how best to handle it. While awake, they respond as best they can to crises of the moment. During sleep, with the external world shut off, they can further integrate information taken in during wakefulness, thus reprogramming their behavior.

Dewan's paper, which appeared in an Air Force research journal, proved influential, serving as catalyst for a number of related theories. In his recent *Landscapes of the Night,* Christopher Evans offers one version. During sleep, he writes, "the great software files of the brain become open and available for revision in the light of changes that have taken place as the result of the horde of new experiences which occur every day.

"What are dreams?" he asks. "Well, they are the programs being run."

Which programs? Social adaptation programs, for one. In an insight he credits to Cambridge University psychologist Nicholas Humphrey, Evans pictures dreams as "dress rehearsals for events we can expect, hope for or fear in everyday life. Situations present themselves in which the dreamer is an actor, playing a part, coping with the often strange twists of the plot, keeping abreast of the unfolding drama."

In perhaps the most ambitious synthesis of disparate strands of evidence yet, Jonathan Winson outlines a theory of dreaming that seeks to reconcile the latest neurobiological research with information processing models and the insights of Freud. His view, outlined in the book *Brain and Psyche* and based in part on research in his own laboratory at Rockefeller University in New York, sees a sausage-shaped piece of the brain known as the hippocampus as a kind of neural gatekeeper for information from the world outside. Closed during non-REM sleep, when the cortex is largely quiet, the gate opens during REM sleep and wakefulness, both periods of high arousal. For Winson, here is the physiological expresion of "off-line processing."

That processing has immense survival value, helping to integrate experience and consolidate memory. "In man," Winson writes, "dreams are a window on the neural process whereby, from early childhood on, strategies for behavior are being set down, modified or consulted."

In support of his theory, Winson turns to a stunning variety of evidence—among which the echidna again figures. To him, the echidna is nature's last attempt at doing without dreaming. The complex task of associating memories with new experience and formulating new behavior—which more "modern" mammals do through dreams— the echidna does while wide awake and busy with the outside world. That's what demands the animal's enormous prefrontal cortex. "Should the organization of man's brain have been similar to the echidna's," Winson observes, "he might have needed a wheelbarrow to carry it around."

Unlike many neurobiologists, Winson does not wholly reject Freud, but rather demands of his own theory that it explain such Freudian staples as dream distortion, repression and transference. Indeed, he interprets as the Freudian unconscious just those mechanisms involved in REM sleep.

Winson was not the first to point out that Freud's psychological theories mirrored the neurophysiology of his day—which was then in its infancy. So of course he was wrong much of the time. Today's understanding of the nervous system and its workings, says Winson, suggests that dreams are not the work of some mysterious "censor" at all, but simply a by-product of evolution.

For example, why do dreams invariably reduce the abstractions of waking human thought to visual form? Because, says Winson, dreaming is "a vestige of our animal past," which had no place for abstractions. "Language and abstract concepts derived therefrom," he writes, "played no part in the lower mammalian brain." Does the dream have some abstract "work" to do that might, when awake, require one or two sentences to convey? Well, the ancient, evolutionarily bound dream mechanism does it by manipulating images.

So you dream of riding atop a mare. *Mere* is mother in French; Freudians might have a field day with *that* one, and Winson doesn't automatically discount their interpretations. It's just that the image, far from the work of a Freudian dream "censor," is really just "your brain trying to do its job within the limitations of evolution. . . . Your unconscious mind is

searching for something to represent the abstraction."

For Winson, the Freudian unconscious is real. Dreams, he writes, "are the statements, wishes, hopes and fears of the unconscious personality" of the dreamer. "Despite their apparent dramatic swings from the bizarre to the logical and matter of fact, the content of dreams day by day forms a cohesive pattern and reflects each individual's unconscious strategy for survival."

Whatever their neurobiological function, dreams evoke a sense of freedom from the grey constraints of everyday reality, suggest to us secret knowledge transmitted, mysteries revealed, a suspension of the ordinary rules of daily life.

Into that nighttime freedom, artists and thinkers have sunk deep wells of creativity. The modern black writer Gloria Naylor (*The Women of Brewster Place* and *Linden Hills*) reports that the titles of her first four books all came to her in dreams. A dream inspired Robert Louis Stevenson's *Dr. Jekyll and Mr. Hyde*. A dream of a snake consuming its own tail brought the 19th-century chemist Friedrich August Kekule the long-sought chemical structure of benzene—six carbons arranged in the form of a ring. It was in a dream, too, that Otto Loewi conceived the crucial transmission of nerve impulses. (Reportedly, he forgot the dream when he awoke, then had it again the following night.)

Dreams so proclaim freedom from dreary reason and waking constraint that the word has come to mean, figuratively, a wild fancy, ambition, or vision of the beautifully unlikely. In her song "Woodstock," Joni Mitchell imagined a world at peace: "I dreamed I saw the bombers/Riding shotgun in the sky/Turning into butterflies. . . ." When the sociologist and student of the

adult life cycle Daniel Levinson sought a metaphor for youthful hopes not yet dashed by adult reality, he chose The Dream. Martin Luther King, Jr., expressing his vision of racial harmony at the Lincoln Memorial in 1963, cried, 'I have a dream!"

Wrote Shelley: "Some say that gleams of a remoter world/Visit the soul in sleep." A dream is an opening in the universe, a momentary freedom from the fixed and immutable laws of causality.

For most of a century, Sigmund Freud had a virtual lock on how the thinking world viewed dreams. It was through the filter of his rich and innovative intellect that generations of educated people perceived dreams in particular and the unconscious mind in general.

There are still Freudians to be found, and they can offer a spirited and impressive defense of the master's thinking. But their ranks are thinning. "It's in great decline," says David Cohen of Freudian thinking, "embarrassed by poor results in the lab and in therapy." It's plain, as Graeme Mitchison says, that "the mainstream of psychology no longer holds rigidly" to Freudian orthodoxy.

The old, comforting certainty is gone. At least until new insights or new experimental paradigms come along, anything goes.

In a way, it's sad. We are back in the state of befuddlement and uncertainty that existed on the eve of the publication of The *Interpretation of Dreams*, when Freud could write: "In spite of many thousands of years of effort, the scientific understanding of dreams has made very little advance . . . [In the scientific literature of the time], many stimulating observations are to be found and a quantity of interesting material bearing upon our theme, but little or nothing that touches upon the essential nature of dreams or that offers a final solution of any of their enigmas."

The Unbalanced Brain

Parkinson's disease and Huntington's chorea appear to be at opposite ends of a chemical seesaw in the brain

Patrick L. McGeer

Patrick L. McGeer is a neuroscientist who specializes in research on the chemical messengers of the brain. He is presently on leave from his position as director of Kinsmen Laboratory of Neurological Research at the University of British Columbia in Vancouver in order to serve as Minister of Education for the Province.

Probing for life on Mars, like the walk on the moon, has been a spectacular achievement. Both were accomplished with enormous efficiency, incredible ingenuity and several billion tax dollars. But there is an inner universe yet to be explored fully that is as diverse and as potentially rewarding as anything that lies amongst the stars.

By the inner universe I mean, of course, the human brain. The master organ of the body, it weighs only three pounds. It looks no more impressive than a lump of molded putty, yet it represents the pinnacle of nature's achievement on this planet. It is the reservoir of all human knowledge and the limiting tool in exploring the outer universe.

As one philosopher put it, "Without the human brain, the drama of the cosmos would be played before empty stalls. There would be no knowledge, because there would be no way of knowing."

The human brain is far larger than that of any other species except some much larger mammals such as killer whales. It is this organ alone which permits humans to demonstrate superiority over other species. The kidneys, the heart, the lungs, and the biceps are as good, or better, in other mammals. But the brain is not a forgiving organ. Neurons do not divide beyond the time of birth. If they die or are destroyed, they are not replaced. We are left with the gloomy prospect of knowing that, at least in terms of numbers of neurons, we are born with all the brains we will ever have. Misuse, disease and age deplete neurons throughout life.

In the past decade much of neuroscience has been focused on the neurotransmitters—the chemical messengers that convey electrical impulses from one neuron to another. These chemicals are released from the so-called boutons—the specialized enlargements of axons—that synthesize and store them in ping-pong ball-like vesicles. The released transmitter travels across a tiny gap called the synaptic cleft and then acts at the

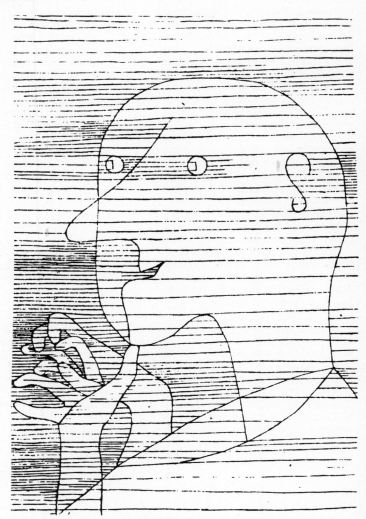

Old Man Figuring, by Paul Klee (1929) The Museum of Modern Art

specific receptor, or "docking site," on the postsynaptic neuron.

This mechanism is unique to the nervous system—not shared by the liver, lung or kidney. And here, within the nervous system, is a way of influencing particular types of neurons: just discover the neurotransmitters which govern those cells having a particular function—like mood or movement—then discover agents that change the action of the transmitter or its receptors. We will then possess the power to modify the function. Other organs, or other neurons within the nervous system not using that particular transmitter, will remain unaffected.

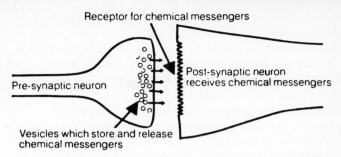

Schematic diagram of a neuronal connection showing role of the chemical messenger or neurotransmitter.

It is a wonderful theory, but neuroscientists are faced with difficult problems. The brain is fantastically complex. The writing diagram is more sophisticated by far than the most advanced computer. There are probably more than 50 billion neurons in the human brain—each with many thousands of nerve endings—and so far only seven established neurotransmitters. A dozen other chemicals are suspected of being neurotransmitters. It is clear that there must be considerable overlap. Therefore, whenever there seems to be a close association between functions that would not ordinarily be thought to be related, one suspects that the common link is the neurotransmitter. Mood and movement are two such physiological functions.

Relationships between mood and movement have been revealed in research on two diseases involving movement—Huntington's chorea and Parkinson's disease—and on two aspects of mood and behavior—tranquilization and schizophrenia. A curious overlap exists and seems to be connected with three main neurotransmitters—dopamine, acetylcholine and GABA (-aminobutyric acid).

Much of the research centers on a functional system in the brain known as the extrapyramidal motor system, which has long been thought to control movement, and on four large neuronal clusters, known as nuclei, in this system. They are the caudate, the putamen, the globus pallidus, and the substantia nigra. The caudate and putamen together are known as the neostriatum, and, with the globus pallidus, they form the basal ganglia. In the accompanying illustration, we see not only these nuclei, but some of the neuronal tracts that link them.

Some 25 years ago, the first psychopharmacological agent, reserpine, was unveiled. Originally employed as an anti-hypertensive agent, its tranquilizing side effects were first regarded as unwanted. However, psychiatry—which so often uses the unwanted crumbs from the table of medicine—found reserpine to be the first really effective treatment for psychosis. Its clinical success spurred tremendous interest in discovering its mechanism of action. It was found that reserpine depleted the body of four amines that were later identified as neurotransmitters. Three of them were chemically classed as catecholamines and one (serotonin) as an indoleamine.

The effect on blood pressure was quickly established as being due to depletion of noradrenaline, the neurotransmitter for sympathetic nerves supplying blood vessels. But it was anybody's guesss, and to some extent still is, what chemical was responsible for the antipsychotic action. In the mid-1950s, clinicians first noticed the relationship between reserpine and movement. It produced in some people a rigidity and poverty of movement strikingly similar to that caused by Parkinson's disease. The common denominator turned out to be dopamine. The Swedish pharmacologist, Arvid Carlsson, showed that one subcortical region of brain—the neostriatum—contained large quantities of dopamine and that reserpine depleted dopamine in the brain. He speculated that dopamine was important to the control of movement, since this extrapyramidal function was believed to be centered in the neostriatum. This was a key discovery around which much subsequent research turned.

A second key discovery was made by Oleh Hornykiewicz, then at the Pharmakologisches Institut der Universitat in Vienna but recently working at the Clarke Institute in Toronto. He showed that Parkinsonian patients also had depleted dopamine in their neostriatum, just as Carlsson had found for reserpine-treated animals.

Why not then build up the dopamine in Parkinson's disease victims by administering its precursor L-DOPA? Several false starts in this direction were made—including one by our own laboratory—all because the doses were too low. But in 1967, George Cotzias of Brookhaven National Laboratory demonstrated L-DOPA's therapeutic effectiveness. It is now the treatment of choice in this disease. Since it was the treatment of disturbance of mood that led to this startling advance in the therapy of a movement disorder, scientists guessed that one of the side effects of high doses of L-DOPA might be the induction of psychosis. And this turned out to be true for 10-15 percent of parkinsonian patients treated with the compound.

Psychiatric medicine provided still further clues. Shortly after the introduction of reserpine, another antipsychotic—chlorpromazine—was discovered. This also occasionally produced parkinsonian side effects. The mechanism of action turned out to be somewhat different—it did not deplete dopamine, but it blocked its receptor sites.

In either case the treatment turned out to be the same—to use the standard anti-parkinsonian agents of the day. These were atropine-like agents which block the action of another chemical believed to be a neurotransmitter—acetylcholine—also known to be present in high concentrations in the neostriatum.

At about that time, I was asked to see a patient who was incorrectly believed to be suffering from drug-induced parkinsonism. Treated with large doses of anticholinergics, she developed a syndrome remarkably similar to another neurological disorder, Huntington's chorea. In many ways, chorea is the clinical opposite of parkinsonism. Parkinsonism produces rigidity and an inability to initiate movement. Victims of chorea find themselves unable to stop movement or control overshooting voluntary actions.

Together with some of my colleagues, I suggested that in the neostriatum, dopamine was balanced by acetylcholine in a simple teeter-totter arrangement. If the dopamine side became light, parkinsonian rigidity developed; if the acetylcholine side became light, chorea activity developed. I was somewhat reticent to publish an analogy with a child's teeter-totter, but the great Canadian neurologist and a good friend, André Barbeau, did so, adding some ideas of his own, and generously acknowledging our studies.

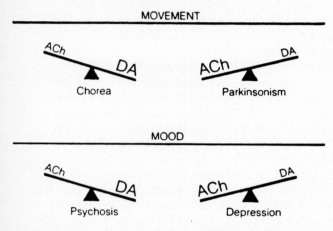

Hypothesized balance between cholinergic (ACh) and dopaminergic (DA) extrapyramidal neurons in the control of movement and mood.

How does this balance come about?

Many years ago pathologists found that the substantia nigra degenerated in Parkinson's disease. There are two of these structures, each about the size of a flattened pea, one on each side of the brain. As people grow older, even normal individuals seem to lose cells in this area. At birth, a normal child has about 400,000 cells in each substantia nigra. With age this number progressively decreases, so that by 65 it is reduced to about 150,000 cells. Those with Parkinson's disease possess even fewer cells as they age—about 60,000-120,000 cells. Physiological breakdown begins to occur when the cell level drops below about 120,000. If we knew why substantia nigra cells die, we might learn how to protect ourselves from Parkinson's disease.

In the mid-sixties, Louis Poirier of the University of Laval in Quebec City and Ted Sourkes of McGill University in Montreal cut the tract between the substantia nigra and the neostriatum in monkeys and showed that a loss of dopamine resulted. Almost simultaneously, a Swedish group, including Arvid Carlsson, developed a highly sophisticated technique which revealed the pathway that dopamine travels between the substantia nigra and the neostriatum. Later, our labora-

tory discovered that there are large numbers of small cholinergic cells within the neostriatum which are connected to the dopamine neurons. The interconnection must be of a fairly massive nature—several hundred contacts per cell at a minimum, which accounts for the strength of the dopaminergic-cholingeric reciprocal relationship.

A third type of neuron in the extrapyramidal system, this one using GABA, was found to connect with dopaminergic cells and to decrease their activity. This connection may be another, as yet unexploited possibility for the treatment of Parkinson's disease. Blocking the inhibition of dopamine cells should speed up their activity and help to correct the deficiency.

No sooner had these connections in the nervous system been discovered than they turned out to be important also in Huntington's chorea. Chorea is a genetic disorder, which emerges in the third to fifth decade of life. It is relentlessly progressive. It exhibits two characteristics—a movement disorder (opposite to that of Parkinson's disease) coupled with a behavioral disorder (which, in some ways, resembles schizophrenia). In advanced stages, the unfortunate victim cannot control movement. The limbs jerk endlessly in a purposeless fashion. Progressive dementia takes hold until death. What is particularly cruel about the disease is that every person with a choreic mother, father, older brother or older sister knows that there is one chance in two that he or she will suffer the same fate. There is no reliable way of determining who has the faulty gene and who is free until the disease strikes. The news that the damaging gene is present comes after children have been born and reared. It is this uncertainty which brings such fear to affected families.

It had long been known that in Huntington's chorea there was a loss of cells in certain regions of the brain. Thomas Perry of the University of British Columbia in Vancouver, Canada, first reported in 1974 that GABA was decreased in the neostriatum and substantia nigra of Huntington's chorea patients. Later that same year, our laboratory, as well as that of Edward Bird and Leslie Iverson of the University of Cambridge, published data indicating that both GABA and acetylcholine cells are lost in various parts of the brain in patients suffering from Huntington's chorea. Pathologists could and did for years see the gross pathological damage (cell loss) in Parkinsonism and chorea. Now biochemists have added a new dimension to this classical pathology by revealing fundamental changes in disease processes which could not be seen by microscopic examination. It is to be hoped that these methods will soon shed new light on other unsolved diseases.

A fourth possible type of neuron which may play a role in these events is a cell known as the glutamatergic neuron, which would use glutamate, a suspected neurotransmitter. For almost 15 years, neurophysiologists have been speculating that glutamatergic neurons exist in the central nervous system. But no definite pathways using glutamate have ever been identified.

Just a few months ago, Joseph Coyle and Robert Schwarcz of Johns Hopkins University, and a group in our laboratory found that the biochemical and pathological events in Huntington's chorea could be reproduced in animals through the injection into the striatum of glutamate or glutamate analogs such as the powerful kainic acid. Although the reason why GABA and acetylcholine cells are destroyed by glutamate is not known, John Olney of Washington University in St. Louis suggests that glutamate may overstimulate the cells and quite literally excite them to death. Some evidence exists that the massive neuronal tract from the cerebral cortex to the striatum may use glutamate as its neurotransmitter.

The model of brain activity that emerges is of a network where dopaminergic cells from the substantia nigra (which appear to be inhibitory) and glutamate cells from the cortex (which would be excitatory) provide input to the striatum. Within the striatum we have acetylcholine and GABA cells, which are excitatory and inhibitory respectively, and which process the information internally. Finally, we have GABA neurons that go out from the regions of the brain known as the caudate-putamen and the globus pallidus. This scheme is incomplete, because it ignores a number of pathways which have already become established anatomically but which, so far, have unknown biochemistry. Moreover, even the connections that so far have been described may need improvement or revision. But at least we now have a start towards a biochemically coded "wiring diagram" of one of the major motor systems of the brain.

We are perhaps many years behind in our knowledge of mental, as opposed to neurological diseases, but the same kinds of clues as there were for movement may be present for mood and behavior. Huntington's chorea is not just a movement disorder. It is much more prominently a behavioral disorder. In fact, many cases not showing the characteristic changes in movement are misdiagnosed as schizophrenia in the very early stages. There is a loss of cortical cells in Huntington's chorea, which has always been cited as the cause of the behavioral disorder. But this may not be the real reason.

While psychosis is a well-known feature of Huntington's chorea, it is almost never seen in Parkinson's

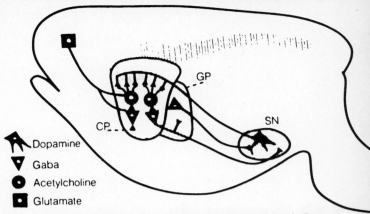

Diagram of major nuclei of the extrapyramidal system in brain (long believed to control movement) with some of the important neurons identified as to the specific chemical each uses as messenger. CP = caudate putamen (neostriatum), GP = globus pallidus, SN = substantia nigra.

disease. Instead, parkinsonians tend to develop depression. Schizophrenia is counteracted by agents which deplete or block dopamine. On the other hand, L-DOPA which leads to excessive dopamine, can sometimes produce psychosis as a side effect. Amphetamine, an agent which preferentially releases dopamine, will produce, in high doses, a paranoid psychosis in normals almost without fail.

On the other side of the teeter-totter, high doses of anticholinergics can produce hallucinations and psychotic behavior. Lower doses do not and are often used to counteract the parkinsonian-like reactions produced by anti-psychotic drugs. As a consequence, it has often been believed that cholinergic mechanisms play no part in psychosis. But in view of the results obtained with higher doses, this may prove to be incorrect.

An understanding of the physiological role probably played by certain elements of the circuit has been supplied by two well-known neurological disorders. Parkinson's disease and Huntington's chorea. In turn, a better understanding and better treatment of Parkinson's disease has emerged. We hope the same may soon take place for Huntington's chorea. Vital clues as to possible underlying physiological disorders of mental diseases have been unearthed. We should be warned away from acceptance of the glib notion that mental disorders are psychogenic diseases related only to environmental circumstances.

Physiology

Are we a piece of machinery that, like the Aeolian harp, passive takes the impression of the passing accident? Or do these workings argue something within us above the trodden clod?

Burns

Physiology, as the dictionary defines it, is a branch of biology dealing with the processes, activities, and phenomena characteristic of life. All organisms, plants, and animals face similar challenges in life: finding food; defending themselves against predators, parasites, and diseases; regulating body composition; coordinating the activities of the various parts of the body; and reproducing, to name a few. How organisms cope with these problems is the province of physiology. This section looks at several aspects of physiological functions as they apply to living systems.

Most physiologists are convinced that research will help achieve a better understanding of normal physiological functions. They believe that by understanding what is normal, we will be better able to deal with what is abnormal and be in a better position to correct it. For example, in "The Aging Body" the author presents the results of recent findings regarding the process of aging in humans. As he points out, aging is a normal process; it cannot be corrected, but it can be understood.

The article "Immune System: Great Mystery Is Solved After Long Quest" explains that immunological defense depends on recognition signals. Failure in this kind of recognition can lead to infection, cancer, or other illnesses.

Breakthroughs in fetal surgery are discussed in "Saving Babies." The author presents some of the difficulties and risks involved in operating on the unborn child.

"Weight Regulation May Start in Our Cells, Not Psyches" investigates the mystery of why some people are overweight and what can be done about it. Recent research shows that obesity problems may be caused by the body's fat cells which can "learn" to counteract repeated dieting.

In "Ergot and the Salem Witchcraft Affair," Mary Matossian presents evidence that the "witches" of Salem may not have been witches at all but victims of nutritional poisoning.

Looking Ahead: Challenge Questions

Is it possible to slow or reverse the aging process of humans through exercise or types of life-style changes? Explain.

What type of substance is nerve growth factor? How does it function?

Unit 5

The Aging Body

After twenty, the decades take their toll. You may get wiser, but your memory dims. Your body parts grow, shrink, disappear. It's a process you can only watch with wonder, so you'd do well to know the wonders to watch for.

John Tierney

John Tierney is a staff writer for Science 82 *magazine.*

There are many gruesome things to be said about a man's body as it creeps past the age of thirty. But first a word about the President's hairline:

"It's a hairline you normally see only on a child or a eunuch," says Dr. Norman Orentreich, the inventor of the hair transplant, who has studied men's scalps for three decades and still has a hard time accounting for Ronald Reagan's hairline. Men typically lose hair around the temples. It's an effect of androgen, a hormone produced in the testicles after puberty, and it seems to happen to all men, even those who otherwise keep a thick head of hair all their lives. Yet here is Ronald Reagan with a straight line of hair above his forehead. "He's not wearing a hairpiece," Orentreich says, "and he hasn't been castrated, so I'd have to assume that he happens to have some sort of rare hereditary variation."

This is a comforting fact. Ronald Reagan's scalp is further proof of what gerontologists have come to realize in the past two decades—that the only absolute rule about the aging process is that it eventually stops. The individual variations are enormous at every age and in every part of the body.

THE CATALOG OF DECAY

that follows is merely a list of the average ravages, most of which will hit you sooner or later but some of which you may escape. The victim is a hypothetical American man—one, say, who works in an office, gets a little exercise, has no serious vices, and doesn't dabble in such exotica as macrobiotic diets or megavitamins. At thirty he's not a bad specimen. A little plumper than he used to be, a little slower, a little balder, yet smarter than ever. Still, his body has just passed its peak. It has started dying a little every day, losing about one percent of its functional capacity every year. Cells are disappearing, tissues are stiffening, chemical reactions are slowing down. By seventy his body temperature will be two degrees lower. He will stand an inch or so shorter and have longer ears.

No one understands why. The most appealing theory for why we disintegrate as we get older was offered in an eighteenth-century treatise called *Hermippus Redivivus, Or, The Sage's Triumph over Old Age and the Grave, Wherein, a Method Is Laid Down for Prolonging the Vigour of Man, Including a Commentary upon an Ancient Inscription, in Which This Great Secret Is Revealed, Supported by Numerous Authorities.* A man aged, according to this theory, because he lost vital particles every time he exhaled. The Great Secret— how to find a new source of particles—was revealed by the discovery of a tomb whose occupant had lived to 115. The fellow managed to live so long, according to the tomb's inscription, WITH THE AID OF THE BREATH OF YOUNG WOMEN. Today's physicians advise jogging.

"Exercise will make you feel fitter, but there's no good evidence that it will make you live longer," says Dr. Jordan Tobin of the National Institute on Aging. The same goes for practically every other prescription. Because scientists don't know what causes cells to break down with age, they can't say that anything causes longevity. They can only note that certain types of men age better—those who have long-lived parents, a satisfying job, and plenty of money. Married men tend to outlive bachelors, with one notable exception: if you look at a chart comparing the average life expectancy of men according to their occupations, it turns out that the best job may well be pope (or at least cardinal). Of course, worrying about statistics like these will only hasten the aging process. So your best strategy may simply be to relax, hope that you have the right genes, and accept peacefully the indignities as they occur. You might also consider taking afternoon naps. They don't seem to have hurt Mr. Reagan.

30 In most ways, he is at his peak—the tallest, strongest, maybe the smartest he's ever been. And yet he can see the first lines on his forehead, he can't hear quite as well as he could, his skull's circumference has even started swelling. And his degeneration has just begun.

40 He's an eighth of an inch shorter than he was ten years ago, and each hair follicle has thinned two microns, but not everything's shrinking: his waist and chest are ballooning. All over, he's begun to feel the weight of time's passage: his stamina is greatly diminished.

50 His eyes have begun to fail him, particularly at close range. He notices quirky changes: his speaking voice has risen from a C to an E-flat, his thumbnails are growing more slowly, and his erections have dipped below the horizontal mark. His waist is as big as it will get.

60 By now he has shrunk a full three quarters of an inch, he has trouble telling certain colors apart, trouble distinguishing between tones, trouble making distinctions among the different foods he tastes. His lungs take in just about half what they could thirty years ago.

70 His heart is pumping less blood, his hearing is worse, vision weakening still; yet if he's made it this far, say the statistics, he'll live another eleven years. And if he has the right attitude, he will look back with awe at the wonders that have made him what he has become.

Weight

He loses a bit of his body each day, yet the body just gets bigger. The reason is fat. He's not burning up enough food—both because he's not as active as he used to be and because his basal metabolism (the rate at which the resting body converts food into energy) is slowing down about 3 percent every decade. So while muscle and other tissue is dying, accumulated fat is taking up more of his body. That's the case until middle age, after which his weight levels off and then slowly declines: he starts losing more tissue than he gains in fat.

Age 20: 165 pounds, 15% of it fat
Age 30: 175 pounds
Age 40: 182 pounds
Age 50: 184 pounds
Age 60: 184 pounds
Age 70: 178 pounds, 30% of it fat

Nails

Aged nails grow more slowly, which makes for easy grooming but weathered nails. Measured in millimeters per week, his thumbnail grows:

Age 20: 0.94 millimeters
Age 30: 0.83 millimeters
Age 40: 0.80 millimeters
Age 50: 0.77 millimeters
Age 60: 0.71 millimeters
Age 70: 0.60 millimeters

Stamina

The weakening of our man's heart, lungs, and muscles means that there's less oxygen coming in and that the heart is slower in dispersing it through the bloodstream to the muscles. A healthy seventy-year-old man can still run a marathon if he trains properly, but it will take him at least an hour longer than it did at thirty. The best measure of our man's limits is the work rate, which measures how many pounds he can turn with a weighted crank in a minute and still have his heartbeat return to normal after two minutes of rest:

Age 30: 1,110 pounds
Age 40: 1,020 pounds
Age 50: 950 pounds
Age 60: 870 pounds
Age 70: 800 pounds

Skin

A middle-aged man makes his own wrinkles: the lines on his face are drawn from repeated facial expressions, which is why the surly have more furrows on their brows. But an old man's wrinkling happens automatically—the inside of his skin loses water, and nearby molecules bind to one another, making for a stiffer, less elastic skin structure. Meanwhile, the skin itself thins and spreads out, much like a piece of dough that's been stretched. The result is a baggy suit, with the skin too large for the body. This is especially troublesome in places like the jaw, where the bone is shrinking as the skin is expanding.

Age 30: Lines in forehead are present
Age 40: Lines from other facial expressions show up, especially crow's-feet (from squinting) and arcs linking the nostrils to the sides of the mouth (from smiling)
Age 50: Lines are more pronounced; skin begins to loosen and sag in the middle of the cheek
Age 60: Excess skin and fat deposits etch bags under the eyes
Age 70: Face wrinkles everywhere; skin is rougher and has lost its uniform color—he can see a variety of shades in his face

Eyes

The lens of the eye steadily hardens throughout life and begins to cause problems for a man in his early forties. By then the lens is too big for the eye muscles to focus properly on close objects. Eventually this can cause cataracts, but the odds are that our man will die before that happens. The amount of light reaching the retina steadily declines with age (perhaps because the pupil shrinks), which means that our man will have trouble seeing in the dark; he will need especially bright light to read.

Age 30: 20/20 vision; reads without glasses
Age 50: 20/20 for distance vision, but needs glasses to read; a less elastic eye lens makes him more sensitive to glare; his depth perception is beginning to get worse
Age 60: 20/25 vision; a less elastic, yellower lens filters out some shorter wavelengths of light, making it harder for him to distinguish between blues and greens
Age 70: 20/30 vision; peripheral vision is diminished; night vision is worse, and his eyes take longer to adjust to the dark

Flab

With increased weight comes flab—or, to be precise, an increased subscapular skinfold, which is the best index of flab. This skinfold, measured by pinching the skin beneath the shoulder blade and determining the distance the skin can be stretched, is twelve millimeters wide at age twenty, fourteen millimeters at thirty, and remains about sixteen millimeters after forty. Unfortunately the flab also extends below the shoulder—conspicuously to the waist and chest.

	Waist	Chest
Age 20:	33 inches	36 inches
Age 30:	36 inches	39 inches
Age 40:	39 inches	40 inches
Age 50:	40 inches	41 inches
Age 60:	39 inches	41 inches
Age 70:	39 inches	41 inches

Height

A man is able to withstand gravity only so long. As his muscles weaken, his back slumps. And as the disks between the bones of his spine deteriorate, those bones move closer together. The result: The inexorably shrinking man.

Age 30: 5'10"
Age 40: 5'9⅞"
Age 50: 5'9⅝"
Age 60: 5'9¼"
Age 70: 5'8⅞"

Reflexes

His reflexes slow down, for which his brain is probably more guilty than his nerves. The speed at which signals travel along his nerve fibers declines only 2 percent each decade, which is a relatively minor deterioration compared with other changes in the body. The real slowdown happens because the brain takes longer to process information, make decisions, and dispatch signals. If a man looking at numbers flashing on a screen is told to press a button whenever he sees two consecutive even or odd numbers, this is how long it takes him to react:

Age 30: 0.88 seconds
Age 50: 0.90 seconds
Age 60: 0.92 seconds
Age 70: 0.95 seconds

Teeth

Eating slowly files down a tooth, but not enough to make any significant difference to anyone under the age of two hundred. The problem is keeping the tooth, and it is one problem a man can control. Despite the fact that the amount of enamel on the surface will decrease with age and the layer of dentin underneath will become more translucent, most tooth and gum decay is a result of neglect and disease. The average seventy-year-old man today has lost a third of his teeth; because of fluoridated water and better dental care, his descendants should fare better.

Age 30: 2 teeth missing
Age 50: 7 teeth missing
Age 60: 8 teeth missing
Age 70: 10 teeth missing

Hair

Men actually do get hairier with age, but, alas, not where it does them good. Hair grows in the ears, in the nostrils, and sometimes on the back. Eyebrow hairs tend to get longer and more noticeable. As for the top of the head, there are different hormones at work. Balding usually begins at the temples, producing a widow's peak that recedes with age. Next hit is the monk's spot, that circle on the back of the head—it keeps growing until it meets the receding widow's peak, leaving the top of the head bare. Men bald at different rates, of course, and some never bald at all. Still, if our man is going to bald, he will—no amount of scalp massaging will help. Although there are marked differences among men in the rate at which hair falls out or turns gray, there does seem to be a consistent pattern in the way individual hairs thin. A man's hairs are thickest at about twenty; after that each hair shrinks, and by seventy his hairs are as fine as they were when he was a baby. The diameter of a single hair (measured in microns—millionths of a meter) changes like this:

Age 20: 101 microns
Age 30: 98 microns
Age 40: 96 microns
Age 50: 94 microns
Age 60: 86 microns
Age 70: 80 microns

Muscles and Strength

At thirty, about seventy of his 175 pounds are muscle. Over the next four decades he loses ten pounds of that muscle as cells stop reproducing and die. His shoulders narrow an inch. Connective tissue replaces fiber, causing his muscles to become stiffer and to tense, relax, and heal more slowly. The remaining muscle grows weaker as the fiber becomes frayed, jumbled, and riddled with deposits of waste material. His strength peaks at about thirty and then steadily diminishes. The muscles in his hands perform as follows, as measured by the amount of force that can be exerted by the right (if that's his dominant hand) and left grip:

	Right Hand	Left Hand
Age 30:	99 pounds	64 pounds
Age 40:	97 pounds	62 pounds
Age 50:	92 pounds	58 pounds
Age 60:	86 pounds	48 pounds
Age 70:	80 pounds	42 pounds

Head

His features become more distinguished, which is a kind way of saying that they get bigger. Because of the cartilage that begins to accumulate after age thirty, by the time he is seventy his nose has grown a half inch wider and another half inch longer, his earlobes have fattened, and his ears themselves have grown a quarter inch longer. Overall, his head's circumference increases a quarter inch every decade, and not because of his brain, which is shrinking. His head is fatter apparently because, unlike most other bones in the body, the skull seems to thicken with age.

Mouth

He tastes less. When he's thirty, each tiny elevation on his tongue (called a papilla) has 245 taste buds. By the time he's seventy, each has only eighty-eight left. His mouth gets drier as the mucous membrane secretes less. His voice begins to quaver, apparently because he loses some control over his vocal cords. He talks more slowly, and his pitch rises as the cords stiffen and vibrate at a higher frequency: after fifty his speaking voice rises about 25 hertz (cycles per second), from a C to an E-flat (in the octave below middle C).

Now for the Good News...

"No wise man ever wished to be younger," said Jonathan Swift, and some gerontologists today might even agree with him. There's no denying that a seventy-year-old body doesn't work as well as it once did, but it's also true that people's fears of aging are greater than they should be. In the course of chronicling decay, researchers have come up with some reassuring findings:

▷ A man becomes less sensitive to pain after the age of sixty. It takes longer for an old man to notice a disturbing stimulus, thus he can endure greater levels of pain without complaining. This is probably due to the degeneration of his nerve fibers and to the decline in the central nervous system's ability to process sensory information.

▷ A man sweats less with age; his sweat glands gradually begin drying up.

▷ Fat isn't as bad as they say—it may even be good for you. Because thin people have always been thought to outlive the obese, doctors have urged patients to conform to a chart of "desirable weights" 15 percent lighter than the men's average. Yet recent studies show that a man ten to twenty pounds above the desirable weight is likely to outlive the virtuous man who follows the chart. "There clearly is something strange going on," says Dr. Reubin Andres of the National Institute on Aging. It could be that the mildly overweight have more reserve capacity to survive illnesses, but no one knows for sure.

▷ Until the age of sixty-five, a smoker is at least twice as likely as a nonsmoker to suffer a heart attack, but after sixty-five, the odds change. The risks are about even for nonsmokers and for men who smoke less than a pack a day. And a man who smokes more than a pack a day is three times *less* likely than a nonsmoker to have a heart attack. This doesn't mean you should take up smoking in your old age, only that a heavy smoker who survives past sixty-five is a hardy fellow.

▷ A nearsighted youth may be able to put away his glasses when he reaches middle age. The thickening of his eyes' lenses cures nearsightedness.

▷ It may be only a small victory in the battle of the sexes, but a man's skin ages about ten years more slowly than a woman's—his skin has more oil, so it's slower to dry out. Shaving also helps: The mild scraping of a razor strips away dead skin and leaves the face smoother. It's not unlike epidermabrasion, a beauty treatment that skin specialists frequently administer to women with more elaborate instruments.

▷ A man's sexual decline isn't usually traumatic. When Clyde Martin, one of the authors of the Kinsey report, asked old men how they'd react to the discovery of a safe drug that could restore their youthful sexual vigor, most said they wouldn't bother taking it. Martin found that men who were most sexually active in their youth showed the least decline in activity during old age; it was the less active men who dropped off most drastically and brought down the average. The conclusion: Old men can continue with sex if they're interested (most men in their seventies can still produce sperm); it's just that most of them aren't. "They bow out gracefully," reported Martin. "They have other interests." Or, as an aged Sophocles said when he was asked about his love life, "Peace, most gladly have I escaped the thing of which you speak; I feel as if I had escaped from a mad and furious master."

▷ If a man reaches seventy, the odds are that he'll live to see eighty. Diseases wreak far more havoc than normal aging; physically, a healthy seventy-year-old has more in common with a healthy thirty-year-old than he does with an ill man of his own age. Seventy is the average life expectancy for a man, but that's only because accidents and disorders kill so many before then. A man who survives until his seventieth birthday, according to actuarial tables, will live eleven more years.

Lungs

As the muscles that operate the lungs weaken and the tissues in the chest cage stiffen, the lungs can't expand the way they used to. A deep breath isn't as deep as it once was. The maximum amount of air he can take into his lungs:

AGE 30: 6.0 quarts
AGE 40: 5.4 quarts
AGE 50: 4.5 quarts
AGE 60: 3.6 quarts
AGE 70: 3.0 quarts

Brain

His brain shrinks as it loses billions of neurons. The cell loss varies in different parts of the brain—the region that controls head posture, for instance, doesn't seem to lose any, while the region that controls sleep stages is hit especially hard (which helps explain why he sleeps about two hours less at night). If IQ tests are any measure, then his intelligence declines. In order to make the average IQ at every age be 100, the test scores are automatically adjusted according to age. If these adjustments weren't made, the average score would be:

AGE 20: 110   AGE 50: 100
AGE 30: 111   AGE 60: 93
AGE 40: 106   AGE 70: 83

This may only mean that old people are out of practice at taking standardized tests; beyond that, a slight loss of memory is probably the most noticeable change after fifty, though even that is more a matter of faulty retrieval than of lost information. If an old man and a young man each try to memorize a list of words and then are given clues to each of the words, the old man recalls them as well as the young man. But without clues, the old man has a harder time remembering what was on the list.

AGE 20: 14 of 24 words recalled
AGE 30: 13 words recalled
AGE 40: 11 words recalled
AGE 50: 10 words recalled
AGE 60: 9 words recalled
AGE 70: 7 words recalled

Can Youth Spring Eternal?

The more realistic question might be "Can aging be slowed?" and, as you would expect, there are two schools of thought about that. Those scientists who believe that an inner clock automatically shuts down each cell at a predetermined time naturally suggest that it's impossible to lengthen the life-span radically. The less deterministically inclined think that a cell dies because of gradual processes that wear it down and that therefore man has the power to somehow slow the decay. They just haven't figured out how.

All of which means that science offers little help for your aging cells at the moment. But whether or not you can ever extend their little cell lives, you can certainly make their stay in your body more comfortable by following these simple health guidelines:

▷ **EXERCISE:** Studies show that regular exercisers outlive their sedentary peers. But it's a chicken-and-egg problem: Are they healthy because they exercise, or do they exercise because they're healthy to begin with? A recent study found that monkeys reduce their risk of heart disease if they work out on a treadmill regularly. This wasn't absolute proof—monkeys aren't men—but it was probably the most compelling evidence yet for the life-lengthening advantages of jogging. And even if exercise doesn't actually prolong life, it produces other rewards—firmer muscles, less fat, stronger lungs, better circulation—worth aspiring to in their own right.

▷ **SUN:** Unless he's a nudist, a man's youngest-looking skin is on his buttocks. That's because exposure to the sun's ultraviolet rays roughly doubles the havoc normal aging wreaks. Protecting your skin from the sun will help keep it from drying and stiffening, and help you avoid brown "age spots" or "liver spots," which can be signs of overexposed skin.

▷ **DIET:** It's probably a good idea to go easy on fats, but nobody really knows what makes the ideal diet. The fact that life expectancy is almost the same in Japan, France, and the United States suggests that vastly different diets can produce similar results. The old rules probably still apply: Keep your food tray balanced, and Eat your vegetables.

▷ **TOBACCO:** It should be avoided, of course. A cigarette smoker's lung capacity is usually equal to that of a nonsmoker ten to fifteen years older.

▷ **ALCOHOL:** Surprisingly, it's probably better to drink a little than not at all. A recent study found mortality 50 percent higher among both teetotalers and heavy drinkers (three to five drinks a day) than among light drinkers (one or two drinks a day). Apparently, moderate amounts of alcohol increase the blood's supply of high-density lipoprotein, which in turn reduces the risk of heart disease.

▷ **VITAMINS:** As you age your stomach takes longer to digest food, but that doesn't really matter. The body can still extract all the nutrients it needs, and a seventy-year-old man who eats a balanced diet shouldn't require any special vitamin supplements. Some gerontologists think we should take extra vitamin E because it stops "free radicals," which the doctors believe cause aging, from forming. But leading gerontologist Dr. Nathan Shock, among others, doesn't think much of this advice: "Taking vitamin E probably won't hurt you," says Shock. "But the main effect is probably just going to be an increase in the profits of the pharmaceutical companies."

For now, there are few other reliable pieces of advice. When asked for any great secrets to be gleaned from all the studies of the aging corpus, Dr. Jordan Tobin of the National Institute on Aging turns his palms heavenward. "I guess the best general rule is to practice moderation in the way you live," he says. "Well, it's probably a good idea not to let yourself get extremely overweight. Don't drink and drive. And wear seat belts."

Bones and Joints

His bones lose calcium. That's bad for the bones and also for the nearby blood vessels, where the lost calcium can accumulate, clogging up the works. His bones become more brittle and slower to heal. Relatively few men suffer rheumatoid arthritis, but after sixty, chances are good that our man will develop a less serious condition called degenerative arthritis. Years of flexing have worn down and loosened cartilage around the joints; the presence of this stray cartilage, coupled with depleted lubricating fluid in the joints, makes for a slower-moving, stiffer man. Movement is further restricted by ligaments that contract and harden with age. The hardened ligaments are more liable to tear.

Sex

By seventy he has found new activities for the nighttime, and he's all but stopped daydreaming about sex. Just why a man's sex drive declines is unclear—lower levels of sex hormones may be a factor, but psychological changes and the general loss of vitality in the body are probably more important. With age, the testes sag and the penis takes longer to become erect, longer to reach orgasm, longer to recover. The orgasm itself is shorter.

ANGLE OF ERECTION
AGE 20: 10% above horizontal
AGE 30: 20% above horizontal
AGE 40: Slightly above horizontal
AGE 50: Slightly below horizontal
AGE 70: 25% below horizontal

FREQUENCY OF AWAKING WITH ERECTION
AGE 20: 6 mornings per month
AGE 30: 7 mornings per month
AGE 50: 5 mornings per month
AGE 70: 2 mornings per month

FREQUENCY OF ORGASMS
AGE 20: 104 per year (49 solo)
AGE 30: 121 per year (10 solo)
AGE 40: 84 per year (8 solo)
AGE 50: 52 per year (2 solo)
AGE 60: 35 per year (4 solo)
AGE 70: 22 per year (8 solo)

Heart

His resting heartbeat stays about the same all his life, but the beats get weaker as his heart muscles deteriorate. As a result, his aged heart pumps less blood with each beat. The decline in blood flow is more marked during exercise, because his pulse can no longer rise as high as it used to.

BLOOD PUMPED BY THE RESTING HEART	MAXIMUM HEARTBEAT DURING EXERCISE
AGE 30: 3.6 quarts per minute	**AGE 30:** 200 beats per minute
AGE 40: 3.4 quarts per minute	**AGE 40:** 182 beats per minute
AGE 50: 3.2 quarts per minute	**AGE 50:** 171 beats per minute
AGE 60: 2.9 quarts per minute	**AGE 60:** 159 beats per minute
AGE 70: 2.6 quarts per minute	**AGE 70:** 150 beats per minute

Heart disease is the most common cause of death in men over forty years old and is responsible for more than half the deaths of men over sixty. As the level of cholesterol in the blood increases with age, the cholesterol accumulates on the artery walls, which are themselves thickening. The net effect is to clog the arteries, increasing the pressure of the blood against the arterial walls, which in turn forces the heart to work harder to pump blood and makes strokes and heart attacks more likely.

AGE 20: 180 milligrams cholesterol; 122/76 blood pressure
AGE 30: 200; 125/76
AGE 40: 220; 129/81
AGE 50: 230; 134/83
AGE 60: 230; 140/83
AGE 70: 225; 148/81

Kidneys

At seventy his kidney can filter waste out of blood only half as fast as it could when he was thirty. He also has to urinate more frequently because his bladder's capacity declines from two cupfuls at age thirty to one cupful at seventy.

Ears

Over the years, things like a good stereo just don't seem as important anymore—a man can't hear the highest notes no matter how well they're reproduced. A child can hear sounds reaching as high as 20,000 hertz, but in early adulthood the range starts decreasing. This seems to be a direct result of a breakdown of cells in the corti, the organ in the inner ear that transforms the vibrations picked up by the outer ear into nerve impulses, as well as of deteriorating nerve fibers. Fortunately, hearing diminishes least in the range of everyday human speech (below 4,000 hertz)—the average old man can hear conversations fairly well. To the young, an old man often seems deafer than he really is simply because he's not paying attention (perhaps with very good reason).

AGE 30: Has trouble hearing above 15,000 hertz (a cricket's chirp)
AGE 50: Can't hear above 12,000 hertz (a "silent" dog whistle)
AGE 60: Can't hear above 10,000 hertz (upper range of a robin's singing); has trouble distinguishing among tones in range he can hear
AGE 70: Misses some words in normal conversation; can't hear above 6,000 hertz (high notes on a pipe organ)

GROW, NERVES, GROW

How can severed nerve cells be encouraged to reestablish functional connections?

JULIE ANN MILLER

In an episode of Hugh Lofting's *The Story of Dr. Dolittle,* a resourceful band of monkeys spans a river by creating a "living bridge," each one's arms holding the legs of the one before it. Another kind of living bridge may be the key to promoting repair of damaged spinal cords and brains. In this case, the bridge is built of implanted tracts of cells derived from the more peripheral regions of the nervous system.

Injury to the brain or spinal cord — the central nervous system, or CNS — tends to produce permanent damage in people and other mammals. If nerve cells are killed, they cannot be replaced. Even when the cell bodies are spared, if the injury severs some of the millions of long fibers that carry signals, connections are lost and a person may suffer loss of sensory input or motor control. The cell body does not regenerate those thin cellular extensions.

In amphibia and fish, some parts of the brain and spinal cord *will* regenerate the long fibers, which are called axons. Even within the human body, the peripheral nervous system (PNS), which runs between the spinal cord and the muscles and sense organs, can regrow damaged processes. Medical researchers are making progress in encouraging this regrowth after PNS damage (SN: 9/21/85, p. 183). But the failure of nerve cells in the CNS to regenerate cut processes continues to frustrate physicians searching for treatment for the people — almost half a million each year in the United States alone — who suffer from head and spinal cord injuries.

Since the 1920s some scientists have suspected that the source of the problem lies not in the capabilities of the nerve cells but in an inhospitable environment. More recently, Albert Aguayo and his colleagues at McGill University in Montreal, Quebec, have demonstrated that, when placed in the environment of peripheral nerve cells, even CNS nerve cells can regrow axons (SN: 12/5/81, p. 363). Now Aguayo is taking advantage of this observation to direct the growth of rat CNS neurons, developing techniques that may be useful for direct repair of CNS injury and for transplanting brain tissue.

"Once you dress up like a peripheral nerve cell, you behave like one — within certain limits," Aguayo told a recent science writers' workshop at the Cold Spring Harbor (N.Y.) Laboratory.

The technique may be thought of as an environment transplant, Aguayo says. In his experiments, Aguayo transplants into an injured rat brain or spinal cord a segment of a peripheral nerve, the sciatic nerve of the leg. The nerve cell components rapidly degenerate, but the supporting cells, called the Schwann cells, divide and remain aligned in columns, which are surrounded by a continuous tube of material called basal lamina.

"The graft is not just a bridge," Aguayo says, "but a biologically active system." The Schwann cells secrete molecules that stimulate nerve growth. One of these substances is nerve growth factor, a molecule that causes axonal growth under a variety of conditions.

A search for other growth-stimulating substances is under way in several laboratories, including that of Eric Shooter at Stanford University. Shooter and his colleagues have identified several candidates. One is a protein that accumulates in peripheral nerves but not in central nerves. It appears to bind fat molecules spilling out of damaged nerves and provide material for new construction. Another candidate is a molecule found in the axon's growing tip, both in young animals and in regenerating nerve cells. The scientists expect that ultimately at least several proteins will be found to be involved in the natural switching on of nerve growth.

"We don't know yet what is crucial," Aguayo says. "But for the first time we are getting information about the molecular events that take place." The PNS graft may also provide a surface — the basal lamina — conducive to regeneration.

Absence of inhibitory factors may be another crucial factor in the PNS environment. Hans Thoenen and Martin Schwab of the Max Planck Institute for Psychiatry in Martinsried, West Germany, propose that the CNS contains substances that inhibit nerve outgrowth. These substances may be necessary in the uninjured adult nervous system to prevent axon growth and rearrangement of connections, processes that could be detrimental to brain function.

Whatever the underlying mechanism, Aguayo has clearly demonstrated that a peripheral nerve environment can trigger regrowth of CNS axons. In a typical experiment, he makes an incision into the rat's brain or spinal cord, thereby cutting many nerve cell axons. One end of a peripheral nerve segment, the PNS graft, is inserted into the incision. Some of the damaged nerve cells then regenerate axons that grow into the PNS graft (see cover and p. 195).

The scientists, thus far, have no control over which damaged nerve cells enter the graft. Aguayo says, "It's like ice fishing. We make bait and wait for something to take it."

Those axons that do enter the graft may grow to span even greater distances in the graft than they normally span in the brain. Severed axons in the retina, for example, can grow 2 to 3 centimeters, twice the length of the normal optic nerve. Spinal cord nerve cells have sent axons through a 7-centimeter graft.

The scientists find normal electrical activity in some of these axons, both spontaneous activity and response to stimulation. Axons in a graft to the retina have demonstrated, in response to light on the retina, electrical activity that is indistinguishable from that of axons in a normal optic nerve. And the axons that enter grafts to the part of the brainstem that receives signals from the rat's whiskers give an electrical response when the rat's whisker is tweaked.

"We assume that many cells don't respond, but now we know some cells do respond normally," Aguayo says. "We are pleased by the fact that some of these cells communicate in appropriate language. But we don't know yet if there is communication with any target cells."

Although it is too early to conclude that all CNS cells can send fibers into PNS grafts, such growth has been demonstrated in dozens of areas of the brain and spinal cord. Aguayo has shown that the axon regrowth arises from cells of different sizes, geometry and chemistry. However, cell types differ in what fraction of injured cells will populate a graft. This variability "provides but a hint of the multiplicity of influences that shape neuronal responses to injury in the adult mammalian brain," Aguayo says.

While Aguayo has demonstrated that axons can grow through a PNS graft, he does not yet know whether they can establish functional connections at the other end. Such connections might be necessary to the longevity of the axons as well as to restoration of brain function, Aguayo and his colleagues reported at the recent meeting in Dallas of the Society for Neuroscience.

Practical applications of this technique might someday include using a PNS graft as a bridge to span a damaged area of the spinal cord or brain. Aguayo has already begun studies

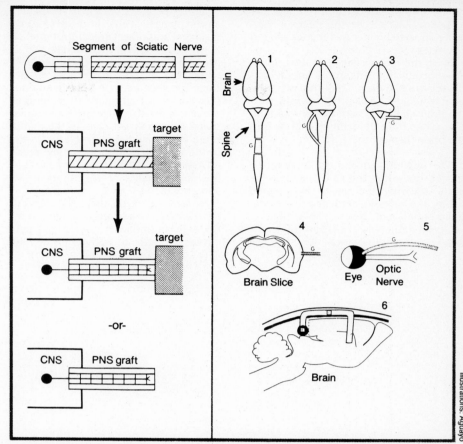

Illustrations: Aguayo

In a segment cut from a peripheral nerve (left), the nerve axons degenerate (dashed line), but the surrounding Schwann cells (striped area) remain aligned in columns. In the transplantation experiments, researchers insert one end of the segment into the central nervous system (CNS). Axons from the CNS then grow through the graft, whether or not it is connected to a target tissue. At right, nerve axons grow through segments of peripheral nerve grafted into the central nervous system in many ways. The graft (G) may link cut ends of the spinal cord (1), bridge widely separated neural areas (2) or provide a conduit from the spinal cord (3), brain (4) or retina (5). The graft may also channel axons from a reservoir of transplanted fetal nerve cells into a chosen site in the adult brain (6).

in which a PNS graft, instead of an optic nerve, connects a rat's retina to its brain. The researchers drill a hole through the skull above the site that normally receives retinal input. They lay the graft on the top of the skull and thread it through the hole. "The axons are led by this curious course back to their target," Aguayo says.

He finds that the axons from the retina spread radially as they leave the graft and grow about 1 millimeter to the appropriate brain area, the tectum. He has observed growth cones, the specialized axon tips characteristic of growing axons during embryonic development. "We are presently looking to see if the axons eventually culminate in differentiated synapses [the specialized junction across which nerve cells communicate with chemical or electrical signals]," Aguayo says.

This technique also may open up new possibilities for brain cell transplants.

"We can use this strategy to establish new interactions between distant neurons," Aguayo says.

If fetal cells are implanted into an adult brain to replace damaged cells, the CNS environment discourages axonal growth. A PNS graft can encourage such growth as well as guide the axons across long distances. This procedure can provide not only a pathway but also a reservoir of fetal cells. Instead of implanting the cells into the brain, risking further brain damage, researchers can place them in a pouch constructed of peripheral nerve sheath and implant the pouch at a convenient location outside the brain. The graft then leads the axons to the appropriate brain site.

This approach is being attempted by Aguayo, working with Fred Gage of the University of California at San Diego and Anders Björklund of the University of Lund in Sweden. In a rat's brain, they destroyed connections that supply the chemical dopamine to the area called the striatum. This surgery produces a characteristic behavior—the rat goes around in circles (SN: 11/20/82, p. 325).

Next, the researchers put fetal dopamine-producing cells into a pouch connected to a PNS graft. They implanted the pouch outside the back of the brain, with the graft making a path to the striatum. The axons from the fetal cells grew into the brain through the graft.

The implant reduced the turning behavior. But when the scientists cut the graft, severing the newly established axons, the behavior returned to the previous abnormal level. "This demonstrates that there is a new pathway and a new source of innervation," Aguayo says.

In this instance, normal function is restored by nerve cells simply providing a chemical. Important questions remain as to whether regenerated axons will be useful in those many parts of the brain where the specific pattern of connections between cells underlies function. Can the axons, all jumbled in the graft, recognize their appropriate connections? Do the target cells in the mature brain still show the characteristics that in the embryo guided incoming axons? As Aguayo asks, "Will the axons smell out the determinants of specificity in the adult organ?"

Aguayo's work on the optic nerve fibers of rats is addressing these questions. In amphibia, he notes, cut optic nerves are able to regenerate and the axons make sufficiently accurate connections to restore the animal's sight. Therefore it is possible, but not yet demonstrated, that with the aid of a PNS graft, a sensory system with all its specific connections could be restored in a mammal.

Aguayo likes to quote an early neuroanatomist who foreshadowed these developments. Santiago Ramon y Cajal, a Spanish scientist, wrote in 1928, ". . . if experimental neurology is someday to supply artificially the deficiencies in question, it must accomplish these two objects: It must give to the sprouts, by means of adequate alimentation, a vigorous capacity for growth; and place in front of the disoriented nerve cones . . . , specific orienting substances."

Immune System: Great Mystery Is Solved After Long Quest

Harold M. Schmeck Jr.

A mystery that has puzzled biologists for more than a decade has been solved, and a new chapter has opened in the understanding of the human immune system.

The result could be greater knowledge and perhaps better treatment of important diseases, including cancer, but the immediate excitement is over the new discovery itself.

What had eluded scientists for so long was an understanding of how a certain cell vital to the body's defense against illness—the so-called T cell—can recognize invading enemies such as viruses and bacteria. It was known that it must have a recognition site called a receptor on its surface, but the nature of that receptor had long escaped detection. Now, however, the chemical and physical structure of the receptor and its genes have been determined.

Immunity depends on recognition signals. Defense cells in the body must always be able to tell self from alien, friend from foe. Failure in this kind of recognition can lead to infection, cancer or other illnesses.

Among the most crucial cells of the immune defense system are the types called T cells and B cells. In their various ways they act as organizers, sentries, assassins or guided missiles to deal with invasions and subversions of many kinds.

To insure that their attacks are properly directed and their actions properly controlled, the chemical configurations of the cells' receptors act as keys in a lock-and-key recognition system. If something on the surface of a virus fits the receptor on the surface of a B cell, for example, the cell starts a train of events that produces a swarm of protective antibodies against the virus.

The identity of B cell receptors has been known for a long time. Those recognition sites are actually antibodies that constitute a normal part of the cell surface. But until recently the nature of the T cell receptor was a mystery.

Since B cell receptors are antibodies, it seemed logical that T cell receptors might be antibodies also, but all efforts to prove this failed. It was also possible that the T cell receptor was a distinct kind of substance, but if so it could not be found.

So difficult and fraught with disappointments has the search been over the years, that the T cell receptor has been called "the Holy Grail of immunology."

But within roughly the past year the nature of the T cell receptor was discovered independently by two teams of scientists.

One was led by Dr. Mark M. Davis, now of Stanford, and Dr. Stephen M. Hedrick of the University of California at San Diego. They did much of the work while at the National Institute of Allergy and Infectious Diseases in Bethesda, Md. The other team was led by Dr. Tak W. Mak and colleagues at The Ontario Cancer Institute and the University of Toronto in Canada.

The group in the United States identified the T cell receptors of mouse cells, while the group in Toronto worked with human cells. Chemically the two are sufficiently similar to make it clear that they are variants of the same thing: a molecule related to an antibody and put together by the cell in much the same complex way that antibodies are generated but nevertheless distinct from an antibody.

Somewhat earlier, three other groups had found evidence that the T cell recep-

tor consists of two linked chains of amino acids, the building blocks of proteins. They are known as the alpha chain and the beta chain. The leaders in this research were Dr. Ellis Reinherz of the Dana-Farber Cancer Institute, Boston; Dr. James Allison of the University of Texas, Smithville, and Dr. Philippa Marrack and Dr. John Kappler of the National Jewish Hospital and Research Center, Denver.

The T cell receptor has been called "Holy Grail of immunology."

Beta Chain Identified

The groups led by Dr. Hedrick and Dr. Davis and by Dr. Mak have worked out in detail the chemistry and genetics of the beta chain. Their work is considered an important advance in understanding of the T cell. It is considered virtually certain that the alpha chain will be reported soon. Indeed, widespread rumor among immunologists indicates that Dr. Susumu Tonegawa and colleagues at the Massachusetts Institute of Technology have already done so.

Dr. Davis made a preliminary report on his group's achievement at a scientific meeting in Japan last summer, but the definitive reports by his group and by the scientists in Canada were made in March in the journal Nature.

"It is a problem that immunologists

have been plagued with for a generation and now it is solved," said Dr. William Paul of the national institute in Bethesda. He said the discovery of the T cell receptor was important intellectually because the problem was so difficult.

It was also important, he said, because of what it will probably allow scientists to discover concerning the actions of T cells.

Third, Dr. Paul said, it may prove important because greater knowledge of the details of T cell function may make it possible to manipulate the human immune defense system in many ways that have not previously been possible for the treatment and prevention of disease.

Separate but Interlocking

The functions of T cells and B cells are separate but interlocking. Better understanding of the T cell receptor may make those relationships easier to understand and control.

In an important sense it is all a matter of shapes.

The body generates millions of different antibody types, each capable of recognizing a different chemical configuration or shape. There are so many different configurations of the coupling sites of antibodies, in fact, that one or another can be found that will match a shape on the surface of any invading microbe or virus.

Once the antibody on the surface of a B cell encounters its matching shape on a foreign invader, the B cell is stimulated to grow and produce new cells called plasma cells, which in turn produce antibodies of that specific identity. These antibodies act like guided missiles circulating throughout the body. When such an antibody encounters a virus, for example, that has a chemical profile that matches the B cell receptor, it attaches to that matching shape and, by doing so, takes the virus particles out of circulation.

The immune defenses preserve a memory of any such incident so that a second attack by the same kind of virus will be

counterattacked so fast and so powerfully that infection never takes hold. Vaccines generate such immunological memories in people who have never encountered the natural virus.

T Cell Roles Are More Complex

T cells play an even more complex set of roles. Some, called helper T cells, bolster the activity of B cells. Others, called suppressor T cells, act to shut off that activity when it has gone far enough. The two are important in keeping the immune defense system in balance. When it gets out of balance the result can be serious disease.

For example, acquired immune deficiency syndrome, AIDS, involves a virtually complete lack of one set of helper T cells known as T-4 cells. Some leukemias and lymphomas, cancers of the blood-forming system, seem to result from wild proliferation of defective B cells in some cases and T cells in others.

Attempts to treat such B cell disease have already been made by finding the precise clone of cells that have gone cancerous and trying to eradicate that clone. Now that the T cell receptor is known, Dr. Hedrick said recently, it may prove possible to do the same for some cancers that involve the T cells.

The defensive actions of T cells also include the production of powerful substances called lymphokines that help bring other immune cells into action in the body's defense. Some T cells, distinct from helpers and suppressors, can be activated to become killers themselves. Such T cells can destroy bacteria, cancer cells or cells infected by viruses.

Unlike B cells, which can recognize and attack free-floating viruses or other particles, the T cell seems to concentrate its attentions exclusively on cells.

'Foreignness' and 'Self'

Dr. Paul said a T cell's receptors usually must recognize both "foreignness" and

"self" on the cells it will attack or regulate.

The natural determinants of "self," known as the major histocompatibility complex, are vitally important to doctors involved in organ transplantation. A patient who needs a kidney transplant, for example, must be guarded with powerful immunosuppressive drugs to keep his own T cells from recognizing the transplanted tissue as foreign and destroying it.

Conceivably, greater knowledge of T cells and the ability to prevent them from functioning by generating antibodies against their receptors might lead to a new means of dealing with transplant rejection, the main obstacle to such operations.

Dr. Davis said the new grasp of the T cell's biology should lead to better knowledge of how many varieties of such defensive cells exist and give important new insights into the regulation of the immune response.

Many difficult diseases including rheumatoid arthritis are caused by what are called autoimmune reactions, in which the body's defenses turn against some of its own tissues. T cells that act abnormally are almost certainly a key factor in such crippling diseases. Better knowledge of T cells and ability to augment some of their clones or eliminate others, might help doctors cope with such disorders.

Someday it might be possible, Dr. Paul said, to design T cells virtually to order, making varieties that would serve as powerful drugs to fight viruses or cancers and for other purposes.

"Whether such things will become feasible, I don't know," Dr. Paul said, "but they could not even have been considered before we knew the structure of the receptor and its genes."

In a commentary on the discoveries, Nature noted that it will be possible to answer for the T cells all the questions that have been answered over the years for the B cells.

"From now on," said the commentary, "things will move fast."

Weight regulation may start in our cells, not psyches

Recent research shows overweight problems may be caused by body's fat cells, which can 'learn' to counteract repeated dieting

Gina Kolata

Gina Kolata, a Science *magazine writer, reported on computer crime in the August 1982* SMITHSONIAN.

Every year, Susan Goldsamt, a Weight Watchers lecturer in Washington, D.C., tells her classes her life story. Her struggle with weight began when she was three years old and discovered bananas—gobs and gobs of them—with which she stuffed herself. She started her first real diet when she was 14 and began what was to be more than a decade of dieting and regaining weight. Her weight, she says, fluctuated 50 to 60 pounds a year. And, with each diet, she found it harder and harder to lose weight at all. Finally, in 1972, her weight soared to a new high of 237 pounds, she joined Weight Watchers for the fifth (and last) time, and after five years and nine months, she reached her goal of 137 pounds and became a Weight Watchers lecturer. But, she says, the years of yo-yo dieting took their toll. Now she can eat only about 1,100 to 1,200 calories a day if she wants to maintain her weight and she keeps her eating habits under rigid control by writing down a menu plan each day and recording every bite of food that goes into her mouth. She has done this for the past seven years. "It's not easy," she says.

Mrs. Goldsamt's story rings all too true to the millions of people who have tried again and again to diet. But there are elements of it that particularly intrigue medical researchers who are beginning to uncover the mystery of why some people are overweight and what can be done about it. There are now hints of scientific evidence, for example, that the signals to overeat may come from the fat cells themselves. Some formerly obese people, who have large numbers of fat cells, may actually be in a state of starvation, as far as their fat cells are concerned. And there is animal research indicating that repeated dieting may result in an altered metabolism that makes it harder and harder to lose weight, even on extremely restricted diets. The new obesity research cannot yet promise cures for overeating or being overweight, but medical scientists are convinced that the study of the biochemical basis of obesity at least has the potential for leading to an eventual treatment.

The thrust of today's obesity research is "a 180-degree change of direction," says Albert Stunkard, a slender, white-haired psychiatrist who heads the Obesity Research Group at the University of Pennsylvania. For years, it was widely believed that obesity resulted from an emotional disorder. According to this psychogenic theory, people are overweight because they use food to relieve anxiety and depression. Then, of course, their obesity makes them even more anxious and depressed, so they end up eating still more. This vicious cycle got started early in life when they were either deprived of the food or oral gratification they craved or were given too much food, as a reward or to comfort them. But, gradually, a very different theory has emerged. As researchers studied how animals regulate their body weight and started to look at the biochemistry of obesity, they began to question the pat psychological explanations. They began to suspect, Stunkard says, that "the emotional disorders so closely linked to obesity are the result, rather than the cause, of obesity. In fact, they may be partly the result of trying to control obesity by dieting."

No such thing as "fat eating behavior"

Many obesity researchers now believe that there is no such thing as a "fat eating behavior." Some fat people eat when they are under stress, for example, but so do some thin people. Some fat people gobble their food or go on eating binges. But thin people do, too. One behavior that does seem to be more characteristic of fat people than of thin people is the night-eating syndrome in which people secretly binge at night and hardly eat during the day. But, scientists suspect, even that syndrome is not innately associated with obesity. Instead, it results from the constant pressure obese people feel to lose weight. After strictly dieting all day, their hunger gets the best of them. Embarrassed, they stuff themselves in secret.

It is possible for obese people to diet and stay thin, particularly for those who are not too fat to begin with, but the new findings about the causes of obesity are leading researchers to be cautious in the advice they give. "I tell people that the only way to lose weight and keep it off is to continually mount a counterforce to the forces making them fat," says Jules Hirsch of Rockefeller University. "I advise them to make small changes in their lifestyle—to eat less and to exercise more—and to realize that any changes they make will have to be maintained forever. I also tell them that the more radical the diet they go on, the less likely it is to succeed."

M. R. C. Greenwood of Vassar College is a bit more circumspect. She remarks that although Hirsch's advice is good, it also is very difficult for most people to make lifelong changes in their diet and exercise patterns. "I tell patients that most of our treatments of obesity are predicated on a view of obesity as a behavioral disorder that can be cured by changing behavior." This view of obesity no longer seems correct, Greenwood remarks, and it is increasingly clear that the behavioral treatments are not usually successful. For those individuals who must lose 20 pounds or more, the chances of reaching a normal weight and staying there are "dismal—less than five percent," Greenwood says. But frequently even losing a small amount of weight and increasing physical fitness a small amount can make obese people much healthier. And for those who are only a little bit overweight,

Greenwood says, "my message is to zealously watch your weight." It is feasible to think of losing five to 15 pounds and keeping it off. So no one should let the pounds pile on and get to a weight where obesity is almost certain to become a way of life.

Stunkard maintains that the new obesity research really is not telling dieters anything they did not already know—dieting and maintaining a lower weight is inordinately difficult. "All the new theories do," he says, "is explain why you have to try harder if the genetic and biochemical dice are loaded against you."

The change of direction in obesity research began when investigators came to appreciate the significance of an observation that has been made by scores of scientists over a period of decades: most animals seem to have stable weights. If they are given all they want to eat, they maintain a constant weight. If they are deprived of food or force-fed, they lose or gain weight, but when they are allowed to eat naturally again, their weight returns to its original level.

Such experiments raised new questions: How do animals maintain their weight? And if maintaining a stable weight is so natural, why do so many people have such difficulty with it? Thirty-four million Americans are at least 20 percent above their ideal weights—the usual definition of obesity—and the recidivism rate in obesity treatments is estimated at two-thirds. It is as high as 95 percent for the morbidly obese who are 100 pounds or 100 percent above their ideal weights.

Even people whose weights are normal frequently must struggle to keep from gaining. This is reflected, notes Judith Rodin, a psychologist at Yale University, in surveys of healthy people of normal weight—women, in particular—in which respondents reveal that they constantly fear getting fat and are always on diets. Eating disorders are pervasive. Bulimia, in which people gorge themselves with enormous quantities of food and then purge themselves by vomiting or taking laxatives, "is strictly a Western phenomenon," says Rodin. It is a direct result, she believes, of the pressures that women in particular feel to be slim—pressures that may not let up from youth to old age.

Susan Wooley, a University of Cincinnati psychologist, remarks that bulimia is becoming epidemic among young women because "it is one way to avoid becoming fat. As the standard of thinness has shrunk, more and more women are uncomfortable with their natural body size and diet until they get so hungry they lose control of their eating." She conducted a survey of 33,000 women for *Glamour* and found that about 20 percent of the respondents who were below age 20 reported that they had made themselves vomit to keep from gaining weight. And when Rodin surveyed 200 women ages 62 to 91, their second most widely stated problem after concern about memory loss was worry about gaining weight.

Anthropologists are puzzled by the emphasis on extreme thinness in our society, and especially extreme thinness for women. It was not always so, of course. In the late 19th century, Lillian Russell, considered a great American beauty, weighed 186 pounds. Even today, in non-Western cultures, fat men are frequently those of wealth and power, and fatness in women is associated with fertility. In Nigeria, relates anthropologist Claire Cassidy of the University of Maryland, young girls of the Ibo and Ibibio tribes once entered fattening houses when they reached puberty in order to gain weight before marriage. There they were prohibited from working and emerged later, "mountains of flesh" and fit for wedlock. Cassidy recalls that a few years ago, when she was doing fieldwork in Central America, she began to her dismay, to gain weight. But, she says, "the fatter I got, the more they admired me. 'Finally,' one woman told me, 'you're starting to look healthy. And furthermore, you are starting to look marriageable.' "

In our culture, fat people, far from being viewed as powerful or fertile, are scorned. One study found that children as young as six years old describe silhouettes of an obese child as "lazy," "dirty," "stupid" and "ugly." In a survey of employers by Daphne Roe and Kathleen Eickwort, 16 percent said they would not hire an obese woman under any circumstances.

So, to no one's surprise, many people feel extraordinary pressure to keep their weight down. The great puzzle is why staying thin should be so difficult to do. At Rockefeller University, obesity researchers Hirsch, Irving Faust and Rudolph Leibel believe an animal's—or a person's—stable weight, sometimes called its set point, has something to do with the fat cells in its body. The signals to overeat, they conclude, may come from the fat cells themselves. The number of fat cells in an animal's body is not necessarily fixed, says Faust, although once people or animals acquire fat cells they never lose them. But what does seem to be closely regulated is how large the fat cells become. If scientists destroy parts of a rat's brain that regulate its eating behavior, the rat will become grossly obese. As a consequence, its fat cells grow to four to five times their normal size. Or, says Faust, "we can push the fat-cell size in any animal to a bare minimum by depriving it of food. Knowing this, it is interesting to observe that if you leave an animal or a person alone, the fat cells will stay constant in size. It suggests to us that there is some sort of regulation in effect. We feel that there are signals between the fat tissue and the central nervous system." Since obese people tend to have fat cells that are two to two-and-a-half times larger than those of normal people, Faust and Hirsch hypothesize that the signals regulating fat-cell size may be perturbed in these people.

Faust's first test of the fat-cell hypothesis was to look at rats that have twice as many fat cells as usual. These rats regulated their eating so that their fat cells were the same size as those of normal rats. But since they had twice as many fat cells, they were twice as fat.

Next, Faust removed fat from young rats so that they had only half the normal number of fat cells. These rats, too, ate just enough to keep their fat cells of normal size, but since they had only half as many fat cells as they normally would, they were only half as fat as their littermates.

The next step was to look at people. Hirsch and Leibel began with a group of women and one man who belong to Overeaters Anonymous. All are of normal weight, but each was formerly obese, weighing more than 200 pounds. But although they look normal, their body chemistries are deranged. Their fat cells are tiny and they look like people with anorexia nervosa. The women do not menstruate, their pulse rates are low—about 50 to 60 beats a minute rather than the normal rate of 70 to 80 beats a minute—their blood pressures are low, they are always cold and they burn about 25 percent fewer calories than would be expected on the basis of their weights and heights. Not only do they look biochemically like people who are starving, they act like it. "They are always thinking about food," says Hirsch, "and many are in the diet business." They may become Weight Watchers lecturers, for example.

Analyzing every calorie for a year

Hirsch and Leibel posed a question: Would people of normal weight who had never been fat have similar biochemical changes if they were to lose weight? And would they have more or less the opposite biochemistries if they gained weight? So far, they have carefully studied one volunteer—"It took one year and was very difficult," says Hirsch—and they are halfway through similar studies with a second volunteer.

The volunteer was a young student who agreed to live in a small room in a ward at Rockefeller University for the duration of the experiment. When the experiment began, he was of normal weight for his height —about 190 pounds—and needed 2,850 calories a day to maintain his weight. Of these calories, 183 were used strictly to burn up the food he ate. In the first part of the experiment, he gained about 20 pounds, all of which ended up around his waist. He found it extremely difficult to gain and keep on the extra pounds and felt terribly uncomfortable carrying the excess weight around. When Hirsch and his colleagues measured his caloric needs, they found he now required 4,620 calories a day simply to maintain his weight and that 569 of these calories were used to burn up his food. It was as though a furnace was being stoked. He was burning up his calories at a furious rate, particularly just after meals.

The next step for the young man was to diet until his weight was 20 percent below normal. This time, his metabolism slowed down after meals. He needed 2,871 calories a day to maintain his lower weight but, of those calories, only 109 were used for burning food.

What these numbers show is that as the body gains weight, it burns up excess food at an increasing rate, but when it loses weight, it uses relatively few calories to consume food. In other words, it adjusts its metabolism to counteract any deviation from its ideal weight. The student was reacting like those laboratory animals that keep returning to a stable weight. And the second volunteer, so far, is behaving biochemically just like the first.

In his search to understand why people and animals seem driven to maintain particular weights, Leibel developed a method for measuring the propensity of fat cells to accumulate fat or to break fat down. And in doing so, he came upon an intriguing result. People who cannot spot-reduce—women who can't get rid of their large hips, for example, may have fat cells in their problem areas that simply will not release their fat.

Leibel is looking at small molecules on the surfaces of fat cells called alpha and beta receptors. The alpha receptors stimulate fat accumulation and the beta receptors stimulate fat breakdown. Human fat cells have both of these receptors on their surfaces, but they vary in which kind of receptor is predominant. Leibel and his colleagues found that women tend to have fat cells on their hips and thighs that have predominantly alpha receptors, which means they tend to accumulate fat. This may help explain why so many women have fat that accumulates on their hips and thighs, Hirsch says, and the receptor study in general shows why different people have different fat distributions—the fat goes where the fat cells have mostly alpha receptors.

The alpha and beta receptor story may explain the dilemma of some of the Rockefeller University group's patients. One woman who was pear-shaped at 170 pounds lost 20 pounds, but looked worse than when she began her dieting and exercise. The fat cells on her hips and thighs had mostly alpha receptors and simply did not release any fat as she lost weight elsewhere. A potbellied male patient had a similar problem. The fat cells of his abdomen had too many alpha receptors and so when he dieted he lost weight everywhere except in his paunch.

It would be wonderful if this understanding of receptors could lead to a "skinny pill" that forces recalcitrant cells to burn fat. But Hirsch and his colleagues think that is not forthcoming. They do have drugs, however, that, in the laboratory at least, make isolated fat cells burn up fat. These are drugs that block alpha receptors or stimulate beta receptors. Hirsch envisions someday using drugs such as these to help dieters lose weight where they want to lose it.

Since most overweight people diet not just once but over and over again, investigators are beginning to wonder whether dieting itself can be self-defeating. Is it better not to diet at all, they ask, than to diet periodically? At the University of Pennsylvania, Kelly Brownell is coordinating a multi-university study that asks what metabolic changes occur when people constantly gain and lose weight—the so-called yo-yo syndrome. He explains: "We got interested in this problem about two years ago when we noticed that many obese people in our clinic had been on many diets in the past. They tended not to do well in our weight-loss programs. Maybe, we thought, their past dieting history caused them to do poorly."

Learning the lessons of yo-yo dieting

In collaboration with Greenwood at Vassar College, Brownell's group carried out experiments with rats that clearly showed that animals who were yo-yo dieting got very food-efficient. The second time they were forced to lose weight, they lost it at only half the rate they did on their first diet, even though they had exactly the same caloric intake. And, after their second diet when they were allowed to regain the weight they had lost, they put on the weight at three times the rate that they did the first time they regained weight. "It appears that the animals were defending their body weight," Brownell says.

George Blackburn of the Harvard Medical School is now looking at data from obese people to see if the animal findings might hold true in humans as well. He runs a metabolic ward in which very obese people come to live in an isolated environment and lose weight on a carefully controlled, very low-calorie diet. Blackburn has noticed that a number of people keep returning to the ward twice, three times, even five times to lose weight. And each time they come, they lose weight more slowly than they did the previous time. So he strongly suspects that these people, like the laboratory rats, have learned to be very efficient users of food as a result of yo-yo dieting.

If there is a common theme to the stories emerging from these laboratories and hospitals, it is the fat cell as culprit. Researchers are only beginning to grasp the metabolic nature of these cells. They can't yet say why some people have more than others, or where these extra fat cells come from. But they hope that, with a better understanding of these cells, the signals to overeat may be circumvented. Some researchers postulate that there may be periods of life when fat cells are particularly likely to be deposited. In rats, the first three weeks of life are critical—the young animals develop excess fat cells at that time if they are overfed. No one knows whether humans have a similar critical period or, if so, when it may be.

So, for the time being, the new studies of obesity cannot provide much help to people who want to lose weight. But for Susan Goldsamt, at least, the very fact that researchers are looking seriously at the biochemical basis of obesity is comforting. It is so much more satisfying, she says, than hearing over and over again that she was obese because she was weak-willed or a glutton, or being told that she was not losing weight on her diets because she was cheating.

Ergot and the Salem Witchcraft Affair

Mary K. Matossian

Educated at Stanford University, Mary Kilbourne Matossian is at present Associate Professor of History at the University of Maryland. Her research fields have included European folklore, family history, and demographic history. For the last few years she has been studying the impact of mold poisoning epidemics on human population trends and social behavior.

The witchcraft affair of 1692 had several peculiar aspects. In terms of the number of people accused and executed, it was the worst outbreak of witch persecution in American history. Accusations of witchcraft were made not only in Salem Village (now Danvers) but also in Andover, Beverly, Boxford, Gloucester, Ipswich, Newbury, Topsfield, and Wenham, all in Massachusetts, and in Fairfield County, Connecticut. The timing of the outbreak was strange, since it occurred 47 years after the last epidemic of witch persecution in England. No one has been able to prove why it occurred in 1692, and not some other year, or why it happened in Essex County, Massachusetts, and Fairfield County, Connecticut, and not in other counties.

In 1976 psychologist Linnda Caporael proposed an interesting solution to the problem of why various physical and mental symptoms appeared only in certain communities at certain times (1). She suggested that those who displayed symptoms of "bewitchment" in 1692 were actually suffering from a disease known as convulsive ergotism. The main causal factor in this disease is a substance called ergot, the sclerotia of the fungus *Claviceps purpurea*, which usually grows on rye (Fig. 1). Ergot is more likely to occur on rye grown on low, moist, shaded land, especially if the land is newly cultivated. The development of ergot is favored by a severely cold winter followed by a cool, moist growing season: the cold winter weakens the rye plant, and the spring moisture promotes the growth of the fungus.

People develop ergotism after eating rye contaminated by ergot. Children and teenagers are more vulnerable to ergotism than adults because they ingest more food per unit of body weight; consequently, they may ingest more poison per unit of body weight. Made up of four groups of alkaloids, ergot produces a variety of symptoms. Diagnosis may be difficult because many symptoms are not present in all cases.

According to current medical thinking, the symptoms of early and mild convulsive ergotism are a slight giddiness, a feeling of frontal pressure in the head, fatigue, depression, nausea with or without vomiting, and pains in the limbs and lumbar region that make walking difficult (2). In more severe cases the symptoms are formication (a feeling that ants are crawling under the skin), coldness of the extremities, muscle twitching, and tonic spasms of the limbs, tongue, and facial muscles. Sometimes there is renal spasm and urine stoppage. In the most severe cases the patient has epileptiform convulsions and, between fits, a ravenous appetite. He may lie as if dead for six to eight hours and afterward suffer from anesthesia of the skin, paralysis of the lower limbs, jerking arms, delirium, and loss of speech. He may die on the third day

after the onset of symptoms. Animals suffering from convulsive ergotism may behave wildly, make loud, distressed noises, stop lactating, and die.

Caporael matched the symptoms and their epidemiology in 1692 with those in the above model. She was severely criticized by psychologists N. K. Spanos and Jack Gottlieb on the ground that the facts of the case fit the model very imperfectly (3). I have concluded, after examining the Salem court transcript, the ecological situation, and recent literature on ergotism, that this objection is not as valid as originally perceived.

Previous attempts to explain the witchcraft affair of 1692 have been unsatisfactory. The work of historians Paul Boyer and Stephen Nissenbaum, for example, has been concerned with the social reactions to the symptoms of bewitchment, rather than the origin of the symptoms (4). Other historians have attributed the outbreak to the tendency to make scapegoats of certain members of a community; although this is a widespread and chronic phenomenon, it is insufficient to explain the unique aspects of the case. New Englanders believed in witchcraft both before and after 1692, yet in no other year was there such severe persecution of witches.

The suggestion that the afflicted teenage girls in Salem Village were feigning their symptoms or, as Spanos and Gottlieb suggested, role-playing in the presence of social cues cannot explain the symptoms of the animal victims or of the other human victims who were apparently not stimulated by social cues. The suggestion made by an English professor, Chadwick Hansen, that the be-

Reprinted by permission, *American Scientist* 70:355 (1982), "Ergot and the Salem Witchcraft Affair," by Mary K. Matossian.

Figure 1. The three ears of rye on the left of this early nineteenth-century drawing contain ergot, the sclerotia of the fungus *Claviceps purpurea.* Ergotism, a type of food poisoning caused by eating rye contaminated with ergot, produces a variety of symptoms similar to those commonly associated with bewitchment, such as pinching sensations and convulsions. The first two ears of rye contain a great number of ergots, more detailed examples of which are shown in the detached ergots on the drawing. The third ear, of stout rye, contains only one large ergot. The ear at the right, which also bears one ergot, is wheat, a grain rarely contaminated by the fungus. (From ref. *18.*)

witched were suffering from hysteria is also unsatisfactory (*5*). People in the afflicted communities may have been hysterical in the sense that they were excited and anxious, but such psychological stimuli alone have not been shown to be capable of producing an epidemic of convulsions, hallucinations, and sensory disturbances in any case in which a diagnosis of ergotism or other food poisoning was seriously considered and then ruled out (*6*).

Symptoms in 1692

In Essex County, Massachusetts, 24 of 30 victims of "bewitchment" in 1692 suffered from convulsions and the sensations of being pinched, pricked,

or bitten. According to English folk tradition, these were the most common specific symptoms of a condition called "bewitchment" (*7*). Hence, they were the symptoms most often mentioned in the court records, for the intent of the court proceedings was to prove "witchcraft," not to present a thorough medical case history.

Some of the other symptoms of "bewitchment" mentioned in the court record, like the most common symptoms, may also occur in cases of ergotism. These include temporary blindness, deafness, and speechlessness; burning sensations; seeing visions like a "ball of fire" or a "multitude in white glittering robes"; and the sensation of flying through the air "out of body." Three

girls said they felt as if they were being torn to pieces and all their bones were being pulled out of joint. Some victims reported feeling "sick to the stomach" or "weak," having half of the right hand and part of the face swollen and painful, being "lame," or suffering from a temporary, painful urine stoppage. Three people and several cows died.

The Salem court record does not mention certain symptoms often associated with mild or early cases of ergotism, such as headache, nausea, diarrhea, dizziness, chills, sweating, livid or jaundiced skin, and the ravenous appetite likely to appear between fits. If these symptoms were present, they may not have been reported because they were not commonly associated with bewitchment. Nor does the court record establish whether or not the victims suffered relapses or how the cases ended.

Social cues in the courtroom may have stimulated some of the hallucinations, but such stimulation does not disprove a diagnosis of ergotism. Ergot is the source of lysergic acid diethylamide (LSD), which some mycologists believe can occur in a natural state (*8*). People under the influence of this compound tend to be highly suggestible. They may see formed images—for instance, of people, animals, or religious scenes—whether their eyes are open or closed (*9*). These hallucinations can take place in the presence or absence of social cues.

Symptoms similar to those mentioned in the Salem court record also appeared between May and September of 1692 in Fairfield County, Connecticut. A 17-year-old girl, Catherine Branch, suffered from epileptiform fits, pinching and pricking sensations, hallucinations, and spells of laughing and crying. On 28 October she died, after accusing two women of bewitching her. John Barlow, aged 24, reported that he could not speak or sit up and that daylight seemed to prevail even at night. He had pain in his feet and legs (*10*). These symptoms also suggest a diagnosis of ergotism.

Epidemiology

The victims of bewitchment in Essex County were mainly children and teenagers. Seven infants or young children are known to have developed symptoms or died. According to recent findings, nursing infants can develop ergotism from drinking their mother's milk (2).

Spanos and Gottlieb, citing the court record, asserted that the proportion of children among the victims in 1692 was less than that in a typical ergotism epidemic. However, in a recent epidemic of ergotism in Ethiopia, the ages of the victims were not much different from those in the Essex County epidemic of 1692: more than 80% of the Ethiopian victims were aged 5–34 (11).

There can be no doubt that rye was cultivated in Salem Village and in many other parts of Essex County in the late seventeenth century (12). The animal cases could have resulted from ingestion of wild grasses such as wild rye or cord grass, some of which in Essex County were also liable to ergot infection (13).

The first symptoms of bewitchment appeared in Salem Village in December 1691. Beginning about 18 April 1692, the pace of accusation increased. It slowed in June and then reached a peak between July and September. Exactly when the symptoms terminated is unknown. After 12 October 1692 there were no more trials for witchcraft, by order of the governor of Massachusetts. However, during the winter of 1692–93 in the area around Boston and Salem there were religious revivals, during which people saw visions (14).

If rye harvested in the summer of 1691 was responsible for the epidemic, why did no one exhibit any symptoms before December of that year? In the ergotism epidemics of continental Europe the first symptoms usually appeared in July or August, immediately after the rye harvest. But these episodes occurred in communities heavily dependent on rye as a staple crop and among people so poor that they had to begin eating the new rye crop immediately after the harvest. The situation was otherwise in New England. The diary of Zaccheus Collins, a resident of the Salem area during the epidemic, and probate inventories show that the rye crop often lay unthreshed in the barns until November or December if other food was abundant (15). Since ergot can remain chemically stable in storage for

up to 18 months, stored rye might have been responsible for the symptoms of December 1691.

But if people normally delayed threshing rye until winter, why was there a peak of convulsive symptoms in the summer of 1692? Such a peak might be expected in time of food scarcity: was this the case in 1692?

Unfortunately the usual sources of information about food supply, government records, are missing for 1692, but data from tree rings indicate that in 1690, 1691, and 1692, the growing season in eastern New England was cooler than average. Diarists in Boston recorded that the winters of 1690–91 and 1691–92 were very cold (16). Since rye is a crop that flourishes in cold weather when other crops fail, people may have been more dependent on rye and therefore may have begun consuming it earlier in the year. In coastal areas, such as Essex and Fairfield counties, cool conditions are usually also moist; ergot grows more rapidly in moist weather.

In several other years for which tree rings indicate especially cool weather, there were epidemics of convulsions. The most widespread epidemic in New England occurred in 1741. In 1795 a Salem epidemic, labeled "nervous fever," killed at least 33 persons (17).

The growth of population in Salem Village provided an incentive for local farmers to utilize their swampy, sandy, marginal land. This land, if drained, was better suited to the cultivation of rye than other cereal crops. But this was the very type of land in which rye was most likely to be infected with ergot (18). All 22 of the Salem households affected in 1692 were located on or at the edge of soils ideally suited to rye cultivation: moist, acid, sandy loams. Of the households, 16 were close to riverbanks or swamps and 15 were in areas shaded by adjacent hills. No part of Essex County is more than 129 m above sea level. As in Essex County, in southern Fairfield County, Connecticut, the predominant soil type was fine sandy loam, elevations were low, and the population was expanding (19).

Beginning in the 1590s, the common people of England began to eat wheat instead of rye bread. The settlers in New England also pre-

ferred wheat bread but, troubled by wheat rust, in the 1660s they began to substitute the planting of rye for wheat. This dietary shift may explain why the witchcraft affair of 1692 occurred 47 years after the last epidemic of witch persecution in England (20).

Although the limitations of surviving records make certainty impossible, the balance of the available evidence suggests that the witchcraft accusations of 1692 were prompted by an epidemic of ergotism. The witchcraft affair, therefore, may have been part of a largely unrecognized American health problem.

References

1. L. R. Caporael. 1976. Ergotism: The Satan loosed in Salem? *Science* 192:21–26.
2. B. Berde and H. O. Schild, eds. 1978. *Ergot Alkaloids and Related Compounds.* Springer.
3. N. P. Spanos and J. Gottlieb. 1976. Ergotism and the Salem Village witch trials. *Science* 194:1390–94.
4. P. Boyer and S. Nissenbaum. 1974. *Salem Village Possessed.* Harvard Univ. Press.
5. C. Hansen. 1969. *Witchcraft at Salem.* Braziller.
6. H. Mersky. 1979. *The Analysis of Hysteria.* London: Baillière Tindall.
   F. Sirois. 1974. *Epidemic Hysteria.* Copenhagen: Munksgaard.
   M. Gross. 1979. Pseudo-epilepsy: A study of adolescent hysteria. *Am. J. Psychiatry* 136:210–13.
7. P. Boyer and S. Nissenbaum, eds. 1977. *The Salem Witchcraft Papers.* New York: Da Capo Press.
   A. MacFarlane. 1970. *Witchcraft in Tudor and Stuart England.* London: Routledge & Kegan Paul.
   C. L. Ewen. 1933. *Witchcraft and Demonianism.* London: Frederick Muller.
   R. Scot. 1584. *The Discoverie of Witchcraft.* Reprinted 1973. Totowa, NJ: Rowman and Littlefield.
8. R. Emerson. 1973. Mycological relevance in the nineteen seventies. *Trans. Brit. Mycological Soc.* 60:363–87.
9. A. Hoffer and H. Osmond. 1967. *The Hallucinogens.* Academic Press.
   R. C. De Bold and R. C. Leaf, eds. 1967. *LSD, Man and Society.* Wesleyan Univ. Press.
   R. Blum. 1964. *Utopiates.* Atherton.
   M. Tarshis. 1972. *The LSD Controversy.* Thomas.
10. Wyllys Papers, Connecticut State Library, 23–29.
    J. M. Taylor. 1908. *The Witchcraft Delusion in Colonial Connecticut, 1647–1697.* Stratford, CT: J. E. Edward.
11. T. Demeke, Y. Kidane, and E. Wuhib. 1979. Ergotism, a report on an epidemic, 1977–78. *Ethiopian Med. J.* 17:107–14.

5. PHYSIOLOGY

12. *Danvers Historical Collections.* 1926. Danvers, MA: Danvers Historical Society. Diary of Josiah Green. 1866, 1869, and 1900. *Essex Institute Historical Collections* 8:215–24, 10:73–104, and 36:323–30.

13. L. N. Eleutherius. 1974. Claviceps purpurea on Spartina in coastal marshes. *Mycologia* 66:978–86.
   S. K. Harris. 1975. *The Flora of Essex County, Massachusetts.* Salem, MA: Peabody Museum.

14. G. L. Burr, ed. 1914. *Narratives of the Witchcraft Cases, 1648–1706.* Scribners.
   C. Mather. N.d. *Diary.* Ungar.

15. Diary of Zaccheus Collins, Essex Institute, Salem.

A. L. Cummings. 1964. *Rural Household Inventories, 1675–1775.* Boston: Society for the Preservation of New England Antiquities.

16. Diary of Lawrence Hammond. 1891–92. *Mass. Hist. Soc. Proc.,* ser. 2, 7:160.
   S. Sewall. 1972. *Diary.* Arno Press.
   E. De Witt and M. Ames, eds. 1978. *Tree-Ring Chronology of Eastern North America.* Tucson: Univ. of Arizona Press.
   H. C. Fritts and L. Conkey. Pers. comm. 1981.

17. Holyoke Papers and Salem Bills of Mortality, Essex Institute, Salem.
   M. K. Matossian. 1982. Religious revivals and ergotism in America. *Clio Medica* 16:185–92.

18. O. Prescott, Jr. 1813. *A Dissertation on the Natural History and Medicinal Effects of the Secale Cornutam, or Ergot.* Boston: Cummings and Hilliard.

19. M. F. Morgan. 1930. *The Soils of Connecticut.* Bull. 320. Conn. Agriculture Station.

20. W. Harrison. 1587. *The Description of England.* Reprinted 1968. Cornell Univ. Press.
   A. B. Appleby. 1979. Diet in sixteenth century England: Sources, problems and possibilities. In *Health, Medicine and Mortality in the Sixteenth Century,* ed. C. Webster, pp. 97–116. Cambridge Univ. Press.

Saving Babies

The risks, dilemmas and rewards of fetal surgery

B.D. Colen

B.D. COLEN *is the Pulitzer Prize-winning science editor of Newsday (New York). He also writes the "Hot Seat" column for* HEALTH, *which appears every other month.*

between the moment of conception and the moment of delivery, so much can go wrong that it truly is a miracle that so much goes right. Until the development of prenatal diagnosis, women rarely, if ever, knew when potentially devastating anomalies were occurring. Now, imaging technologies, particularly ultrasonography, have provided a window into the uterus. Once a problem is diagnosed, though, there is still little that can be done for the "defective" fetus. Only one surgical fetal intervention, interuterine fetal blood transfusion, is now considered standard therapy. In fact, while some of the slightly more than 100 surgical attempts worldwide to correct fetal defects have received a good deal of media attention, most human fetal treatments are still more conceptual than factual.

Hard Questions

Like any risky, experimental medical technique, fetal surgery raises serious moral and ethical dilemmas. If the treatment is not completely successful, the physician may have saved a horribly handicapped child for a lifetime of institutional existence.In addition, the physician is treating the fetus as a patient, thus endowing it with human value and rights, even though society has collectively agreed it has no rights and may be killed upon the request of the woman carrying it.

Before considering these two issues though, listen to the stories of Michael Skinner and Nicholas Hannan.

Twins At Risk

Rosa Skinner had no premonitions or worries about the fetus she was carrying in January 1981. But due to her "advanced maternal age" of 41, she was scheduled for amniocentesis. The first thing Skinner learned was that she was carrying twins. "You're kidding!" she remembers telling the ultrasound technician at the University of California's Moffit Hospital in San Francisco. The next thing she learned—from her doctors, Mitchell Golbus, MD, and Roy Filly, MD—was

that the abdomen of one of the two fetuses was overloaded with fluid. If the two fetuses were carried to term, the doctors said, one might be healthy. "But they said, 'Don't paint the room and don't buy baby clothes, because you're going to abort them.'"

Through the winter and into the early spring, Skinner worried and the doctors observed the fetus with ultrasound. Then in April, Dr. Filly, senior ultrasonagrapher at Moffit, discovered that urine was backing up into the kidneys of the male fetus and its organs were being forced up into the chest cavity, compressing the lungs. "They said that the female fetus was in danger, too," Skinner recalls.

It was then that Skinner learned how fortuitous it was she had decided to come to San Francisco for her amniocentesis. For Drs. Golbus and Filly are two of the three co-directors of the pioneering Fetal Treatment Program at UCSF, which has been centered on the treatment of *hydronephrosis,* the condition threatening the male twin. "[Golbus] explained that he could try to insert a catheter into the fetal bladder and empty it into the amniotic sac," Skinner says. "He also said it had never been done before. Right then my husband and I looked at each other and said it should be done. The doctors were giving me a chance—not just a chance to save one baby, because by saving one baby, I was saving the other, too," she says.

A Pioneering Procedure

Three days later as Rosa Skinner lay on an operating table under mild sedation, a catheter was inserted through her abdomen into the fetus. She recalls feeling "this tremendous rush of water, just a *whoosh!* across my abdomen." While the physicians managed to drain the urine, the instruments they had wouldn't work properly with the catheter, so the procedure had to be concluded without the installation of a semi-permanent drain. "It was so disappointing," Skinner recalls. "My husband and I were thinking how sad it would be to see the fetus die for something that would be so easy to correct." Then a firm agreed to manufacture the necessary equipment. So once again, the doctors inserted a catheter through Rosa Skinner's abdomen into the fetus.

On May 10, 1981, Mother's Day, the Skinner twins were born about five to six weeks premature, a common occurrence for twins. Mary was completely normal. Michael was not.

Michael was operated on the day after his delivery and his ureters (the tubes connecting the kidneys and bladder) were cut and led outside his body. At the same time, a hernia in his diaphragm was repaired and the organs that had been forced up against his lungs were repositioned.

During his first year, Michael had respiratory difficulties caused by the compression of his lungs in utero. But by the time he returned to Moffitt a year later for an operation to reconnect his ureters, he was chattering away. And at two years he was walking, talking and climbing onto and into every available object. "I didn't make any promises to God," Skinner says. "I just prayed to God to help me and . . . it's a miracle. His kidneys are fine—there's no damage. His heart's fine and his lungs are fine. Everything is fine."

A Different Problem

Sara Hannan was in the 18th week of

If treatment is not successful, doctors may have saved a handicapped child for an institutional existence.

her second pregnancy when she began to sense that something wasn't quite right with the fetus. "He was very active," she recalls, "I could tell he was kind of in pain." An ultrasound test revealed that Sara was carrying a fetus suffering from *hydrocephalus*, a build-up of fluid in the *ventricals*, or normally empty chambers of the brain, which can cause brain damage before birth. After birth, if a shunt, or drainage tube, is not surgically implanted, the head can swell, causing severe retardation and eventual death.

At the University of Colorado Medical Center in Denver, surgeon William Clewell, MD, told the Hannans about an experimental surgical procedure that could relieve the pressure on the brain and prevent further brain damage. He said that normally, if a fetus with hydrocephalus were not aborted, it would be observed with ultrasound to gauge how much of the brain was being destroyed and a shunt would be implanted shortly after birth.

"The doctors painted a really bleak picture for us," says Sara Hannan. "They said there are kids with this condition who are diapered when they are in their 30s. But they didn't push the surgery. They let us think about termination first. But we don't believe God makes mistakes. We couldn't consider termination."

Risky Business

In late December, Hannan, "sedated but very much awake," lay on an operating room table surrounded by "eighteen to twenty doctors at one time," as her husband describes it. "They were all trying to look at this little eight-inch screen," he says. Not only were the team members watching the screen, they were also watching a doll, which one of the physicians held in a position identical to that assumed by the fetus. Thus Clewell and his asso-

ciates were able to relate to a three-dimensional object as they inserted a needle containing a shunt through Hannan's abdomen and into the skull of the fetus. The following day an ultrasound examination showed the ventricals greatly reduced as the fluid had drained into the amniotic sac. But in a few weeks, it became apparent that the fetus had pulled the shunt out, because the ventricals were enlarged even farther than they had been prior to the procedure. So a second attempt was made to insert a shunt. "They about gave up the second time because they couldn't get him to hold still," Sara Hannan recalls. Despite the fetus's kicking, the surgeons managed to insert the second shunt in a better position. And in late February 1982, Nicholas Hannan was delivered by caesarean section.

Premature delivery at 32 weeks did not end Nicholas's problems. The four-pound, three-ounce baby's lungs were compressed because he was suffering from a *diaphramatic hernia*, a condition requiring surgical correction. He also had a hole in one lung.

As of six months, however, all the correctable problems were corrected. "He's developing," the surgeon says. "You can't say much in the first months of life, but we have a follow-up program to at least age six. In the long run, that's the most important part of our program"—to see if prenatal surgical intervention pays off in improved outcome for the babies involved.

In Nicholas Hannan's case, the point is debatable. His mother says at age three, he's "doing excellently as far as we're concerned" but "still has a long way to go." Developmental specialists describe the three-year-old as being on about a nine-month level. Additionally, Nicholas has twice undergone surgery to correct his diaphramatic hernia, has had a double hernia repaired and his shunt replaced and had required a surgically implanted feeding tube because at a year and a half he weighed only 15 pounds.

"We have a faith in God that we believe that he's going to be healed," says Sara Hannan. "I'm very pleased. I know that if he hadn't had that surgery before birth, he'd be a lot worse."

Which Babies Can Be Saved?

With the possibility of saving a child who would only face a lifetime of pain and difficulty, why attempt fetal therapy in the first place? Why not simply put our medical efforts and resources into treating those patients who are already born, and leave parents with their current choice: either abort the "defective" fetus or attempt to correct its defect after birth? Because, says Michael Harrison, MD (one of the fetal specialists at UCSF who treated Rosa Skinner, and co-author of the definitive text on fetal surgery, *The Unborn Patient*), the medical issues in fetal therapy are similar to those in any other area of medicine. "Our problem in medicine at any age" is finding a condition too late to treat it, he says, "and many kids can only be salvaged before birth. But the biggest danger of what we do is that we will fix kids halfway. That is a potential tragedy of the first order."

The Rights of the Fetus

Consider, once again, the examples of Nicholas Hannan and Michael Skinner. As fetuses at the time they were surgically treated, not only would they not normally be considered "patients," they would not even be considered "persons" in the eyes of the law. They lie in a truly schizophrenic social, legal and medical no-mans land. On one hand, in order to protect the "rights" of fetuses and even human embryos, the National Institutes of Health have issued strict guidelines limiting research and experimentation involving embryos and fetuses. On the other hand, society legally sanctions abortion for whatever reason the pregnant woman deems appropriate.

New York University philosophy professor William Ruddick, though, doesn't see any difficulty in saving a fetus on one floor of a hospital while aborting a fetus on another.

"People have already reached moral opinions and adopted a terminology appropriate to it [abortion]," he says. "If you believe for metaphysical reasons that the fetus is a spiritual equal, you will talk about mother and child from day one. If you don't, then you will hold off talking about a child until the birth.

> ## *"They didn't push the surgery. They let us think about termination first. But we don't believe God makes mistakes."*

"The problem about the options that medicine creates is that they create obligations, too," he says. "The serious debate, I think, is about women: whether motherhood is a natural and obligatory situation in women."

Ruddick is referring to a question that concerns many others: If we have the ability to provide unquestionably beneficial fetal therapy, are we *obligated* to provide it? How do we balance the conflicting rights of the mother and the fetus?

John Robertson, a law professor at the University of Texas at Austin and specialist in the rights of the handicapped, notes that conflict arises not in a case when a woman wants to abort a defective fetus, but rather, when she wants to continue with the pregnancy without having fetal treatment. "We are talking about the right of the child to be born healthy," he says. "Ethically and legally the interests of the child should take priority."

The Rights of the Mother

On the other hand, NYU's Ruddick views the dilemma from the viewpoint of the mother, stating that "women should have the liberty to define the obligations they are willing to undertake. The fact that they started out on a project that has gone seriously bad doesn't commit them to continuing that project come what may.

"In the case of a pregnant woman with a fetus diagnosed as seriously imperiled, there is only one set of interests, and that is not because the fetus counts as nothing. It is rather that the interests of a woman wanting to have a child prevail," Ruddick says.

Bioethicist John C. Fletcher is even more troubled by the possibility that society might adopt Robertson's view of conflicting maternal-fetal rights. "The mother's interest comes first whenever you have to go through her body to treat the fetus," Fletcher says. "We should not violate or coerce women against their will. Then we would have a health police state to monitor pregnancies. Women should not be exempt from the consent form. Pregnancy is not a jail sentence."

Medicine

"Thou know'st 'tis common,—all that live must die, passing through nature to eternity."

William Shakespeare

"All men," said Hippocrates, the father of medicine, "ought to be acquainted with the medical art. I believe that knowledge of medicine is the sister and companion of wisdom." Clearly, everyone still believes in society's right to medical knowledge. Today, discoveries in biology find their way into medicine even though they may be based on organisms very different from ourselves. Since there is a unity of life found at the molecular level, even information gained from the study of noninfectious bacteria may aid in our understanding of how the human system functions. The selections in this unit consider some recent medical discoveries.

To most people, pain is a transient nuisance. It may be nothing more than a tension headache or a pulled muscle. For others, however, pain is a chronic disease, a debilitating affliction that rules its victims' bodies. The article "Pain, Many Causes, Fewer Cures" explores recent findings regarding the cause of and the techniques used for treating intense pain.

The Acquired Immune Deficiency Syndrome (AIDS) epidemic which began in 1981 has reached its sixth year. What started as thirty-one mysterious cases of healthy young men dying of a rare cancer and pneumonia now stands conservatively at an estimated 22,548 Americans dead due to AIDS. As many as a million and half more Americans are thought to be infected with the AIDS virus. No preventive medicine except prudence is anywhere in sight, and neither is a cure. In 1984 the causative agent of this syndrome was isolated from human blood and the infectious process characterized. Where the virus came from and how long it has existed are explored in "The Natural History of AIDS." Another mystery surrounding the AIDS epidemic is why some people who have been infected by the AIDS virus live for years—if not their whole lifetimes—without developing the syndrome. "What Triggers AIDS?" sets forth a hypothesis about possible cofactors necessary to promote a symptomless AIDS infection into the disease state.

Millions of people suffer from various allergies, the symptoms of which can often drive the sufferer to use drastic measures to acquire partial relief. The author of "Racing Toward the Last Sneeze" reports that new and better therapy will be available soon to aid allergy victims.

Researchers are close to finding a cure for genetic diseases caused by simple errors in the genetic code. The articles "The Gene Doctors" and "Mapping the Genes, Inside and Out" examine the possibility of removing the defective gene as potential therapy. Besides the medical benefits, many scientists are eager to sequence the entire human genetic code and look inside the genes to their protein products. Determining the nucleotide sequence of genes may not only eradicate genetic disease but could also provide biologists with a tool to attack some fundamental biological questions.

Looking Ahead: Challenge Questions

What are the possible cofactors that are thought to trigger an AIDS infected individual into the actual disease state?

How have recent advances in chemistry, molecular biology, and genetic engineering contributed to the fight against human disease?

What are the promising new advances that have occurred in the treatment of allergy symptoms?

PAIN
Many Causes, Fewer Cures

Stephen Budiansky

Stephen Budiansky is a free-lance writer living in Clarksburg, Md.

Pain is no one's stranger. Whether it comes as a headache, a warning of injury, or a symptom of any of countless diseases, pain is a universal ill that people experience. The encounter is usually brief. But for an estimated 70 million Americans, pain is a permanent state that remains long after its original purpose.

According to John J. Bonica, the recent past president of the International Association for the Study of Pain, three-quarters of a billion workdays are lost each year because of chronic pain; the annual economic dent in terms of health-care costs alone is $20 billion. Yet, until quite recently, pain was the orphan child of both medical research and clinical practice. Its elusiveness, its subjectivity and its dependence on individual personality, environment and past experience with pain made its scientific study seem nearly impossible. From a clinical perspective, the intractability of chronic pain, and perhaps the tendency of physicians to view pain as a symptom rather than an ailment in itself, formed a similar obstacle to progress.

But now, the study and treatment of pain are beginning to be adopted by the medical world. Remarkable discoveries—for instance, the body's ability to produce its own morphinelike pain reliever—are giving pain research a more scientific footing. Clinical practice is keeping pace, with the introduction of new, more effective drugs, the establishment of clinics devoted exclusively to pain therapy and a new willingness to try such unconventional treatments as acupuncture and biofeedback.

The Body's Opiates

Perhaps nothing has done more for the recent upsurge in pain research than the discovery of endorphins and enkephalins, chemicals produced within the body that act like morphine—a strong painkiller—in blocking the transmission of pain. Their existence was hinted at in 1971 when scientists discovered that nerve cells carry on their surfaces special sites specifically designed to receive morphine and similar molecules. When a morphine molecule plugs into one of these receptors, the nerve, which normally acts as a relay station for painful stimuli, ceases its activity.

The discovery of these opiate receptors, as they are known, was at once intriguing and puzzling. Since morphine is a product solely of plants, it seemed to be a remarkable and very unlikely feat of biochemical evolution to have built such receptors into animals—that is, if they were tuned strictly to morphine. Researchers thus immediately suspected that they were instead designed to receive substances made within the body.

The answer came in 1975 when Hans W. Kosterlitz of the University of Aberdeen in Scotland announced that he had isolated what would turn out to be the first of an entire series of opiates produced within the body. He called them enkephalins (from the Greek words meaning "in the head"); later studies revealed a second class of naturally produced opiates, the endorphins (a contraction of "endogenous morphines").

Although the discovery of the endorphin–enkephalin system has greatly complicated the picture of how pain works, it has also, according to Bonica, provided a scientific way of examining that picture. "These findings have opened a whole new area for future research, which is very exciting because there is some promise of being able to control pain by noninvasive techniques, without cutting nerves and without injecting drugs. Of course the intriguing and fascinating question is 'how can we activate these systems that reduce pain or suppress pain?' "

Circumstantial evidence points to stress as one way to stimulate the release of endorphins within the body. For example, Bonica says, "There are many, many anecdotal reports of soldiers in the battlefield or athletes on the playing field who get severe injuries and actually feel no pain for a long period of time."

Recently, more concrete evidence has emerged pointing to stimulation of the painful spot (by vibration or rubbing, for example) as a way to get the endorphin–enkephalin pain-relief system going. Maria Fitzgerald of University College, London, and Clifford J. Woolf of the Middlesex Hospital Medical School, London, reported in September at the triennial World Congress on Pain held in Edinburgh, Scotland, that when nerves not associated with pain are stimulated in rats, a fast message is sent to the spinal cord, resulting in the release of enkephalins. The enkephalins, in turn, switch off the spinal cord "relays" responsible for transmitting to the brain pain signals coming from the same area. (This study is one of several that have provided evidence for the conceptual "gate theory" of pain; see box, page 146.)

Recent evidence also shows that endogenous opiates are responsible for the pain relief associated with acupuncture and the related technique of electrical nerve stimulation.

Whether the emotional factors that have long been known to affect the perception of pain are also mediated by the endorphin–enkephalin system remains to be seen. But the growing conviction is that it, or possibly some other chemical system, does in fact underlie the psychological element. The placebo effect—the ability of a sugar pill to induce powerful pain relief—is then not purely psychologic; rather it's a psychologic triggering of a very real, chemical pain-relief mechanism.

Providing Pain Relief

But as powerful a pain reliever as the endorphin–enkephalin system is, and as clever as researchers eventually may be in finding ways to trigger it, there will always be a certain need for pain-relieving, or "analgesic," drugs.

Just as the body is capable of relieving pain, so too is it capable of promoting pain. Whenever tissues are injured or irritated, a wide variety of chemicals is released; some cause redness and swelling, while others, notably prostaglandins, render the tissues more sensitive to pain. A familiar example is the pain that even gentle rubbing of sunburned skin can cause.

As long as the pain is mild, it serves a useful function in bringing attention to the injury and preventing further accidental damage. But when the pain is severe, as in bone fractures, or when it is persistent, as in backaches, it can come to be as debilitating as the injury itself. And because inflammation is at least partly responsible for nearly every sort of pain, fighting pain usually means fighting a bodily mechanism—prostaglandin production—that is actually geared to enhance pain.

The use of drugs in the battle follows two divergent strategies. Drugs similar to morphine act on the central nervous system, blocking the reception or the relaying of pain signals; aspirinlike drugs, on the other hand, act on the peripheral nerves, reducing the initial transmission of the pain signals.

According to William Beaver of the Georgetown University Schools of Medicine and Dentistry, research has focused lopsidedly on the morphine drugs. "In 1900 we had aspirin and a number of other peripherally acting analgesics. We also had some opium derivatives: morphine, codeine and heroin. The interesting thing that happened in the intervening 80 years is that most all of the effort to develop better analgesics has been based on modifications of the morphine type of drug, and really, very little has been done up until the last 10 years or so in terms of making a 'super aspirin.' Morphine was felt to be very effective but to have certain drawbacks, most notably the potential for addiction. So people felt the best starting point to make a better analgesic was to take the one that was the strongest and just get rid of some of the undesirable properties."

But "just getting rid of" the addictive power of morphine has proved difficult, if not impossible. Heroin, acclaimed around 1910 to be nonaddicting morphine, was one notable failure. Yet the belief was ingrained in researchers that the aspirin class of analgesics would not lead anywhere. Though aspirin is actually quite effective, Beaver says, "it was common, it was cheap, and although it was very much used, it was a case of familiarity breeds contempt."

But the aspirin-type drugs had some other benefits, in particular their effectiveness against inflammation—a valuable asset in the treatment of rheumatoid arthritis. That, in fact, is what finally helped to refocus attention on them. In the course of searching for new antiarthritis drugs, Beaver says, the drug companies, in effect, rediscovered the link between antiinflammation and analgesia among these compounds and began sifting through them not only for arthritis drugs but for general analgesics as well.

The effort was spurred on by the discovery that the analgesic effect of these peripherally acting compounds was due to their ability to block prostaglandin production. "This," Beaver says, "gave a theoretical basis for the search for new compounds."

That search has already begun to pay off. Zomax (generically known as zomepirac sodium), a drug developed by McNeil Pharmaceutical in 1969 and recently tested in clinical trials, has been described as a "super aspirin." According to Beaver, "It turned out to be more effective than aspirin. Well, there just hasn't been anything more effective than aspirin unless you went to strong, injectable narcotics."

Tests in patients with post-operative pain found Zomax in fact to be as effective as morphine. In addition, it is apparently nonaddictive and retains its effectiveness even in repeated doses. The side effects of Zomax are very similar to those of aspirin—chiefly gastrointestinal upset—and, so far, appear to be no more severe than those of aspirin.

Zomax was approved for prescription-only use in the U.S. last year; but though it actually may be safer than aspirin, it is not likely to be sold over the counter—ever. Edward C. Huskisson of St. Bartholomew's Hospital in London says that physicians and regulators have become much more cautious in recent years. "I think that, undoubtedly, if aspirin were to be brought along now and somebody said, 'please, can we sell this over the counter?' the answer would clearly be no."

Living with Pain

Many physicians feel, though, that given the general overuse of pain-relieving drugs, the caution is more than justified. That's particularly true for sufferers of chronic pain, for whom prolonged use of drugs often ends up doing more harm than good.

"Although a person may have started taking pills for pain," says Gerald M. Aronoff of the Boston Pain Unit at Massachusetts Rehabilitation Hospital, "gradually that person learns, 'after taking this I feel better.' So the person tends to gradually develop a set of behaviors geared toward taking pills and forgets that it was originally only because of severe pain."

The Boston Pain Unit, like several hundred other pain clinics that have opened in the past few years, is aimed at helping chronic pain sufferers who have not been relieved by drugs or more radical treatment such as surgery. Aronoff says the effort is focused on finding ways for the patients to take control of their pain and, essentially, to learn to live with it. "Obviously, if we can do something to remove the source of pain, we try to do that," he says. "But for the majority of these people that is not possible, and so we are talking about approaches which can help them be more active, be less depressed with the pain, and get on with their lives and not only focus on their pain."

To that end, pain clinics are trying some of the more unconventional treatments, especially those that patients can apply themselves and that are noninvasive (treatments applied outside the body). Transcutaneous nerve stimulation, the application of electrical pulses to the skin

The Gate Theory: Regulating the Flow of Pain

"The discovery of endorphins and enkephalins reveals how the gate may work," explains Ronald Melzack, who with Patrick D. Wall proposed the gate theory of pain in 1965. They put forth the idea that instead of traveling uninterrupted from the peripheral nerves to the brain, pain signals are instead controlled by a "gate" in the spinal cord, and that the gate can open or close in response to other processes in the body.

Experiments like those of Fitzgerald and Woolf (see page 145) are beginning to reveal details of the gate's wiring system. Pain nerve fibers, called C-fibers, are the smallest of all nerve fibers, and they carry nerve signals most slowly. The C-fibers enter the spinal cord and there terminate on thousands of different nerve cells. The more rapidly conducting A-fibers, which activate nerve cells to pass on information concerning such low-threshold events as touch, pressure and hair movement, have been shown to be the control lever for the gate. They also terminate in the spinal cord.

The link between the action of these types of fibers and the gate is the body's chemical, enkephalin.

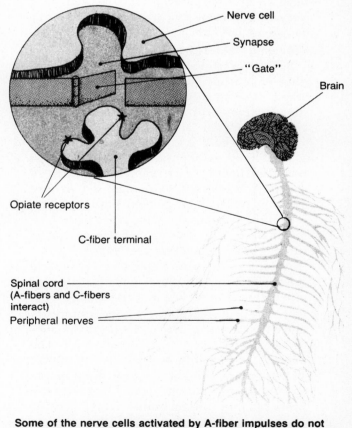

Nerve cell
Synapse
"Gate"
Brain
Opiate receptors
C-fiber terminal
Spinal cord (A-fibers and C-fibers interact)
Peripheral nerves

Some of the nerve cells activated by A-fiber impulses do not transmit information but act on the C-fiber terminals instead, inhibiting passage of a pain signal. This inhibition of C-fibers by A-fibers, mediated by the peptide enkephalin compounds, occurs totally within the spinal cord and does not involve messages descending from the brain.

Pains and Needles

Traditionally, Chinese acupuncture was a complete system of medicine. Needles as long as 23 cm were thrust into the body and twisted; depending on the site, this was supposed to elicit results ranging from cancer cure to restoration of hearing in the deaf.

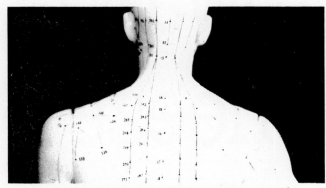

A plastic model shows various points of the body used for acupuncture anesthesia and therapy.

One of the most publicized modern uses of acupuncture in China is in providing anesthesia for surgery. But according to John J. Bonica, a member of the first American medical mission to witness acupuncture treatment in China, it is not likely to come into widespread use. "I think it has very little place, even in China, where only about three percent of all operations have been with acupuncture anesthesia and in a highly selected group of patients."

On the other hand, he says, "It's quite clear that acupuncture does something—certainly physiologically; but undoubtedly it also involves some psychological processes. I can tell you I was very impressed by seeing people with their chests open, looking at me and drinking orange juice. They had some narcotic on board, but not a sufficient amount by Western standards to permit us to operate on our patients."

Lorenz K.Y. Ng, chief of the Pain Studies Program at the National Institute on Drug Abuse, uses acupuncture clinically. He points to his own studies and those of others that have shown acupuncture to activate the endogenous opiate system, and to what he sees as the clinical fact that it works. "You can say it's just a placebo effect, but even a placebo has to have an underlying physiological mechanism."

Ng's study of rats found elevated levels of endorphins in the cerebrospinal fluid of rats subjected to electro-acupuncture—the application of electric pulses via a needle inserted into the rat's ears. Other studies have shown that electroacupuncture does in fact have an effect different from the related—and fairly well accepted—technique of transcutaneous nerve stimulation. This technique of stimulating an area by applying electric pulses to the surface of the skin does not appear to activate the endogenous opiates.

around the affected area, is an example. Aronoff says it has proved effective in about one-third to one-half of the patients at the Boston Pain Unit who try it. Like the related techniques of acupuncture and electroacupuncture, transcutaneous nerve stimulation appears to work according to the gate theory of pain—by closing the "gate" on the spinal-cord pain relays. (The role of acupuncture in clinical practice remains uncertain, however; see box, this page.)

Aronoff says that a much simpler yet effective treatment is application of ice to the area. "In fact, ice tends to be much more effective than heat for chronic pain. It's incredibly safe, and it's also desired because patients can be taught to do something for themselves and get more control over their own pain."

Another important way for patients to gain control is to realize that although the pain itself is very real, it often is a result of tension that has psychologic origins. Aronoff says that biofeedback, the method of translating physiological factors such as muscle tension into a visible or audio signal that the patient can observe, is one very useful tool in teaching patients to recognize when their tension increases and what they can do to control it.

Aronoff believes that this work holds a valuable lesson for everyone in dealing with everyday bouts with pain, such as headache. "Most of the headache sufferers in this country have tension headaches. It doesn't mean they are not real; they can be excruciating. But the main way to prevent it is not by taking a lot of pills but to try to resolve the stress."

Suggested Readings

(1) Fagerhaugh, Shizuko Y., and Strauss, Anselm, "Politics of Pain Management: Staff–Patient Interaction," Addison–Wesley, Menlo Park, CA, 1977.
(2) Hassett, James, "Acupuncture Is Proving Its Points," *Psychology Today*, **14**, 81–85 (December 1980).
(3) Neal, Helen, "The Politics of Pain," McGraw-Hill Book Company, New York, NY, 1978.
(4) Wall, P. D., "The Gate Control Theory of Pain Mechanisms: A Re-examination and Re-statement," *Brain*, **101**, 1–18 (1978).

Racing Toward the Last Sneeze

Dianne Hales

Dianne Hales, who never suffers from any allergies, is the author of The Complete Book of Sleep *(Addison-Wesley, 1981) and* New Hope for Problem Pregnancies *(Harper & Row).*

Last August, Robert Appler of 8644 Oak Road, Baltimore, mowed the lawn, went on picnics and slept with the windows open. Commonplace events in an unremarkable life? Not during the ragweed season. And not for a man who'd been a virtual prisoner of pollen over many summers—barricaded inside, nose raw from wiping, eyes streaming, throat aching from a constant cough.

"At school, my paper would be wet from my sneezes by the end of a test," Appler recalls. "At work, I'd go through a box of tissues during a meeting. Since I'm also allergic to dust and mold, my symptoms never went away entirely. It was like having a bad cold that lasted for 30 years."

Skeptical of the usual allergy treatments, Appler sniffled through season after season of his discontent until he learned of an experiment at the Johns Hopkins Allergic Diseases Center in Baltimore. From spring to late fall in 1980 and again in 1981, he received biweekly injections of a "conjugated allergen," a chemical compound attached to ragweed extract. According to his daily diaries of sneezes and wheezes, Appler breathed easier than he had ever done before.

Philip Norman, M.D., the director of the center, is encouraged but cautious. Month by month, allergy victims such as Robert Appler report dramatic results, but Dr. Norman does not yet have enough objective data based on blood analysis to back up the subjective reports.

"Right now," he says, "we can only offer hope, not deliverance. But if things work out, we could be talking in terms of treating the cause and not just the symptoms."

Even if this particular clinical trial doesn't lead to the ultimate antiallergy treatment, the theory behind the therapy marks an enormous advance. For the first time, investigators are trying to locate the on/off switch for allergy in humans in order to turn off the allergic response before it starts. As part of a grander effort to get to the source of allergy, this experiment may herald the beginning of the end of chronic discomfort for the 35 million Americans with allergies.

Within 10 to 20 years, predicts Michael Kaliner, M.D., head of the allergic diseases section of the National Institute of Allergy and Infectious Diseases (NIAID), allergy may become a preventable disease. It may go the way of polio and smallpox.

Such optimism, echoed by allergists and immunologists around the country, is new in the history of a disorder that has mystified medicine. Only in the last century did science begin to work on the basic concept of allergy as a strategic error in the body's defense system. The term *allergy* itself, derived from the Greek for an altered or unusual reaction, goes back just 75 years. And until almost yesterday, allergy was a clinical version of Rubik's cube—fascinating and frustrating, a puzzle that stubbornly resisted quick solution.

By its very nature, allergy seems capricious and contradictory. It attacks an equal percentage of men and women, the newborn and the aged, the rich and the poor. Its possible causes read like an inventory of creation, including life's pleasures (foods and flowers), perils (poison ivy and insect stings) and inescapable realities (mold and dust), as well as such manmade potions as deodorants and detergents. "Everything under the sun, and even the sun itself," says one dermatologist, can provoke some reaction in some susceptible soul.

And the reactions can be as varied as Job's godsent afflictions: itching (from eczema and other skin allergies); nasal congestion, eye irritation and coughing (from allergic rhinitis, more commonly misnamed "hay fever"); wheezes and gasps (from allergic asthma); hives, internal and exter-

Our Costliest Ailment

These days, $25,000 can buy a turbocharged Porsche 924, serve as down payment on a three-bedroom house or buy two years in any Ivy League college. But for the William Randalls, $25,000 barely covered the costs of their allergies over the past 10 years.

Bill Sr. had to quit his job as a baker after developing a flour-dust allergy. His wife, Shirley, sensitive to house dust, ragweed, mold and dog dander, can't work because of her frequent asthma attacks and ear infections. Son Billy misses two to three days of school a month because of allergic rashes, sinus headaches and breathing difficulties. Little Judy, four years old, has had wheezing spells since she was one. The Randalls spend $200 to $250 a month for allergy outlays on doctors, prescriptions, over-the-counter antihistamines and nasal sprays. Their health plan, like most insurance policies, doesn't cover these expenses. And their indirect costs—such as air conditioning, filtering and special foods—aren't even tax deductible.

The Randalls are a fictitious family, but the problems are real. An estimated 15% of Americans suffer from allergies. These sneezing, sniffling, scratching citizens spend more than $1 billion a year on allergy care. The economy loses another $800 million in lost workdays and lowered productivity. Allergy is the No. 1 chronic disease that causes school absenteeism, and the commonest chronic complaint in industrialized nations.

When you seek help, a standard diagnostic evaluation might run $100 to $250, including extra skin tests. Many allergists use a new approach called RAST (for radioallergosorbent technique) that detects allergies by laboratory analysis of a blood sample. It replaces the tedious, at times painful, process of skin testing, but it costs more ($10 to $30 for each substance tested) and is less precise.

If the allergen is a weed, grass or tree pollen, and drugs fail to bring relief, then desensitization shots may be the only option. Each biweekly injection of a specific allergen costs $3 to $10, with annual resupply of the extract running about $60.

Some impatient patients bet their hopes and dollars on offbeat treatments that range from the unproven (such as desensitization for food and chemical allergies) to the bizarre (urine injections or Chihuahuas). "Quackery thrives in the treatment of allergy," says immunologist Lawrence Lichtenstein, M.D., Ph.D., of Johns Hopkins. "Everyone has a theory or a cure."

Allergists realize that until they learn to offer more effective therapies, the brisk business in pseudoscientific treatments will continue. For now, Dr. Lichtenstein offers his own alternative therapy for seasonal allergies: Spend the ragweed season in ragweed-free Paris. It may be cheaper.

---

nal (from drug or chemical allergies); vomiting and diarrhea (from food allergies); sudden collapse (from anaphylaxis—the direst, deadliest allergic response).

Heredity looks like a coconspirator, not the sole cause of allergy. Stress gets part of the blame, although allergy is not, as some contend, a psychosomatic disease in which "repressed weeping" results in hives and rashes. Some children outgrow their allergies; some adults try to outrun theirs, only to end up acquiring new ones on their new home ground. Victims seldom die of allergies; they just as seldom recover.

Because allergy has been so elusive for so long, we envision the doctors who treat it as sleuths, gathering clues and tracking down culprit irritants. But the new researchers—many of them immunologists intrigued by the enigmas of allergy—are looking beyond the whos, whats, whens and wheres of allergy and wondering *why*: Why does a walk in the woods bring smiles to one person and sneezes to another? Why does one family live contentedly with a menagerie of pets, while another wheezes collectively the minute a poodle strolls through the yard? Why does one of six Americans react, sometimes violently, to substances as soft as goose down and as smooth as silk?

Among the first scientists to ask why, was a German physician named Carl Prausnitz. Why, he wondered in 1921, could he eat fish with impunity, while a colleague could not? He injected himself with his friend's blood. The next day, he injected fish extract at the same spot. His skin reddened in the classic "wheal and flare" of allergy. (The wheal is the area showing a reaction; the flare is the flesh swelling up red in reaction.) Something in his friend's blood, which Prausnitz dubbed "reagin," had made him temporarily allergic to fish.

It took 45 years before a husband-and-wife team of immunologists—Drs. Kimishige and Teruko Ishizaka, now at Johns Hopkins—learned what goes on under the flared skin. What triggers allergy, they discovered, is an excessive amount of a formerly unrecognized group of proteins in the immune system, which they called Immunoglobulin E (or IgE) antibodies. Unlike other immunoglobulins, IgE does not destroy invading bacteria and viruses. A pseudofighter, it marshals the body's forces against harmless particles such as pollen.

Some researchers believe that IgE may actually be an evolutionary holdover that protects man against such parasites as flukes and hookworms. Natives of regions where parasites still are common have higher IgE levels but a lower incidence of asthma and other allergic diseases. And in animals and humans, IgE increases when some parasitic diseases strike.

With no parasites to fight, IgE wages pointless war on innocent particles. After too close or too frequent encounters with a particular allergen, individuals with a genetic predisposition to allergy may speed up production of IgE antibodies. By the millions, these misguided soldier-molecules swarm through the bloodstream. Many make their way to the mast cells that line the mucous membranes of the air passages and digestive tract. These cells are miniature cannons, loaded with noxious chemicals called mediators. Five hundred thousand IgE antibodies may attach themselves to a single mast cell, like so many fuses waiting to be lit. The next time the victim inhales or ingests the allergen, it sparks, and the mast cell explodes. Out spew the mediators that set off the sneezes, sniffles and spasms. Like an artillery barrage, the firing continues as long as the enemy allergen is in range.

This is where today's allergy treatments come in. They intervene in this attack sequence either by diverting the body's protective forces or by mopping up after the damage is done. Tomorrow's therapies may be more direct. Rather than fighting powerful chemicals with powerful drugs, immunologists are trying to disarm allergy. They want to remove the firing pin (IgE) that sets off the mast cell explosion or to prevent the release of the explosives (the mediators).

Mystery Allergens: Case Histories that Might Have

Soybean Formula Puts Mom in Shock

A pediatrician advised one young mother to use a soybean-based formula for her new baby to prevent possible allergic reactions. Curious, she tasted it and fell into anaphylactic shock, the allergic reaction that can be deadly. Tests with small amounts of soybean and other bean extracts later showed that she was sensitive to almost every known bean.

Her reaction was probably a food allergy, easily verified by increased levels of Immunoglobulin E (IgE)—the chemical essence of allergy—in her blood after exposure to bean extract. But many allergists believe that such reactions to food are rarer than most physicians or patients suspect, particularly in children.

Charles May, M.D., and Allen Bock, M.D., allergists at the National Jewish Hospital in Denver, reinforce professional skepticism about food allergy. Dr. Bock gave capsules containing various foods to 290 children between the ages of 3 and 16, all with an impressive history of adverse reaction to a food. In 60% of the cases, there was no indication of allergy. For the 35% of the children who were clearly allergic, the most offensive foods were peanuts, eggs, milk and soybeans. The food incriminated least often: chocolate.

Pediatricians fear that parents who worry about food allergies may let kids be undernourished. A mother who believes a child is allergic to dairy products may ban all milk products from the diet and cause a calcium deficiency. Many discomforts that plague children after they drink milk—diarrhea, gas, bloating, vomiting—may stem from an entirely different cause: an abnormally low level of the intestinal enzyme lactase, which breaks down the principal sugar in milk, lactose. This problem is commonest in American blacks and Indians, and in families with Latin American, Oriental and Mediterranean roots.

After years of food debates, the National Institute of Allergy and Infectious Diseases has undertaken a monumental research program. "We're assuming nothing and testing everything," says one investigator. "It's the only way we're going to find out whether food allergies are just a figment of mothers' imaginations."

The Case of the Broken Wine Cork

The prospect of a business lunch or a romantic dinner used to delight a San Diego woman—until she began to develop rashes, itches and asthma attacks while eating in restaurants. She sought help at the Scripps Clinic in La Jolla, but a complete allergy workup, even elimination of suspected foods from her diet, failed to turn up any clues.

A few years later, a family friend gave a dinner party in her honor. He opened some wine that he had bottled himself and, ignoring a few bits of cork that had fallen into the glass, offered her the ceremonial first sip. Minutes later, she collapsed.

Rushed to the hospital and resuscitated, the woman asked her friend if the wine contained anything unusual. No, he said, but he *had* treated the cork with a solution of potassium metabisulfite, a common preservative. Many restaurants, he explained, spray this chemical on lettuce in salad bars to prevent wilting or on avocado dip to keep it looking appetizingly green.

Convinced that she'd found the missing clue to her restaurant allergy, the woman showed up at the Scripps Clinic with a bottle of the preservative in her hand. Allergist Donald D. Stevenson, M.D., tested her with a tiny amount. "That was it, bingo!" he recalls. Her skin reddened and rose in the classic "wheal and flare" that signifies an allergic reaction.

Since then, Dr. Stevenson has traced several other mysterious allergies to metabisulfites. The only "treatment" is some simple advice: Ask if the restaurant uses the preservative before taking a single sip or bite.

The clinical trial of conjugated allergens at Johns Hopkins is the pioneer attempt to control allergy by controlling IgE in humans. David Katz, M.D., and an immunology research team at the Scripps Clinic and Research Foundation in La Jolla, California, developed one of the complex chemicals, called DGL, involved in the study. Dr. Katz first tested this compound by linking it chemically to penicillin and showing that injections of the mixture prevented allergic reactions to the drug in mice. Katz's colleague, Fu-Tong Liu, Ph.D., tacked a ragweed extract onto DGL. Once again, this conjugate eliminated an allergic response. Similar compounds, including polyethylene glycol (just plain old antifreeze), also stopped IgE production, although how is not yet clear. If measurements of IgE in the volunteers' blood come up with the evidence to support the researchers' enthusiasm, conjugated allergens may be a true scientific breakthrough.

But they aren't the ultimate solution. For this treatment to work, the allergen that produces symptoms must be identified, and an extract of it linked to a chemical like DGL. But many people are allergic to rare substances, and for more the source of their sneezes can't be identified.

Katz has a better idea: Control the universal factors that suppress or enhance IgE production. He has identified such modulating molecules in mice and in human tissue culture. Dr. Kimishige Ishizaka has pinpointed similar, but chemically distinctive, substances in rat tissue. By manipulating these factors, Katz has raised IgE levels in mice, thereby creating allergy, or lowered levels again, thereby eliminating allergy. He estimates that clinical trials are five years away, if all goes smoothly.

And then? If he's right, "we would have an effective treatment for IgE-mediated diseases, regardless of whether they're caused by a single allergen, or multiple allergens, or dust, or pollen, or whatever." But it's far too early to speculate whether, in men as in mice, a single injection might turn or the IgE spigot once and for all.

Other allergists take a different approach. They concentrate on the mediators—the chemical ammunition of allergy. The best known of these molecules of misery is histamine. Antihistamines, the most widely used allergy drugs, counter the debilitating effects of histamine, but antihistamines are only part of the solution, because histamine is only part of the problem. In the last two years, scientists have identified two other mediators, each potentially more potent than histamine. One, Slow Reacting Substance (SRS), came out of 34 years of research at Sweden's Karolinska Institute and at Washington University in St. Louis. It consists of several "leukotrienes," unique combinations of peptides and fatty acids.

The other mediator, AGEPC, emerged from the labs of the University of Texas in San Antonio. Tested in minute amounts, AGEPC induced reactions 100 times more powerful than those caused by histamine. Allergists

Stumped Sherlock Holmes

Christmas Tree Makes People Cry

"Other families just get gifts every Christmas," one mother in a Dallas suburb complained to friends. "Mine gets colds." After she discovered that holiday flu and sniffles had hit several neighbors, an allergist confirmed her suspicions: The Christmas trees they had chopped down in a nearby woods probably made them sick.

The tree is the mountain cedar, a member of the juniper family native to many Southwestern states. Next to ragweed, it is the commonest cause of pollen allergy in the Sun Belt. The male tree, which releases its pollen in the winter months, makes the trouble, particularly if cut down while still green and taken indoors to pollinate.

Mountain cedar wafts woe at millions who never bring it into their homes. When the winds are strong, dust storms of pollen from mountain cedar groves spread multiple miseries of hay fever and head colds, burning and itching eyes, runny noses, sneezes and asthma attacks to people miles away.

Timothy Sullivan, M.D., head of the allergy division at the University of Texas, Southwestern Medical School in Dallas, is trying to find out why so many people react to mountain cedar.

"Pine pollen is in the air a lot," he says, "but practically no one is allergic to it." He and his research team have pinpointed what may be the primary irritant in mountain cedar—a protein-sugar molecule—and are busily hunting for ways to fight it.

Body Reacts Against Effort

"Me exercise? I'm allergic to it." It's a line worthy of W.C. Fields, but for some people it's absolutely true. Jogging, sprinting, basketball, soccer, tennis, even dancing can leave well-conditioned athletes breathless, swollen and covered with hives.

A varied group of Boston and Albany athletes, age 11 to 45, suffered life-threatening allergic reactions in workouts. All itched. Most had breathing problems — some deadly — and swelling of the face, palms and soles. A few suffered days of extreme nausea or severe headaches.

Some researchers speculate that a strenuous workout may release noxious histamines. Preventing the attacks is easy—avoid vigorous exercise. All of the reported victims refused to follow that recommendation. Some now use antihistamines regularly or before exercising, but they only reduce the severity of attacks.

High-Risk Babies

If you have allergies, your children may also suffer. To help them:

Breast-feed for at least six months. Doctors disagree on the protective merits of nursing, but breast milk is the best possible nutrient and, unlike milk-based formulas, it's easy for newborns to digest.

Feed no eggs, fish, milk products or citrus juices during the first year.

Buy no woolen clothes or blankets.
Don't get a pet. If you already love a cat or dog like your child, don't let the pet sleep indoors.
Damp-mop often to stop dust, mold.
Don't smoke. Tobacco smoke does not seem to provoke allergies, but does worsen asthmatic pains.
Keep shoes in closed closets (they may smuggle pollen indoors).
Divide your child's wardrobe into indoor and outdoor clothes and keep them in separate closets.

Research at the University of California, San Francisco, shows that children are most vulnerable after the onset of a cold or flu. IgE levels rise dramatically, multiplying the chance of an allergic reaction later on.

Prevention may ultimately be as simple as protection against such once-dread diseases as diphtheria and smallpox. An anti-IgE vaccine could, in theory, eliminate the susceptibility of high-risk infants once and for all.

Quicker, Better Desensitizers

In 1911, a London physician named Leonard Noon reported a successful new treatment for hayfever—"prophylactic inoculation" with watered-down extracts of grass pollens. His theory was simple: Periodic injections with gradually increased doses might build up tolerance, or immunity, to pollen. Although Noon's techniques were rather crude, his therapy—the first allergy shots—worked well enough and often enough to become a mainstay of allergy treatment.

Seventy-one years later, allergists annually inject hundreds of the more than 1,500 chemical extracts into several million patients. The basic concept is the same, but there are some important improvements in desensitization. "For a long time, there was an aura of black magic around allergy shots," says James Wedner, M.D., of Washington University in St. Louis. "To other physicians, they seemed as scientific as shaking maracas or throwing bones." But yesterday's medical Merlins now view themselves as clinical immunologists who try to outmaneuver the body's natural defenders.

Until the new breakthroughs get well tested, today's improved desensitization shots are the allergic's best hope. For selected allergies—particularly ragweed, tree and grass pollen reactions—allergy shots can give 65% to 80% of patients relief from their symptoms. They work mainly by stimulating production of the protective defensive protein, Immunoglobulin G (IgG), which combines with the allergen and prevents stimulation of the immunoglobulin that would otherwise trigger an allergic response.

The new shots literally save lives if you are one of 850,000 Americans who go into shock when stung by a bee (40 deaths a year). Survivors of anaphylactic reactions to stings describe the experience as terrifying. "I saw the yellow jacket set down on my arm and bite me," recalls one woman. "And I was surprised that I didn't feel anything. Then my whole body began to swell. I was wheezing because I couldn't breathe. I had this terrible, ominous sense of dread. And then I fainted." These symptoms, caused by a sudden flood of histamine, can be reversed by a quick injection of adrenalin.

Most desensitization shots are for problems that are much more common and far less hazardous. The most frequent target is ragweed pollen, which researchers have broken down into component irritants to make diagnosis more precise and treatment more specific. The conventional approach is a series of up to 60 injections over three to five years.

But several new techniques now make pollen desensitization more efficient and effective. Exposing the allergen extract to formaldehyde causes molecules to form cross-links, swelling the molecular size of the allergen and shrinking its ability to irritate.

Another approach, developed by allergists at Northwestern University, modifies the allergen with chemical polymers that speed production of protective IgG antibodies and lessen the likelihood of side effects. In clinical trials in Chicago, St. Louis, Memphis and Boston, injections 40 times the normal size have proved effective in only 15 weeks.

Other allergists get good results with "rush immunotherapy," a stepped-up program of one to four allergy shots in a single day. Still other investigators are testing nasal intake of medication.

Practicing allergists may be enthusiastic about such refinements, but researchers see desensitization therapy as a therapy of last resort. It works for only a few specific allergies. With the coming of new drugs, such as steroid inhalants, they feel that many people, particularly those hit by pollen, may get as much relief without desensitization's considerable investment in time and money.

hope that, as they synthesize more mediators, they will be able to develop treatments that will stop them before they strike.

"The importance of the mediators cannot be overstated," says Robert Goldstein, M.D., chief of the Allergy and Clinical Immunology Branch of the National Institute of Allergy and Infectious Diseases. "The more allergists know about them," he adds, "the more we have to work with to find effective therapies."

No one can predict which of these avenues of investigation will turn into dead ends and which will become expressways to new insight, better treatment—and solutions. But the researchers feel they're close—very close—to something important. "We don't need to know much more than we know now to find an IgE regulator or drugs that will control the mediators," says Dr. Goldstein. "It could be several years before we get there, but I don't think it'll be decades."

And after that? The scientists emphasize that new answers often lead only to new questions, but the odds are that, at the least, sufferers will have some relief. Our old allergic bugaboos may never be the same again. At best, allergy will become extinct—a disease wiped out by safe, effective therapy. Then, the huddled, harried masses yearning to breathe free may look forward, at last, to a future of no more tears, no more sneezes, no more allergy.

THE GENE DOCTORS

SCIENTISTS ARE ON THE VERGE OF CURING LIFE'S CRUELEST DISEASES

More than anything else, Tara Dew would like a puppy. That's not unusual for a nine-year-old girl. But for Tara, a pet could be fatal. She is a victim of ADA deficiency, the inherited immunological disease that killed David, the famous "Bubble Boy." Although her disease is not as severe as David's, a stray virus picked up from any animal or person could mean death. As a result, Tara cannot go to school. She must stay in her home in Bellevue, Neb., listening via a speakerphone to classes going on in a local schoolroom. Her few playmates must be carefully screened for sniffles and chicken pox.

And once a month she must endure lengthy blood transfusions to give her some small but crucial resistance to disease.

Tara's malady might be helped by a bone-marrow transplant. The new marrow would produce the missing enzyme—adenosine deaminase, or ADA—from which her disease takes its name. But the transplant is risky, and it could even be fatal. So even though Tara is living on borrowed time—few ADA children live past the age of two—Tara's parents are holding out for a startling new development called gene therapy. Instead of implanting marrow

cells from a donor, doctors would go directly to the root of Tara's problem and replace the defective gene in her own marrow.

GREEN LIGHT. Just a decade ago, the prospect of manipulating human genes was almost inconceivable. Scientists were just beginning to decipher the tangle of deoxyribonucleic acid, or DNA, that carries the hereditary coding for organisms. But the pace of the research has been astonishing. In the late 1970s, they first put human genes into bacteria, turning them into factories for useful proteins. Then, in 1982, researchers altered the genes of a higher animal.

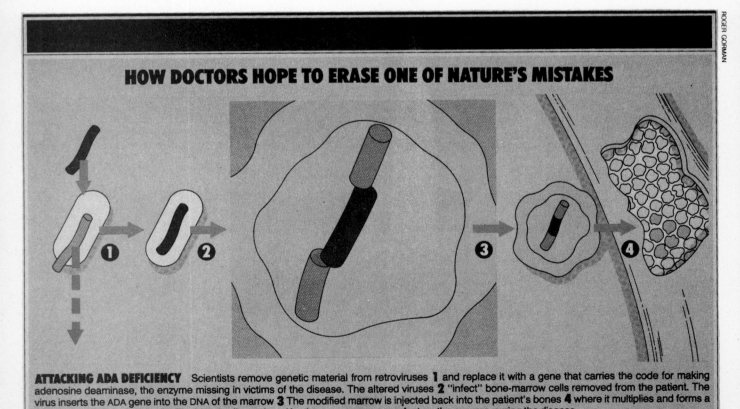

ROGER GORMAN

HOW DOCTORS HOPE TO ERASE ONE OF NATURE'S MISTAKES

ATTACKING ADA DEFICIENCY Scientists remove genetic material from retroviruses **1** and replace it with a gene that carries the code for making adenosine deaminase, the enzyme missing in victims of the disease. The altered viruses **2** "infect" bone-marrow cells removed from the patient. The virus inserts the ADA gene into the DNA of the marrow **3** The modified marrow is injected back into the patient's bones **4** where it multiplies and forms a colony of healthy marrow cells. The blood cells produced by the new marrow manufacture the enzyme, curing the disease

Scientists at the University of Pennsylvania created a huge "Supermouse" by transplanting a gene for rat growth hormone into a mouse embyro.

Now scientists are about to take the next profound step—altering the genetics of humans. Research teams have already performed dry runs of the gene therapy procedure on mice, and they are reporting encouraging results in experiments on monkeys. In September, the National Institutes of Health, which oversee the safety of gene-splicing, gave researchers the green light to seek approval for the first human experiments. Although it will be years before gene therapy becomes widely used, those tests could begin within months.

Tinkering with the genetics of living organisms has been controversial since the beginning. In 1975, at a landmark meeting in Pacific Grove, Calif., concerned scientists debated whether gene-splicing research should be stopped. The immense potential benefits persuaded them to proceed—under stringent safety guidelines. The step to altering human genetics will certainly heighten the debate. But a new BUSINESS WEEK/Harris Poll indicates that most Americans support gene therapy. And if the upcoming experiments succeed, gene therapy may be the first medical tool to cure the thousands struck each year by one of 3,500 genetic diseases.

These hereditary maladies extract an enormous toll. Untouched by the great strides in medicine, genetic diseases afflict as many as 1 out of every 20 newborns. They cause half of all miscarriages, nearly half of all infant deaths, and 80% of mental retardation. What's more, researchers now know that many diseases once thought to be caused solely by environmental factors are also heavily influenced by at least one gene. Such disorders include heart disease, diabetes, many forms of cancer, alcoholism, and even clinical depression.

Inherited diseases inflict a stupendous economic burden as well. The costs for all genetic diseases run into the billions—not taking into account lost productivity. Cystic fibrosis, the most common lethal genetic disorder, alone costs victims and taxpayers an estimated $200 million annually in fees for hospitalization, drugs, and other care.

Such debilitating and ultimately fatal genetic diseases are caused by deceptively simple errors in the genetic code that is written in the DNA of every living cell. The coding is expressed by four chemical subunits, called nucleotides, of the DNA molecule. These nucleotides—guanine, cytosine, adenine, and thymine—combine like the dots and dashes in Morse code to form the genes that determine physical characteristics, such

as sex and eye color, and control every facet of metabolism, from resisting infections to digestion and thinking.

Just one misspelling in the DNA code—an adenine instead of a guanine—and the entire process is thrown off. That, in fact, is one error that can cause Tara Dew's illness, ADA deficiency, which afflicts about 100 Americans each year. And its very simplicity has led many genetic researchers to choose that disorder as the likely candidate for the first human gene-therapy trials. "There aren't that many genetic diseases that fall into this class," says Dr. Stuart H. Orkin, a researcher at Children's Hospital in Boston. "With the more common genetic diseases, we're talking another level of sophistication."

Gene therapy won't reach hospitals for years, but the first experiments on humans are near

---

Orkin and other scientists identified the faulty ADA gene and found a way to insert the correct version of it into mouse cells, which then began producing the missing enzyme. To do this, they turned to one of nature's most efficient invaders of cells: the virus. Since viruses steal into cells and insert their own genetic material into the host's DNA, scientists use them to smuggle in new and beneficial genes.

RACING TO TRY. In a series of key experiments begun in 1983, Richard C. Mulligan of Massachusetts Institute of Technology's Whitehead Institute extracted most of the genetic material from a type of virus called a retrovirus and substituted the healthy ADA gene. Some scientists were concerned that the retrovirus might run wild and cause harmful infections or cancer. It did not. "We've done enough animal experiments not to be fearful," says Dr. Philip W. Kantoff, a member of a research group headed by Dr. W. French Anderson of the National Heart, Lung & Blood Institute.

The next step is inserting a healthy ADA gene into the DNA of a disease victim and getting it to produce ADA. Anderson's team is leading the race to try it. The group is now completing a series of tests on monkeys and will probably begin searching for a suitable human candidate within a year. For the initial experiment, they will try to find an ADA baby less than one year old and otherwise healthy. "There is a small window

of time before the child begins to deteriorate," says Kantoff.

Once the scientists select a candidate, they will extract bone marrow cells with a syringe and infect them with the modified retrovirus. The cells that contain the new ADA gene will then be injected back into the patient's bone. With luck, the cells will multiply and produce healthy blood cells. They should provide enough of the missing enzyme to give the patient a functioning immune system.

If the ADA gene therapy works, scientists will be able to move on to more complicated inherited diseases. One disorder that many have selected as their next target is Lesch-Nyhan syndrome. Like ADA, it is caused by a single missing enzyme. And it results in gout, mental retardation, and a bizarre compulsion to rip and bite one's own flesh.

ON-OFF SWITCH. Conquering Lesch-Nyhan promises to be difficult. For one thing, the missing enzyme—which bears the tongue-twisting name hypoxanthine guanine phosphoribosyl transferase (HPRT)—is used by the brain and nervous system. Researchers are not certain it will be possible to inject new genes into brain tissue. What's more, the body may require HPRT in precise quantities. Simply coaxing some cells to produce it may not solve the problem.

There is evidence, however, that it just might work. At the University of Minnesota, several patients with a multiple-enzyme deficiency who received experimental tranplants of normal bone marrow showed marked progress. Apparently, the enzymes reached the liver and spleen, where they are normally used. "It gives one more hope that gene therapy could correct defective genes at different, distant sites," says Katherine W. Klinger, a senior scientist with Integrated Genetics Inc.

Moreover, researchers believe that one day they will be able to regulate the production of such enzymes by "switching" genes on and off. The exact mechanism by which genes are turned on or off is still poorly understood. Some genes, for example, play a critical role during early fetal development and then, when they are no longer needed, mysteriously stop functioning. Researchers recently discovered one piece of this puzzle: a genetic chemical called antisense-RNA. Its job is to join seamlessly with a strand of genetic material and seal it off, thereby preventing it from functioning. If antisense-RNA for the HPRT gene can be made in the laboratory, it could be used to regulate closely the production of the enzyme.

HIGH RISK. Figuring out how genes are regulated could provide a powerful tool against cancer. So far researchers have identified about 25 genes that apparently help to create tumors in at least 20% of cancer cases. These so-called proto-oncogenes are thought to regulate normal cell growth during the development of the embryo, then switch off. But they may be triggered later in life by, say, chemicals in cigarette smoke. They then cause the wild growth of tumors. Gene therapists may be able to use antisense-RNA to turn them off again.

Gene therapy may even become an effective tool against the nation's leading killer: heart disease. Dr. Jan Breslow, head of the genetics and metabolism laboratory at Rockefeller University, has identified eight genes that produce a group of substances called apolipoproteins, which help carry cholesterol through the bloodstream. When some of these genes are defective, patients have a high risk of heart attacks. Breslow is already transplanting healthy apolipoprotein genes into mice to see if the genes will produce the proper proteins. "The question is," says Breslow, "If we replace bad genes with good genes, can we affect susceptibility to heart attack?"

If critics of gene therapy have their way, Breslow and other researchers may never find out. Moral questions about gene therapy were raised by non-scientists as early as 1980, when the general secretaries of the National Council of Churches, the Synagogue Council of America, and the U.S. Catholic Conference asked President Carter in an open letter: "Who shall control genetic experimentation and its results, which could have untold implications for human survival? Who will benefit, and who will bear any adverse consequences, directly or indirectly?"

Even as that letter was being written, Dr. Martin Cline of the University of California-Los Angeles School of Medicine was taking matters into his own hands. Without proper approval from the human-safety committees involved, Cline tried to use a retrovirus to cure a young girl in Israel and another in Italy of a severe form of hereditary anemia, called thalassemia major.

Cline was censured, and his transgression sparked an ethical debate that has now come to focus on a central question: Is it wrong to perform gene therapy when its effects will be passed on to future generations? While the planned experiments on ADA would affect only somatic, or body, cells, and changes would not be passed on to a patient's offspring, gene therapy performed on germ cells—sperm or egg cells—would affect future generations.

WATCHDOG. For now, most ethicists oppose germ-cell therapy but have concluded that somatic-cell therapy is morally sound. "It's easily acceptable," says John Fletcher, an NIH bioethicist. "It's

'YOU CAN'T THINK, BUT YOU KNOW YOU'RE DYING'

'First, you look fidgety; maybe you have a few tics," says Nancy S. Wexler, quietly. "Then while you're setting the table, you drop some silverware. Or you can't balance your checkbook anymore. In a couple of years, you're flailing every part of your body, nonstop. You can't think—but you know you're dying."

Wexler is describing Huntington's disease, a genetic disorder of the nerves that first shows its symptoms at about age 40. She has just reached that age, and there is a 50% chance she has inherited the disease.

Wexler first learned of Huntington's in 1968, when her mother was diagnosed as having the invariably fatal disorder. A clinical psychologist, Wexler reacted with what she calls "implosion therapy" and what a layman might call dedication and courage. She sought out other Huntington's families and formed a committee to fight the disease. In 1976 she headed up a congressional commission that led to a nationwide network of research centers. She soon began annual research trips to three tiny villages on the shore of Venezuela's Lake Maracaibo. They are home to the world's largest Huntington's family—5,000 cases have been documented.

Today, Wexler commutes between Columbia University in New York, where she studies her data from Venezuela, and Los Angeles, where she is president of the Hereditary Disease Foundation, which was founded by her father. "I can't tolerate not doing anything," she says. "I want to explore the possibility of finding the Huntington's gene and fixing it."

Thanks in part to her efforts, researchers have taken a giant step toward finding the culprit gene by isolating a bit of genetic material that signals its presence. That "marker" will soon make possible a diagnostic test that will tell potential victims of the disease whether they will get it. But until gene therapy is developed to cure Huntington's, she warns, taking the test "will be a phenomenal gamble. You may find that you don't have the disease. Or you may find you have it—and destroy your life before you ever show any symptoms." Many young victims, she fears, might commit suicide.

LIKE A FAMILY. As for herself, Wexler does not fear the symptoms of Huntington's as much as the loneliness and sense of emptiness that go with wondering whether she might have the disease. "Just being at risk," she says, "you feel cut off and special in a way you don't want to be."

Her fight against Huntington's, however, has brought Wexler both friendship and fulfillment. "My colleagues and I are like a family, working together to solve this problem," she says. "The more satisfied I am with my life, the less anxious I am about losing it."

just one step beyond bone-marrow transplant." But the case is far from closed. "Some people think if we take this first step, we'll inevitably go on to germ-cell therapy," says LeRoy Walters, head of a group that reports on gene therapy to the NIH's Recombinant-DNA Advisory Committee (RAC). "It's not politically wise of me to say so, but they're right."

That would suit Joseph Fletcher, an ethicist at the Episcopal Divinity School in Cambridge, Mass. Fletcher argues that research on germ-cell gene therapy should move forward as rapidly as possible. Somatic-cell gene therapy, he points out, allows its beneficiaries to pass on their deadly trait, thereby "polluting" the human gene pool.

To make sure that scientists move cautiously, the federal government has already created one watchdog and is forming another. The RAC group has published a set of ethical guidelines that gene doctors must follow in setting up experiments. These provide protection for human subjects of somatic-cell gene therapy—and forbid germ-cell therapy altogether. And on Nov. 4, Congress approved legislation that would create a permanent Biomedical Ethics Board to keep Congress informed about the research and its ethical implications. "The ground rules we have to live by are now firm," says the NIH's Anderson.

Even when the technical obstacles are overcome and gene therapy takes its place alongside surgery and drugs, the ethical dilemmas will remain. But it may all come down to people like Tara Dew, who must endure little indignities that other children don't—like having to put on gloves to pet a puppy. Says Tara's mother: "Anything that will make Tara well, I'm for." In the end, the ethical debate may well give way to this choice of life over death.

By Roger Schulman in New York, with Teresa Carson in Mira Mesa, James R. Norman in Houston, Lois Therrien in Boston, and bureau reports

'AS LONG AS IT WON'T KILL HIM, WE'RE WILLING TO TRY ANYTHING'

Bob and Felice Weiner live in a tidy home in Mira Mesa, a suburb of San Diego. A huge portrait of four smiling young boys adorns the family-room wall. One of them, Craig, is now 22, and he lies in a nearby room strapped to his bed, a victim of Lesch-Nyhan syndrome. As a result of that genetic disorder, Craig is retarded, his body is wracked with palsy, and he has a compulsion to tear and bite at his own flesh. "The doctors told me he'd be dead by the time he was five," says Felice, smiling. "But we still have him."

Craig was born the same year Dr. William L. Nyhan and his colleague Michael Lesch, then at Johns Hopkins University in Baltimore, identified this rare inherited disease. It affects one out of every 100,000 births, and doctors can do little to treat the disorder, which results from the absence of an enzyme called HPRT. But Lesch-Nyhan is high on the list of candidates for gene therapy. The first attempts to cure the disease may be less than two years away.

Like many other Lesch-Nyhan children, Craig seemed normal at first. "When he was about three months old, I felt he wasn't progressing well," says Felice. From then on, small steps forward became triumphs: "It was a thrill to see him drink from a straw when he was five."

Caring for Craig is a formidable task—especially for Felice and Bob, who are themselves confined to wheelchairs. Felice suffered a stroke 15 years ago that severely restricts her walking, and Bob had both legs amputated last year due to circulatory problems. "It hasn't been an easy job," Felice admits. "Many nights I would sit in the middle of the floor and cry." Even when the couple was healthy, doctors recommended that Craig be placed in an institution. The Weiners refused. "The only one who can take him away from us is God," says Felice.

HOPE. Doctors were concerned that the Weiners' other three sons—one is 26 and two are 25-year-old twins—would grow to resent Craig. But the family has gotten along just fine, insists Felice. "When the boys played baseball, they'd make Craig 'umpire,' " she says.

The Weiners are eager for Craig to become one of the experimental patients to undergo gene therapy. "As long as it won't kill him, we're willing to try anything," says Felice. They have discussed the possibility with doctors, but the family is unsure if Craig will be selected. "We don't know what is next for Craig," says Felice. "But where there's life, there's hope."

'AS A KID, I WANTED TO BE LIKE ALL THE OTHER LITTLE GIRLS'

Chat with Joan C. Finnegan and you'll soon sense her success. At 25, she has a degree from an Ivy League university, a cozy apartment in Cambridge, Mass., and a fast-moving career as a systems analyst at Bank of Boston Corp. She works out every day, has little problem staying away from greasy and fatty foods, and enjoys an active social life. But there's an invisible side to this young executive. "I have what I call a hidden handicap," she says. Finnegan suffers from cystic fibrosis, a genetic disease whose victims rarely live into their late 20s.

CF is the nation's most widespread genetic disease, affecting 30,000 Americans. The disorder causes the body's mucous glands to produce an extremely thick secretion. This can block a pancreatic duct, causing incomplete digestion. More often, the secretions fill the lungs, interfering with breathing and ultimately attracting fatal infections.

Yet to a casual observer, a CF victim like Finnegan might seem like someone who enjoys a healthy diet (greasy foods aggravate the disease) and keeps in shape (frequent exercise breaks up congestion). "Thank goodness it's fashionable for women to exercise now," she laughs.

OUT OF SIGHT. Finnegan's parents adopted her as an infant, knowing she had CF, even though they had already lost a two-year-old daughter to the disease. Almost from the start, Finnegan spent a half hour each morning across her mother's lap, being pounded on the back to help her clear her lungs. Her brother John, also a CF victim, underwent the same daily therapy until he died in 1969 at age 15. "After a while, you feel you've got a death sword hanging over you," says Finnegan. "You wonder when it will drop."

For most of her life, Finnegan kept her disease out of sight. "As a kid, I wanted to be like all the other little girls," she says. Unlike most children, Finnegan was happy to get braces: They hid her discolored teeth, the result of a steady diet of antibiotics.

This year, Finnegan went public. "I finally felt comfortable enough to come out of the closet," she says playfully. She told her boss that she had CF and was assured that her excellent track record was all the company cared about. Not everyone is as understanding. "When I say, 'I have cystic fibrosis,' most people have no idea what I'm talking about," she says. "Sometimes I almost wish I had cancer, just so people would know I'm not contagious."

Currently there is no cure for cystic fibrosis. But researchers recently isolated a gene marker for CF, a distinctive section of DNA that can be used to develop a diagnostic test. Soon they will isolate the gene itself, and eventually gene therapy could be possible. "That would open up a whole new world of optimism," says Finnegan. Meanwhile, she is not brooding. "You've got to put it in a box and get on with your life," she says. "I'm not a disease. I'm a person."

THE GIANT STRIDES IN SPOTTING GENETIC DISORDERS EARLY

For all its promise, gene therapy will not be commonplace for years. But the same revolution in biology that is opening the way for doctors to cure genetic diseases is already helping them identify potential victims of those diseases. New diagnostic tests are making the heartbreaking experience of having a child with a serious genetic disorder more and more avoidable. And they are beginning to provide early warning of genetically linked health problems such as heart attacks.

In the past decade scientists have doubled the number of diseases doctors can spot prenatally. Some 200 can now be detected, and more are coming almost weekly. Routine tests can detect such genetic diseases as sickle-cell anemia, hemophilia, and fatal blood disorders known as thalassemias. And tests for killers such as Huntington's disease and cystic fibrosis are on the way. These are "blasting open new vistas," says Rita Douglas, a genetic counselor at the Uni-

FIVE DISEASES AT THE LEADING EDGE OF GENETIC TESTING

CYSTIC FIBROSIS

The most common of the lethal genetic diseases, it affects one in every 1,800 births in the U.S. Researchers at Integrated Genetics and Collaborative Research recently found a gene marker that could lead to a diagnostic test.

DUCHENNE'S MUSCULAR DYSTROPHY

The most common form of muscular dystrophy, it affects one out of every 3,000 males born in the U.S. Researchers at Children's Hospital in Boston and the Hospital for Sick Children in Toronto recently developed a prenatal test.

HUNTINGTON'S DISEASE

This genetic time bomb strikes its victims in their 40s. Now its potential victims—one in every 10,000 births—know only that they have a family history of it. A gene marker found by researchers at Massachusetts General Hospital could lead to a test within a year.

PHENYLKETONURIA

A genetic enyzme deficiency, PKU affects one in every 12,000 births. Victims' diet must be severely restricted through adolescence to prevent mental retardation. Researchers at Baylor College of Medicine have developed a diagnostic test.

TYPE A HEMOPHILIA

One in every 10,000 males has this disease, that prevents blood from clotting normally. Doctors at Johns Hopkins Hospital in Baltimore recently began testing fetuses with a diagnostic test developed by Genetics Institute in Boston.

---

versity of California at San Francisco's prenatal diagnosis facility.

TELLTALE SIGNS. The new tests are powerful tools for genetic counselors who spend most of their time advising mothers over 35, who are at risk of having babies with chromosome disorders such as Down's Syndrome. Genetic counselors rely mainly on family history and personal data to spot couples who might bear a child with a hereditary disorder. But the new tests can confirm counselors' suspicions—or set prospective parents' anxieties to rest—with an accuracy unheard of only a few years ago.

These gains have been possible because scientists are rapidly cataloging the more than 100,000 human genes. They have developed techniques that allow them to identify the gene and work out its structure. Even before they find the gene itself, they can often identify telltale sequences of DNA, known as gene markers, that indicate whether an aberrant gene is present. Both Lawrence Livermore and Los Alamos national laboratories are working on "gene libraries" that will eventually contain the structure of all the human genes.

While it will be years before that task is completed, the new information is already being put to use. Since both the genes themselves and the markers are unique sequences of DNA, scientists can make chemical probes that latch onto the target DNA.

Until recently, for example, many women known to carry a gene for hemophilia—the bleeding disease that occurs only in males—elected to forgo having children altogether. Or, desperate to be parents, they aborted the fetus if it was male—even though each boy had a 50% chance of being healthy. Now, genetic tests help conselors to determine whether the fetus has the disease. "When I think of the things I used to have to tell parents, I can't believe the progress," says Beth A. Fine, a genetic counselor at Michael Reese Hospital in Chicago.

That progress goes beyond helping couples have healthy children. One new company aims to incorporate the technology into programs directed at promoting good health. Focus Technologies Inc., set up by the Washington venture capital firm of TEI Industries with backing from Equitable Life Assurance Society, plans to market genetic tests for fitness programs.

The startup will package tests developed by such companies as California Biotechnology Inc. and Integrated Genetics Inc. to identify health-risk factors. The company will offer up to 100 tests that can tell a person's risk of getting everything from stomach cancer to heart disease. Focus Technologies will then tailor a nutrition and lifestyle plan to that person's genetic makeup. "Programs today insist that what's good for the general population is what everyone should do," says Nelson Schneider, managing director of TEI's medical ventures. But, he says, "we can find out which part of the population would really be helped by activities such as jogging."

When the pilot battery of tests was given to one healthy executive in his mid-30s, it turned up recommendations that varied greatly from those of his company fitness program—which told him he should avoid stress but otherwise continue to exercise. The gene test turned up a high risk of iron overload—a condition called hemochromatosis—as well as a strong possibility of developing stomach and intestinal cancer, heart disease, glaucoma, and anemia. What to do? Eliminate iron-rich foods and vitamins, reduce alcohol consumption, avoid ulcer medication, and have regular tests for glaucoma and anemia.

Although genetic screening would seem to be much more straightforward than gene therapy, it is actually every bit as controversial. Because parents may seek an abortion if they get bad news, most genetic screening is opposed by the anti-abortion movement. Moreover, the ability to screen for the susceptibility to disease rings warning bells in some. Should companies be permitted to screen workers and move those at risk for getting occupational cancers to low-risk jobs? Could they simply not hire those with certain genes? Could a genetic predisposition be grounds for denying life or medical insurance?

Most doctors, however, believe the benefits of the tests far outweigh the possibility that they will be misused. And genetic counselors are elated. The pictures of smiling babies on Rita Douglas' office wall are testimony to the fact that they can now confidently tell more parents-to-be that their child will be healthy.

By Joan O'C. Hamilton in San Francisco, with Reginald Rhein Jr. in Washington

Mapping the Genes, Inside and Out

SUMMARY: Many scientists are eager to decipher the entire human genetic code and to look inside the genes at their billions of nucleotides. Complete gene identification and nucleotide "sequencing" promise enormous benefits, including prevention or cure of genetic defects and discovery of new drugs patterned on the body's own chemistry. But the magnitude of the task is staggering — and so is the cost.

While Madeleine Bates was in high school and college, her mother was dying slowly of Huntington's disease, becoming increasingly demented as well as physically incapacitated.

Children of parents who have Huntington's chorea stand an even chance of succumbing to the disease, generally during middle age. "For the last 15 years, being at risk . . . has affected every major aspect of my life, including getting divorced, deciding not to have children, not remarrying and working for a large company instead of for myself," Madeleine told a congressional hearing on "Biotechnology: Unlocking the Secrets of Disease."

For her, hope was learning that scientists had found the gene that causes Huntington's. This meant that a test would soon become available which could determine whether she would develop the disease.

The test she hopes to take is a result of a revolution in genetics that could change the face of health care just as surely as antibiotics did in the 1940s. Since genes are the basic units of inheritance, identifying the ones responsible for a variety of genetic defects could lead within the next decade to drugs that could patch the defects or fortify the constitutions of those who are unusually susceptible to age-related diseases such as cancer and dementia. Ultimately, doctors may be able to replace defective genes with normal ones.

How fast all this becomes reality, says Nobel laureate Walter Gilbert of Harvard University, depends on whether the United States embarks on a costly quest for what he calls "the grail of human genetics": the mapping of the human genome, the genetic description of each human being.

There are an estimated 50,000 to 100,000 genes in the human genome. Each gene, in the form of DNA or RNA, is a blueprint for an individual protein. Proteins make up the structural molecules of the body, from skin and bone to eyeball, as well as the thousands of enzymes, the molecular machines that keep the body running.

Some of these enzymes are familiar. Hemoglobin is the blood-cell protein that transports oxygen from lungs to tissues. High-density lipoproteins remove cholesterol from the blood. Whole systems of enzymes turn food into energy. But many enzymes have more obscure roles, and like the genes themselves, most of the body's enzymes have yet to be identified.

The first step of gene mapping is to locate all the genes within their packages, the 23 human chromosomes. Some 1,400 genes have already been mapped, beginning in 1911 with the gene for color blindness. Other genes are being located faster than ever, and already this year approximate locations have been found for the defective ones that cause Alzheimer's disease, cleft palate, cystic fibrosis and, most recently, one that may be involved in the development of brain cancer. But without a national commitment, finishing the map might take decades instead of several years.

The benefits of this first step could be enormous. Most people are silent carriers of genes for at least three genetic disorders — "and that's a conservative estimate," says Dr. George H. Sack Jr. of the Johns Hopkins University School of Medicine. But most genes are like kidneys and lungs: One good one is enough. The danger is that people with the same defective gene might marry, passing a pair of that gene on to their offspring. A complete map would make it possible to screen for any genetic defect.

People also have varying genetic susceptibilities to diseases commonly associated with aging. Says one scientist, "It's quite likely that some people could smoke three packs a day for 80 years and stay healthy, while occasional side-stream smoke could cause cancer in others."

"We will find patterns of genes that are related to heart disease, cancer susceptibility and mental degeneration, and all this is going to happen over the next decade," says Gilbert.

With this information, medical science can investigate the merits of preventive measures for those who are at risk. "Suppose you say, if a person has these markers for hypertension, is it worth putting them on a low-salt diet? You can test them to see if low salt does any good," says Sack.

But for scientists to understand genetic diseases well enough to treat them, the interior of each gene must also be mapped, since that describes the construction of the chain of amino acid building blocks forming that gene's protein.

Sickle-cell anemia is the quintessential genetic disease. A single flaw in the gene for hemoglobin translates into one misplaced building block. That, in turn, distorts the shape of the red blood cell, destroying its ability to transport oxygen.

"Gene sequencing" — determining the interior makeup of a gene — would allow scientists to get down to the business of patching defects, says Sack. An emerging science of protein engineering will enable them to design special drugs that will function like orthopedic braces to make misshapen enzymes function properly.

More exciting — although fraught with ethical problems — is the potential for replacing defective genes with normal ones (**Insight**, Nov. 17). Gene replacement techniques being used on experimental animals are very crude, both because scientists understand only poorly the mechanisms that control the production of enzymes from genes and because they have no way to insert genes in their normal positions within the genome. Knowing the sequence could help solve these problems.

Reprinted by permission from *Insight*/David Holzman (5/11/87, pp. 52-54).

Such information would also help scientists understand cancer and find cures. "One of the keys to understanding cancer is to determine how normal cells and their cancerous counterparts differ from one another," Dr. Leroy Hood and Lloyd Smith write in the Spring 1987 Issues in Science and Technology. If the sequence were known, scientists could look up the differences.

In addition, they might stand to find thousands of potential new drugs within the body's arsenal of protective substances, says Gilbert. Some of today's most promising new drugs, such as interferon and interleukin-2, occur naturally in the human body, but in such minute quantities that they usually are overlooked. Even when they are found, chemical analysis to determine their structure can take years. With sequence in hand, pharmaceutical companies could computer-scan the genome for genes whose protein products resemble existing drugs. They also could look for new drugs, just as they currently screen chemicals for biological activity, but with better results, says Gilbert, because, "They would be going to the master list of chemicals that influence the human body."

Beyond this, the sequence could provide biologists with a tool to attack some fundamental questions. No one really knows, for example, how a body develops from an egg. "If you understood that in detail, you could regrow any part of the human body," says Gilbert. It will take decades — even after the sequence is established — to reach this point, but the benefits could include regrowing damaged limbs or organs.

The sequence might also provide valuable insights into how the brain works. "We don't understand what memory and cognition are at the molecular level, and it is just inconceivable that the genetic code doesn't have something very fundamental to do with that, as well as with feeling," says Dr. Victor A. McKusick of the Johns Hopkins University School of Medicine.

But the task of sequencing the interior of every gene in the genome is daunting indeed. The basic units of the genetic code are molecules called nucleotides, which can be thought of as letters of the genetic alphabet, linked in a spiral pattern described as a double helix. There are 3 billion of them in the genome. If each nucleotide were assigned a single letter, it would take thousands of volumes to record them all. The challenge of organizing this much information poses its own set of problems.

Inevitably, then, not everyone shares Gilbert's enthusiasm for sequencing the genome. The most prominent critic is David Baltimore, Nobel laureate and director of the Whitehead Institute for Biomedical Research, in Cambridge, Mass.

Baltimore fears that a campaign to sequence the genome would change the way

A Matter of Policy

Biotechnology holds great promise for treating genetically transmitted ills. But doctors will be able to diagnose genetic disorders long before they are able to treat them. This disparity, says LeRoy Walters, director of the Center for Bioethics at Georgetown University, will create serious ethical problems.

Chief among these is the issue of how health insurance companies might deal with genetic information on individuals. Those with bad genes might find themselves shut out of the insurance pool, Walters fears.

"Think of the new presymptomatic screening programs set up at Johns Hopkins and Massachusetts General hospitals for Huntington's disease," he says. "In some cases those centers can tell individuals that they have a 95 percent chance of developing Huntington's disease when they reach the age of between 35 and 45. It would certainly be prudent for the individual to have disability insurance."

Naturally, the insurance company would want to know the individual's risk. "We may have to come to a judgment as a society that all insurance companies should play on the level field of not having access to genetic information about individuals," says Walters. "Otherwise, I see serious problems of particular individuals not having access to various types of insurance. If there is anything for which we are not personally responsible, it's the genome that we inherited from our parents."

Another problem arises from the fact that some people's genetic make-up makes them more susceptible than others to chemical hazards of the workplace. Should companies screen potential workers for such predispositions?

One scientist tells the story of a factory that screened applicants to make sure they had an adequate level of an enzyme thought to detoxify a chemical found in that workplace. A certain minority group tended to be deficient in the enzyme, says the scientist. These people were excluded from employment and sued the company. Worse yet, the enzyme was soon discovered to offer no protection from that chemical.

biological research is funded. The decentralized nature of the current system bears similarities to a market economy in that it provides opportunities for many creative projects to gain funding. Individual researchers take their ideas to such agencies as the National Institutes of Health, which fund them on their merits.

Baltimore fears that a megaproject to sequence the genome could turn research funding into political pork. "A senator is unlikely to fight for a $100,000 grant to be awarded in his state, but will certainly fight for a $50 million project," he writes. Sequencing the genome would cost an estimated $300 million to $3 billion.

In addition, Baltimore fears that the cost of sequencing, an amount heretofore unheard of in biological research, would be so great as to cut into the funding of independent research projects, most of which, he says, have considerably more merit than the sequencing project.

But Hood thinks that the project instead could draw new funds to biology. "I think attitudes like Baltimore's will end up hurting the community. I think this is an idea whose time is coming."

Others object that having the complete 3 billion nucleotide sequence is unnecessary to learning how individual genes work; scientists can sequence important genes as they find them. But, argues Sack, "Having to sequence every time you find an interesting gene is like having to build the motor every time you want to drive the car."

The controversy moved into the high-stakes area last year as several government agencies, Congress and the private Howard Hughes Medical Institute of Bethesda, Md., became extraordinarily interested in human gene mapping and sequencing.

It all started with a conference the Department of Energy held in Santa Fe, N.M. Periodic meetings soon followed among other agencies and the department, which became involved in the field because of the effects of radiation on the genome. Various initiatives are taking shape. For example, Energy has earmarked $5 million to map three chromosomes and hopes to go on to map the entire genome.

The president's Domestic Policy Council is developing recommendations for coordinating research and funding of gene mapping and sequencing among the agencies. Although the report is not expected until summer, Dr. Robert Cook-Deegan of the Office of Technology Assessment says that funding is expected for mapping the positions of all genes but not for sequencing the nucleotides in them. "The Domestic Policy Council has approved in principle a document which suggests that gene mapping be funded at $40 million a year, rising to $200 million by 1993 and continuing until $1 billion has been spent."

Gilbert says he is taking the matter of sequencing into his own hands. The founder of one of the early biotechnology companies, Biogen Inc. of Cambridge, Gilbert has formed a new company specifically to sequence the genome over the next 10 years. He says he already has some venture capital toward the $300 million he anticipates the project will cost.

Although many are skeptical of Gilbert's project, they agree that sequencing is inevitable. It won't happen this year or next, but it will happen soon because technology has just begun to automate the process. The cost of sequencing by manual laboratory methods is $1 per nucleotide, or $3 billion for the entire genome. But Hood, of the California Institute of Technology, has invented a machine that he hopes will ultimately cut the cost per nucleotide down to a penny. The Department of Energy, always strong in developing scientific instruments, plans to back this kind of research and development.

Hood is cautionary, though, about the ability of the United States to compete successfully with the Japanese, who have already developed machinery that can sequence for 17 cents per nucleotide.

"It's well known that the U.S. is the best in the world at biotechnology in most of its forms," says Hood. "We are getting to a phase where the next developments are going to depend on money, resources and people, and there is a question of whether the U.S. is going to maintain the commitment to retain the lead. The Japanese have clearly made the commitment. At the rate we are going, we are going to become a second-rate player."

— *David Holzman*

The Natural History of AIDS

The disease may have existed in isolated humans for thousands of years

Matthew Allen Gonda

Matthew Allen Gonda, Ph.D., is head of the Laboratory of Cell and Molecular Structure, Program Resources, Inc., at the National Cancer Institute–Frederick Cancer Research Facility. He has authored numerous scholarly papers on retroviruses, most recently on the AIDS virus.

In 1984, a previously unknown virus was isolated from human blood. Named HTLV-III (human T-cell lymphotropic virus type III), the virus selectively attacked a specific group of white blood cells crucial to the body's immune response. Soon generally recognized as the causative agent of acquired immunodeficiency syndrome, or AIDS, the virus was later discovered to have an affinity for infecting cells of the brain as well.

Although the AIDS disease process has proved to be enormously complex and often baffling—there is no complete parallel for it among the other viral diseases of humans—we have learned a great deal about the virus's molecular biology and structure in a very short time. Structurally and biochemically, HTLV-III belongs to the retrovirus family, a unique subgroup of viruses found not only in humans but also in many animals, from reptiles to primates. Like other viruses, retroviruses don't always cause disease in their hosts.

Also like other viruses, retroviruses are not really living organisms. Lacking the machinery and the energy-generating capabilities to manufacture progeny, they are perhaps best described as infectious chemicals made up of a sticky protein coat encapsulating a genome (the DNA or RNA blueprint for constructing more viruses). Incapable of growth and division on their own, viruses exploit the cells of living organisms to perform these functions for them. Infection occurs when, via highly specific receptors on its protein coat, a virus attaches itself to and penetrates a susceptible cell. Once inside, it is read and reproduced by the host's manufacturing machinery. Sometimes the cell is killed during virus replication; but before its demise, it has released a new generation of viruses into the host's system.

Retroviruses have evolved a particularly effective variation on this parasitic theme. Unusual because their genomes are composed of RNA (in most living things, including most viruses, genomes are composed of DNA), retroviruses also possess a gene for a unique enzyme, reverse transcriptase. When the retrovirus attaches itself to and penetrates a cell, reverse transcriptase transcribes the retrovirus's genetic information from RNA into DNA. The host, often perceiving this new DNA to be its own genetic material, integrates it into its own chromosomes. Once in this new habitat, the retrovirus may be reproduced or it may remain dormant for weeks, months, or even years. The virus stays in the chromosomes for the life of the cell, that is, until the cell has been killed by the infection, eliminated by the immune system, or removed after senescence. The association is permanent; every time host cells reproduce, they also reproduce retrovirus DNA, even in the absence of new virus.

Some mouse and chicken retroviruses have assured themselves of even longer relationships with their hosts. Because in a past event they infected and were integrated into the host's germ cells (that is, sperm and egg or their precursor cells), they are now automatically transmitted to the next generation of host animals without an infectious cycle. There are no known methods of eliminating these so-called endogenous viruses.

Other retroviruses are exogenous—that is, acquired from the outside. The AIDS virus, passed from person to person (or from pregnant woman to fetus) via infected blood or body fluids, is of this type. Exogenous or endogenous, however, retrovirus infections have one feature in common; infected individuals remain infected (though not necessarily ill) for life.

Before the discovery of the AIDS virus,

only two other retroviruses had ever been isolated in human beings. These—the human T-cell leukemia viruses, HTLV-I and II—belong to the oncovirus subfamily of retroviruses, so called because they are oncogenic (tumor producing in their host). Like the AIDS virus, they attack T-4 lymphocytes, the white blood cells that begin the immune reaction. The question therefore arose early as to whether the AIDS virus was also an oncovirus. At first, the idea seemed plausible, because of the properties shared with the the leukemia viruses, the most prominent of which was their affinity for T-cells. In addition, the AIDS virus was suspected of causing Kaposi's sarcoma, a rare cancer of the skin's blood vessels, from which many AIDS victims suffer. Further investigation, however, made it clear that HTLV-III did not directly cause Kaposi's sarcoma. Rather, the tumors were arising opportunistically because of the underlying immune deficiency, just as they do in organ-transplant patients who are given immunosuppressive drugs.

If the AIDS virus was not an oncovirus, what was it? Investigators began to look for similarities in the two other known retrovirus subfamilies—the lentiviruses and the spumiviruses. (There was also the possibility that it belonged to a new group of retroviruses not previously identified.) The spumiviruses, or foamy viruses although they had not been thoroughly studied, were ruled out quickly; they were not known to cause disease, and structurally, they differed sharply from both the leukemia viruses and from the AIDS virus.

Important clues to the identity of the virus were already apparent, however. Most important was that the AIDS virus did not cause cancerous proliferations but instead brought about cell-killing (cytolytic) events. This cytolytic propensity is one of several distinguishing properties of the lentiviruses. Called "slow" viruses for their slow but persistent rate of replication, the lentiviruses eventually induce debilitating diseases, although years may pass between the initial infection and the onset of symptoms. Since the AIDS virus also is associated with the slow evolution of a lethal debilitating disease, this was a second family resemblance. Firmer evidence came from electron microscope pictures; HTLV-III strikingly resembled the visna virus, a lentivirus that infects sheep.

Lentiviruses had been isolated from a variety of ungulates—sheep, goats, horses, cows—that have been closely associated with humans for thousands of years. Visna virus—grouped with maedi

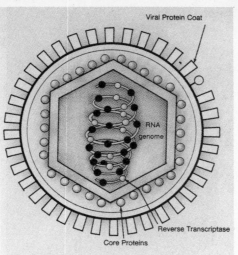

Viral Protein Coat

RNA genome

Reverse Transcriptase

Core Proteins

General retrovirus structure
Joe LeMonnier

and progressive pneumonia viruses, two related retroviruses of sheep—was the first lentivirus to be isolated and the first to be intensively studied. *Visna,* the Icelandic word for "wasting," was the name given to the sheep disease when it suddenly appeared in Iceland in the 1930s. Like the AIDS virus, visna virus induces a complicated disease syndrome. The signs in sheep included lymphadenopathy (infected lymph nodes), encephalitis (brain inflammation), wasting, and susceptibility to infections, the most common of which was an acute pneumonia caused by a bacterium that probably resided in Icelandic sheep populations before visna virus came along.

Lentiviruses that have since been identified in other animals induce a variety of disease syndromes. Caprine (goat) arthritis encephalitis virus, which is genetically very closely related to visna, causes crippling arthritis, paralysis, and encephalitis in goats. Horses are vulnerable to a lentiviral agent called equine infectious anemia virus, which causes intermittent anemia, bouts of fever, and immune-complex glomerulonephritis, an inflammatory disease of the kidneys occurring secondary to the infection. The lentivirus of cows, bovine visnalike virus, also affects the lymph system and causes persistent lymphocytosis, an excessive production of white blood cells.

When observed under an electron microscope, all of the lentiviruses, including the AIDS virus, share a common physical structure. Each infects cells of the immune system, although the specific target cell and the level of interference with the host's immune response differ from species to species. Visna and caprine arthritis encephalitis viruses seem to attack the large white blood cells, the monocytes and macrophages. These cells normally de-

vour foreign bacteria and cellular debris and are a first line of defense against infection. Besides attacking the T-cells, HTLV-III also infects monocytes and macrophages, as well as antibody-producing lymphocytes. Whether other cells of goats and sheep are affected by lentiviruses is not known, since their immune systems have not been as intensively studied as that of humans.

Further analysis of the relationship between the AIDS virus and visna virus awaited direct comparison of their genetic sequences. For if it could be proved that the AIDS virus is genetically related to the lentiviruses, some of the disease's mysterious processes would begin to make sense. DNA hybridization using cloned DNAs of the viruses, an effective way of grossly estimating genetic relatedness, revealed that the AIDS virus and the lentivirus resemble one another even on the very basic level of their DNA sequences. Of the several genetic likenesses investigators saw, the most dramatic was the similarity in the gene for coding reverse transcriptase. This gene, in fact, has changed the least in the evolution of retroviruses, and virologists now depend upon it to determine phylogenetic information for the group. Overall, the AIDS virus and visna virus had significantly more DNA sequences in common than either did with any oncovirus tested, including HTLV-I and HTLV-II.

By this time it was evident that HTLV-III and visna virus were close cousins. But the question of whether HTLV-III was also related to other lentiviruses awaited testing of other representative species. Equine infectious anemia and caprine arthritis encephalitis virus were subsequently cloned and showed an equal amount of likeness with HTLV-III. Clearly, the AIDS virus was a lentivirus.

Final confirmation came from DNA sequencing, which allows nucleotide-by-nucleotide comparisons of the reverse transcriptase gene and the rest of the virus genome. (Each nucleotide represents a single letter of the genetic code.) It demonstrated that the genomes for HTLV-III and lentiviruses were similar in organization and coded for similar sets of genes in the same order and location. This information was important, because sequencing determines how the virus is assembled, how it works, and what it looks like.

In 1985, not long after the structural and genetic studies were reported, another important clinical manifestation of AIDS was recognized. Physicians began to realize that neurological signs and symptoms that they had been seeing in AIDS patients—chronic meningitis, dementia, encephalopathy, loss of motor coordination, and paralysis—were caused directly by the AIDS-virus infection. The findings suggested that HTLV-III was attracted to brain cells as well as to white blood cells. In retrospect, in view of the virus's demonstrated close association with the visna and caprine retroviruses—both of which cause neurological disease—the findings should not have been that surprising.

A great deal was being discovered about HTLV-III, but much of it boded ill for the development of a vaccine. On the one hand, since the virus was exogenous (transmitted from outside), there was an inherent "weakness" in its replicative cycle that could be exploited. Uninfected persons could theoretically be protected via vaccination, as has been done with other horizontally transmitted viral diseases, such as measles, mumps, and smallpox. But the AIDS virus is a retrovirus, and to date an effective vaccine has been made for only one retrovirus, feline leukemia virus, a cancer-causing retrovirus of cats. Although this vaccine has not totally contained the disease—probably because some apparently healthy cats had already been infected—its existence at least raises the possibility that a successful human retrovirus vaccine can also be developed.

Normally, the host immune system counters infection by making protective antibodies that are specially adapted to adhere to and destroy a specific attacking virus. Lentiviruses, however, have developed novel strategies to avoid elimination by the host. Visna virus and equine infectious anemia virus, for example, undergo rapid changes in the gene responsible for their characteristic protein coat. This capacity for rapid change, called antigenic drift, produces variants of the virus that are not recognized by the host's protective antibodies, which were effective in neutralizing the original strain. The variant viruses thus escape destruction and can continue to infect and, sooner or later, to induce a new cycle of disease. (An analogous process takes place in the envelope of the influenza virus and has created a major stumbling block in obtaining a single effective flu vaccine.)

Caprine arthritis encephalitis virus has another means of evading destruction. It evokes a very weak immune response; the antibodies that do respond seem to do so only halfheartedly and do not kill the vi-

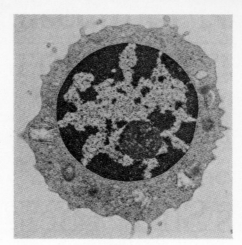

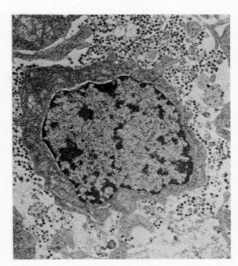

Top: A normal T-4 lymphocyte. Bottom: A T-4 lymphocyte that has been attacked by the AIDS virus.
Kunio Nagashima and Matthew Gonda

rus. The AIDS virus seems to act similarly in this respect, and even though antibodies are present, they do not appear to prevent severe disease or predict survival for the patient. Additionally, the envelope gene of HTLV-III is quite variable, indicating that both of the mechanisms described may be at work.

Effective vaccines have not been made for any lentiviruses, so that producing an AIDS vaccine is no trivial task. Any knowledge gained about lentivirus disease in animals will contribute to the effective control of AIDS.

HTLV-III was presumably introduced into the United States in the 1970s, and AIDS was first recognized clinically in 1981. Although no one knows whether the HTLV-III-induced syndrome is a new disease or where it came from, serologic data are now accumulating to suggest that the virus was in Africa at least a decade before it came to the United States, probably via Haiti. What we don't know is whether the virus was present in humans before the first documented evidence or whether it came from an animal reservoir.

We can only speculate on these possibilities. If the virus was widely present in humans before that time, it must have gone through a genetic change that made it more pathogenic. But there are no data at present to substantiate the coexistence of pathogenic and nonpathogenic forms of the virus. It is hard to believe that the nonpathogenic version could have died out in the few years since giving rise to a pathogenic form.

Another possibility is that there exists a lentivirus family group in animals that resembles the AIDS virus even more closely than already identified lentiviruses and that a virus in this group gave rise to a human variant. A newly isolated virus, called STLV-III (simian T-lymphotropic virus type III), causes an AIDS-like syndrome in the macaque monkey. The simian virus resembles the AIDS virus in growth characteristics and structure, and it is attracted to similar cells. These data suggest that STLV-III may also be a lentivirus. Moreover, the presence of strongly cross-reactive antibodies to STLV-III in the blood of apparently healthy wild African green monkeys suggests that the virus is not disease producing in one species and quite pathogenic when it is transmitted to another—in this case, the macaque.

How close is the relationship between the simian virus and the AIDS virus? There is at least the possibility of a monkey retrovirus giving rise to a human variant. Human leukemia virus type I, for instance, has a correlate in a simian virus (simian leukemia virus type I), to which it is remarkably similar in terms of DNA. There have been no direct comparisons of the AIDS-virus DNA and its simian counterpart, but serological analyses have provided some evidence that the AIDS virus may be closer to STLV-III than it is to other lentiviruses. However, the two are not nearly so closely related as the human leukemia virus type I and the simian leukemia virus type I. Even if STLV-III crossed from monkeys to humans, it is unlikely to have diverged so much in such a short time, to become HTLV-III.

A better analogy may be found in the sudden appearance of visna virus in sheep, first described by Bjorn Sigurdsson, a physician. Before 1933, visna was unknown in Iceland. That year, the government purchased twenty karakul sheep from a farm near Halle, Germany, where a visnalike virus was endemic. (The disease in Germany was less severe than what was later seen in Iceland.) When the sheep arrived in Iceland, they were put into quarantine for several weeks and then distributed to farms scattered all over the country. At least two of the introduced breed apparently carried the infection at the time of quarantine because by 1935, there were outbreaks in two widely separated districts. Until 1939, however, no one realized that the disease was an entirely new entity. The losses were enormous. Between 1939 and 1952 at least 150,000 animals died of the infection.

Between 1949 and 1951 all the sheep in the southern part of the island were destroyed in an attempt to control the virus, and ultimately the disease was brought under control. It is now known that visna can be easily transmitted from ewe to kid during feeding, especially through the virus-laden immune cells found in the colostrum, the fluid secreted by the mammary glands before milk appears. Chance abrasions of the skin or mucous membranes are other possible modes of entry, as are the bites of blood-sucking insects, which are implicated in the spread of equine infectious anemia. It could also be that centuries of isolation made Icelandic sheep particularly susceptible to visna.

If HTLV-III is not a new virus, has not recently jumped into the human species from an animal reservoir, and is not a mutation of a known nonpathogenic virus, what plausible explanation can account for its sudden appearance in humans? Drawing on the parallels with visna, we can make some educated guesses. It is possible that the virus has existed in humans in central Africa for hundreds of thousands of years, but that it resided in an isolated population. Such isolated groups may have coadapted with the virus, lessening the severity of the infection and allowing for mutual coexistence. The persistence of HTLV-III in this scenario may lie in old customs such as scarification and the sharing of needles used for body marking. Parallels exist in the spread of kuru, a slow-acting and fatal viral disease of the nervous system. Kuru is found only in a specific tribe in New Guinea and is spread exclusively by rites associated with cannibalism.

Demographic factors, too, may have a bearing on the spread of AIDS. In the past thirty years, Africa's tribal and geographical boundaries have broken down as individuals moved toward cities for a variety of reasons. Such changes could have brought an infectious agent into contact with previously unexposed populations, both international and local, and the devastating effects of the virus would have been felt more readily, as when visna virus was introduced to Iceland.

Such a pattern of sudden virulence has often been seen when other pathogens have reached unexposed peoples. Examples include the fatal measles epidemic in the Faroe Islands (1781) and the Fiji Islands (1875), and the devastating effects of smallpox on American Indian populations after contact with Europeans in the sixteenth century. Anthropological studies in central Africa may provide further insights into such a scenario.

Fortunately, the AIDS virus, unlike smallpox or measles, is not easily transmitted, and unlike some other retroviruses, it cannot be transmitted from generation to generation. Therefore, even in the absence of a vaccine, we can expect that preventive measures already in place can effectively prevent the spread of the disease.

WHAT TRIGGERS AIDS?

Could other infections or genes boost a symptomless AIDS infection into full-blown AIDS?

JOANNE SILBERNER

One of the great mysteries of AIDS is why some people who have been infected by the AIDS virus go years — if not their lifetimes — without developing the syndrome. Many AIDS researchers believe one or more additional elements, or cofactors, are necessary to turn an AIDS-virus infection into actual disease.

According to the U.S. Public Health Service, about 1 million to 1.5 million people in the United States are infected by human immunodeficiency virus (HIV), and roughly 20 to 30 percent of them will develop AIDS within five years. Who among the infected individuals will get the syndrome and when that will happen are open questions. Finding a cofactor would enable physicians to identify these people and possibly show how to prevent the progression from infection to illness.

Among the many possible cofactors that have been proposed, two of the strongest candidates are the presence of specific, genetically determined proteins in the infected individuals, and exposure to other viruses. If the virus co-infection hypothesis, whose proponents include researchers from the National Institute of Allergy and Infectious Diseases (NIAID) in Bethesda, Md., is true, avoidance of a second virus could be the key to health. But a genetic predisposition, as suggested by researchers at the University of California at San Francisco, would be more difficult to counter.

With most viruses, infection does not always mean a person becomes sick — for example, the majority of people infected with hepatitis B virus or with poliovirus don't develop symptoms. But while cofactors are evidently an element in these and other serious viral infections, there has not been a lot of research into the issue, says epidemiologist Harold Jaffe of the Centers for Disease Control in Atlanta. Questions about cofactors "could be asked for lots of other diseases," he says. The sudden, mysterious and deadly onset of the AIDS epidemic has lent the question "a sense of urgency," he says.

Because many members of the two highest-risk groups, male homosexuals and intravenous drug abusers, have histories of frequent sexually transmitted or blood-borne diseases, some researchers have been investigating whether a second infection can somehow "awaken" the AIDS virus. Recent results from Malcolm Martin and his co-workers at the NIAID provide biological support for the possibility.

Martin, Howard E. Gendelman and their co-workers studied the interaction of HIV and other viral infections in cells growing in culture. To avoid the hazards of working with the entire AIDS virus, they used only a segment of HIV's genetic material, linked to a bacterial gene that directs the construction of an easy-to-test-for enzyme.

Martin and his colleagues introduced the combination genes into a cell line and followed its activity by monitoring the marker enzyme. When they added any one of several viruses that commonly infect people, they found more of the marker enzyme, indicating that the AIDS virus material was much more active. Martin says subsequent experiments using the entire AIDS virus have confirmed the initial results, which were published in the December PROCEEDINGS OF THE NATIONAL ACADEMY OF SCIENCES (Vol.83, No.24).

The viruses, Martin says, could push the AIDS virus in an infected person from a quiet to a lethal stage. "By simultaneous infection, there's a real possibility [of] inducing or activating latent virus."

The viruses used in the experiment are so different from one another that they couldn't possibly all be acting in the same way, he says. Rather than all the viruses producing an identical protein that travels to the AIDS virus and causes it to reproduce, Martin suggests the non-HIV viruses somehow induce the cell itself to stimulate HIV, perhaps by making the cell produce an HIV-stimulating protein.

Several laboratories, including Martin's, are searching for such a protein. Unfortunately, if the infected cell's own protein is responsible, interceding in the process may be difficult. "They [the proteins] are probably there for some important normal function," says Martin. Interrupting that function to keep the proteins from stimulating the AIDS virus could cause other problems. "The more we know," he says, "the less we know."

On the other hand, the cell may also be capable of producing other proteins that inhibit the system, Martin suggests. If so, stimulating those proteins could keep HIV quiet. And whatever the mechanism of action of other viral infections, if they are what's kicking off HIV, avoiding them would be a way to avoid getting AIDS.

While Martin's theory holds that a second infection kicks off AIDS, John Ziegler and Daniel P. Stites of the University of California at San Francisco suggest that the cofactor is a genetic one. They base their theory on the paucity of active AIDS virus found in full-blown, or "frank," disease.

"It's very difficult to find infected lymphocytes [white blood cells] in infected blood," Ziegler says. In frank AIDS, only 1 in 10,000 to 1 in 100,000 lymphocytes show evidence of viral infection.

To explain how so few viruses could

cause such a devastating disease, Ziegler and Stites have suggested that the virus sparks an immune reaction that attacks not only the virus but also the body's own healthy cells (SN: 12/20&27/86, p.388). According to the theory, what controls whether this autoimmune reaction occurs is the degree of similarity between certain immune-system components and HIV itself, and what determines the similarity is genetics.

The AIDS virus attacks and infects the CD4 cell, a type of white cell, at the location where the CD4 normally "docks" with other cells in the immune system. This docking process is a necessary step in a cascade of events that results in the recognition and neutralization of foreign substances.

In order to attach to the CD4 dock, Zeigler and Stites suggest, the virus must in some way "look" like the second set of cells. And this similarity results in the virus affecting the immune system not only by destroying the cell it infects but also by generating antibodies that attack the immune system in two separate ways.

First, antibodies to the virus also attack the cells that normally link up with the CD4 cells, since the virus and the second set of cells have something in common. According to the hypothesis, these antibodies block the interaction of the CD4s and the other cells — even though neither may be infected by the virus. Second, the virus-prompted antibody also triggers the production of other antibodies against both itself and the CD4s, again including those that have not been infected by the AIDS virus. As a result, an entire and vital arm of the immune system is wiped out.

"In this way," says Ziegler, "just a handful of HIV could kick off immune system self-destruction."

Genetics comes into play because the proteins on the immune system cells to which the CD4s attach differ from person to person, and these proteins are inherited. People whose proteins "look" like proteins on the surface of HIV would develop the two sets of antibodies that attack the immune system, and go from infection to full-blown AIDS. People whose proteins differ markedly from the HIV strain would be spared.

If the hypothesis is proven true, it has both positive and negative implications for therapy. The immune self-attack aspect suggests that toning down the immune response could help. Therapeutically, "you'd want to think of ways to remove antibodies to see what happens to patients," says Ziegler. French researchers already have tried damping the immune response with cyclosporine, and a small U.S. trial with cyclosporine began recently.

But it would also throw a wrench into vaccine development. If the part of HIV that is similar to the antigen-presenting cells were used as a vaccine, the antibodies generated against the vaccine material would also be capable of attacking the antigen-presenting cells themselves. Such a vaccine would have the unfortunate result of destroying a normal, necessary arm of the immune system.

Two discoveries would help prove the genetic hypothesis: identifying a single antibody that attacks both HIV and the cells to which the CD4s attach, and the preponderance in AIDS patients of particular classes of proteins on white cells that differ from those in people who are infected but have not developed the syndrome. Collaborators of Ziegler's at UCSF are now in the process of looking for similar classes of proteins among people with AIDS, and there have already been several reports from other laboratories indicating that such clustering exists. Ziegler's collaborators and other U.S. laboratories are also checking an antibody against white cells found in people with AIDS to see if it attacks HIV as well.

"My guess is that everybody who is exposed is capable of being infected, but the progression to illness may well reside in immunogenetic mechanisms," says Ziegler. "Obviously everything isn't going to be explained by genetics. But if it lies there we should be able to find it."

Ziegler's and Martin's theories aren't mutually exclusive — they could each be at work in different people. Nor are genetics and viral infections the only candidates that have been suggested. Ziegler, in fact, has worked with UCSF's Jay Levy on a study showing that some people have a white blood cell capable of suppressing HIV activity. This cell could be producing a protein that counteracts the co-infection effect of Martin's hypothesis.

Other research has pointed to the frequency of AIDS among infected individuals after they were exposed to herpesviruses or hepatitis B. With millions of people infected but not yet showing signs of illness, the problem is more than academic.

But for the moment, what causes infection to develop into AIDS, says Ziegler, "is a biological black box."

The epidemiologic viewpoint

The first clues about the nature of AIDS came from epidemiology. When epidemiologists chronicled the emergence of the syndrome among male homosexuals, intravenous drug abusers and hemophiliacs, their findings suggested that an infectious agent carried by blood or other body fluids was at work. The biologists eventually found a virus.

Epidemiology has offered no solid leads on whether the virus needs a boost from a cofactor in order to cause disease, nor does it provide clues as to what that cofactor might be. But the suspicion of a cofactor is strong enough to have prompted a search for a common behavioral or lifestyle thread among people with AIDS that is absent in healthy infected individuals.

Studies done to date have not identified any causative cofactors. Epidemiologist Harold Jaffe and his colleagues at the Centers for Disease Control in Atlanta, along with researchers from several San Francisco institutions, are conducting an ongoing study of thousands of homosexual men in San Francisco. Information has been collected about the men's behavior, drug use and health. "So far none of these has really been predictive of illness," says Jaffe.

Of 104 men in the study who have gone from being antibody-negative, indicating they had not been exposed to the AIDS virus, to being antibody-positive, 15 percent were stricken with AIDS five years after infection. After seven years, 35 percent of the original group had AIDS, but no common behavior or health factor could be identified in the group that became ill. "The best predictor of illness seems to be the duration of infection," Jaffe says.

While the study argues against a single obvious cofactor, it doesn't rule out cofactors entirely. "It might be that there are a variety of things that slightly increase your chances of being sick," says Jaffe. "Instead of one thing increasing your chances by 50 percent, there might be 50 things that increase your chances by 1 percent, and it's not very likely we'll be able to sort out what those things are."

Proposed cofactors such as genetic proteins or co-infections identified in laboratory research are not necessarily at work in the real world, says Jaffe. "I have no doubt people have already found and will continue to find factors or agents that will activate the infection in the laboratory," he says. "But you don't know if these things are clinically relevant.

"In a sense I think we want to believe there's something else," says Jaffe, "because if there were something else, it would give us a potential intervention. . . . We'd like to believe there is such a thing, but so far nobody's found one."
— *J. Silberner*

The Environment

Go from the creatures thy instruction take: learn from the birds what food the thickets yield; learn from the beasts the physic of the field; thy art of building from the bee receive; learn of the mole to plugh, the worm to weave; learn of the little nautilus to sail, spread the thin oak, and catch the driving gale.

Pope

During the 1980s, the environmental situation predicted by environmental scientists has begun to emerge in a number of ways: population/food imbalances, inflation brought about by energy resource scarcity, acid rain, toxic and hazardous wastes, water shortages, major soil erosion, global atmospheric pollution, and higher rates of plant and animal extinction. Just the past three years have seen drought and famine in Africa, a major environmental chemical accident in Bhopal, India, and a near meltdown of a nuclear power generator in Chernobyl in the Soviet Union. These and other problems have surfaced in spite of the increasing environmental awareness and legislation of recent years. They have resulted, in part, from a curious environmental situation which has favored the short-term expedient approach to problem solving over long-term ecological good sense. People will not be protected against ecological consequences of human actions by remaining ignorant of them. Hopefully, the articles in this section will provide a better understanding of some of the pressing issues at hand.

The growth of population is related to virtually everything else that happens in and to the environment, and the most basic of all the issues dealing with human populations is that of food supply. Several articles in this section address this problem. In the mid-1960s much attention was focused on the projected explosion of the world's population that was supposed to produce massive famine globally by the mid 1970s. This did not occur, but as Paul and Anne Ehrlich point out in "World Population Crisis," it is just a matter of time before it does. They note that although the rate of population growth has slowed somewhat, the absolute numbers of people that will be added to the world's population in the next generation will far exceed those added in the past. Among many scientists there is a generally accepted notion that there are already more humans alive than the earth's resources can tolerate. Increases in the world population necessitate creative new strategies of food production in those parts of the world not previously impacted by the "green revolution." "Beyond the Green Revolution" and "Mimicking Nature" report on the progress of agricultural research and advances that may provide new hope for food production and economic growth for a quarter of the world's population.

As the world's supply of inexpensive petroleum dwindles, it becomes increasingly clear that the energy dilemma is the most serious economic and environmental threat facing the Western world and its high standard of living. In Harry Bacas' article "Energy: A Promise Renewed," he argues that solutions to the energy scarcity problem will be largely technological.

The National Wildlife Federation presents an annual review of the condition of the resource base of the United States. In their "Nineteenth Environmental Quality Index," the editors point out that, in spite of the series of environmental laws passed over the last two decades which have contributed to cleaner air and water, many major resource problems still remain in the United States, one of the most developed nations in the world. The article "Acid Deposition: Trends, Relationships, and Effects" discusses, in a similar fashion, the causes and effects of acid rain on the United States.

Looking Ahead: Challenge Questions

How do poor farming techniques in both the developed and the developing nations of the world contribute to soil erosion? What kinds of strategies might be devised to minimize the effects of cultivation on increasingly fragile soil systems?

How has the traditional view of abundance of US resources contributed to the present resource problems, particularly in terms of the quantity and quality of water?

Unit 7

Energy: A Promise Renewed

Many alternatives to fossil fuels
no longer seem so exotic or expensive.

Harry Bacas

FORTY MILES EAST of San Francisco, following Interstate 580 up and over the grassy slopes of Altamont Pass, the unsuspecting motorist comes upon a sight that would make Don Quixote rub his eyes.

Hundreds of giant windmills, ahead and to the left and right, some like egg-beaters, some with pale arms that reach more than 200 feet in the air, turn endlessly above the grazing cows. They are generating electricity for homes and businesses in the San Joaquin and other California valleys.

The fantastic wind machines represent one of the alternative energy technologies being exploited to augment—and perhaps some day replace—the world's dwindling supply of conventional sources.

Energy experts anticipate that oil, gas, coal and nuclear will be the primary sources of power far into the next century, but research on alternatives continues.

Some alternatives, like wind and geothermal power, are well developed although confined to only a few geographical areas. Some, like photovoltaics, have a wider application and are considered on the verge of a general breakthrough.

One utility company, Southern California Edison, a leader in exploring alternative sources, expects to produce one third of all its new power in the next five years from wind, geothermal, solar, biomass—principally wood—and hydroelectric sources.

Throughout the country an increasing amount of electricity is being generated as a byproduct of other industrial activities. Much of this "cogenerated" power, which Congress in 1978 ordered utilities to buy from its producers, still uses conventional fuels. But one kind of cogeneration—the fastest growing—uses municipal waste to produce electricity.

It is predicted that 100 such plants will be built in the next decade, at a cost of $15 billion and generating revenues of $4 billion.

"This business is really taking off," says David Sokol, president of Ogden-Martin Corporation , one of the largest trash-to-electricity companies.

Ogden finances, builds and operates plants that burn trash and garbage so completely that they produce no dioxins or other air pollutants, create no odors in their neighborhoods, leave a 5 percent residue of ash that is used in road building, and generate industrial grade steam and electricity.

Temperatures of 2,800 degrees destroy harmful dioxins, which poisoned the emissions of earlier incinerators.

Sokol says there is an "overwhelming need" for such plants because city landfills are overflowing and new sites are becoming more expensive.

But even if all 300 million tons of refuse Americans produce each year were converted into electricity, that power would meet less than 5 percent of the nation's needs.

Experts agree that it will take a combination of several technologies to replace the energy now supplied by fossil fuels. Those technologies show varying degrees of progress:

Biomass energy is the oldest and largest alternative technology. It is the conversion of organic material—mostly wood, but also organic material in municipal waste—into heat through burning or the conversion of various crops into gas or liquid fuel. It has nearly doubled in use since 1975 and now supplies more than 5 percent of the nation's energy.

An example of industrial use is Procter & Gamble's $30 million conversion of a plant in Staten Island, N.Y., two years ago, from oil to wood fuel. Another is Burlington (Vt.) Electric Company's completion last year of a $76 million, 62-megawatt wood-fired electric power plant.

Wood also plays a key role in one of the emerging technologies for improved burning of coal, since a little wood, mixed with coal, absorbs 90 percent of coal's sulfur emissions.

Burning municipal waste is the fastest growing use of biomass energy. A third use is production of alcohol fuels from a variety of crops to blend with gasoline as gasohol.

Hydropower is the second largest renewable energy source. Water has been used for centuries to run mills and more recently to produce electricity. Until this year, when it was passed by nuclear power, hydro was second only to coal as a source of electricity. Its growth now is slow because the big rivers have been dammed and future increases will have to come from small projects, most of which face opposition from environmentalists.

Hydro technology is well developed and requires little maintenance. But smaller projects have a higher unit cost because they must be custom-designed to fit the site. Modest but steady growth in small projects is expected.

Wind power has been the fastest growing of all renewable energy sources. The heaviest development has been in California, where more than 8,000 wind turbines, erected in the last three years in windy mountain passes, generate 600 megawatts of electricity, enough to serve 200,000 homes.

Turbines are mostly clustered in "wind farms," and research has aimed at larger and larger machines because with size comes efficiency. But the largest machines have proved vulnerable to damage, and they need frequent repair. Some 20 U.S. companies manufacture large turbines, and 25 companies make smaller wind machines for individual use at locations where utility service is not available.

Although federal tax credits are due to expire at the end of this year, some projections show the industry continuing to grow by as much as 1,000 megawatts a year.

Geothermal energy is derived from the heat inside the earth, which is accessible at some places as steam or geysers. Most of the sites in this country are in the West. Geothermal use has been expanding at more than 10 percent a year, some of it for heating but most for production of electricity.

The largest project is at the Geysers, 90 miles north of San Francisco, where Pacific Gas & Electric Company produces 1,200 megawatts of power, about 10 percent of the utility's total and 6 percent of Northern California's needs.

The Energy Department estimates that by the end of the century U.S. geothermal electric capacity will reach 2,500 to 6,000 megawatts.

Solar thermal energy includes the use of sunlight to heat and cool buildings, heat water and produce electricity. Solar heating of buildings includes simple designs with glass areas facing south and more elaborate systems of solar collectors plus pumps and fans.

The Energy Department estimates that solar heating and cooling will contribute 0.5 percent of the nation's energy by 2000.

Solar hot water heating, already in use in 800,000 residential and industrial systems, typically uses roof-mounted collector plates and mechanical pumps. The Energy Department projects this source will contribute 0.25 percent of our energy by the century's end, 10 times the present amount.

Solar thermal electric power uses curved collector troughs or arrays of reflectors to concentrate the sun's heat and produce the high temperatures needed to generate electricity by steam turbines or hot-air Stirling engines. The largest solar thermal project operating is Southern California Edison's 14-megawatt plant at Daggett, Calif.

Photovoltaics is a different form of solar power. It is based on a technology, developed for space flight, in which sunlight is converted directly into electricity, without any moving parts. A solar cell of thin semiconductor material, usually silicon, when exposed to sunlight yields an electric current. The U.S. industry produced 15 megawatts in photovoltaic systems last year, with sales of about $200 million. Although the technology is still too expensive for widespread use, costs have dropped to one fifth of their former level and efficiency has increased fourfold in a decade. Reliability is already high. Solar cells have a 30-year life.

Photovoltaics are now cost-effective for applications in remote areas that lack access to utility power, but new materials and manufacturing techniques must cut costs further before solar cells can compete with conventional electric utility costs.

Even so, the utility industry considers photovoltaics the likeliest alternative technology to have an early impact on its operations. The Edison Electric Institute, an association of utilities, says: "It is expected that by the early 1990s photovoltaics will be competitive with grid-supplied electricity in some regions."

From Trash To Ash

ILLUSTRATION: OGDEN CORPORATION

This cutaway drawing shows a modern trash-into-energy plant at Harrisburg, Pa. It uses the Martin process, one of the two leading technologies, both developed in Germany. Trucks dump refuse into the pit at left. An overhead crane picks it up in chunks and drops it into the hopper, where it travels to the inclined grate. The burning mass is continuously agitated and subjected to blasts of air from underneath. Air jets above promote further combustion in the exhaust gases. The heat produces superheated steam in boiler chambers above the grate. Then electrostatic scrubbers clean and deodorize the gas before it is released through the chimney. Residual ash is discharged onto the track under the grate. The steam drives electric generating turbines.

World population crisis

Paul R. Ehrlich and Anne H. Ehrlich

Paul R. Ehrlich is Bing Professor of Population Biology at Stanford University and author of The Machinery of Nature *(1986). Anne H. Ehrlich is a senior research associate in biological sciences at Stanford. The Ehrlichs are the authors of* Extinction: The Causes and Consequences of the Disappearance of Species *(1981).*

Rapid population growth, rising competition for resources, and increasing environmental deterioration are intertwined factors in the human predicament that feed the political tensions and conflicts of the late twentieth century.

The following article outlines the dimensions of that predicament and sets the context.

MOST PEOPLE realize that unless something is done soon about the nuclear arms race, civilization may well be destroyed. Understandably, then, a prime focus of attention is on issues that seem immediately related to the possible triggering of a large-scale thermonuclear exchange, such as the deployment of weapons, the adequacy of command and control systems, arms control negotiations in Geneva, and so on.

But those who are concerned with the human predicament should not lose sight of more basic problems that influence nation-states. While we struggle to prevent the nuclear arms race from ending the world with a bang, more subtle global trends may lead to the same end within a century or so, but with a whimper. Moreover, these trends aggravate the conditions that breed conflict, thus increasing the chances that nuclear weapons will be used. They include:

- destruction and dispersion of the one-time bonanza of "capital" (fossil fuels, rich soils, other species, and so forth) that humanity inherited;
- environmental deterioration, including the decay of the systems that provide civilization with the "income" resources that are the only alternative to consuming our capital stock;
- the widening gap between rich and poor nations and related patterns of migration;
- the persistence of economic inequality within nations;
- the rise of ethnic and religious separatism;
- persisting high levels of hunger and unemployment; and
- the relatively slow and geographically spotty progress in ending racism, sexism, and religious prejudice.

All those negative trends are interrelated, and interwoven with them is one of the most basic causes of the human predicament—unprecedented continued growth in human numbers in an already overpopulated world. When the two of us were born in the early 1930s, only two billion people existed; by 1987, five billion will.

This year the population will grow by more than 84 million people—a record number—and each year in the immediate future will see a new record increment. A great deal of attention has been paid to a small decrease in the global growth rate since the early 1960s, a drop from about 2.2 percent annual natural increase to around 1.7 percent. But with that growth rate applied to an ever-larger population base, the absolute annual increase continues to escalate. The ecological systems that support humanity respond primarily to absolute numbers, not rates; those numbers will be a major determinant of how long civilization can be sustained.

EARTH is overpopulated today by a very simple standard: humanity is able to support itself—often none too well, at that—only by consuming its capital. This consumption involves much more than the widely publicized depletion of stocks of fossil fuels and dispersion of other high-grade mineral resources. Much more critical are the erosion of deep, rich agricultural soils, the diminution of our fresh water supply by pollution and mismanagement of groundwater, and the loss of much of the diversity of other life-forms that share the earth with us. All these are intimately involved in providing humans with nourishment from the only significant source of income, the radiant energy of the sun, which, converted by photosynthetic plants into the energy of chemical bonds, supports essentially all life on the planet.

Two crucial points must be remembered. The first is that with today's technology, humanity could not support anything like its current numbers without continually using its nonrenewable resource subsidy. The second is that while exploiting that capital subsidy, civilization is continually degrading the systems that supply its income. Consider only the accelerating extermination of other organisms, which is intimately connected with brute increase in the human population and its exploitation of the planet.

Those organisms are working parts of the ecosystems that provide society with a wide variety of indispensable services, including regulation of the composition of the atmosphere, amelioration of weather, the generation and preservation of soils, the cycling of nutrients essential to agriculture and forestry, disposal of wastes, control of the vast majority of potential crop pests and carriers of human diseases, provision of food from the sea, and maintenance of a vast genetic library, from which humans have already drawn the very basis of civilization, and whose potential has barely been tapped.

All of these services are directly or indirectly involved in providing necessities to humanity derived from our solar income. Ecologists standardly measure that income in terms of net primary productivity. Net primary productivity is the total amount of the energy bound each year by plants in the process of photosynthesis, minus the portion of that chemical energy that the plants themselves must use to run their own life processes. The global net primary productivity can be viewed as the basic food supply for the entire animal world, including *Homo sapiens*, as well as a major source of structural materials, fibers, medicines, and other things of importance to humanity.

The relationship between current human population size and this basic income source is revealed by the answer to a simple question: How much of global net primary productivity is now being coopted by *Homo sapiens*, just one of five to 30 million animal species that completely depend upon it? Humanity not only directly consumes a disproportionate share, but it also reduces production by replacing natural ecosystems with generally less productive, human-managed or disrupted ones. Humanity coopts about 40 percent of terrestrial net primary productivity today, and an additional few percent in aquatic systems.

For technical reasons, it will prove very difficult to increase human exploitation of the oceans significantly, as the decline of per capita yields of food from the sea since 1970 indicates. It is on land that civilization must seek the income to support ever-growing numbers of people. The population is now growing at a rate that, if continued, would double it in about 42 years. Even if *Homo sapiens* could persist after wiping out most of the other animals, population growth clearly would soon carry it past the limits of Earth's short-term human carrying capacity, and a population crash would ensue.

ECONOMIC AND SOCIAL systems also respond to absolute numbers as well as to rates of growth. There is every reason to believe that most aspects of these systems have long since passed the point where economies of scale become diseconomies of scale. Twenty years ago, C.P. Snow, commenting on the declining quality of English telephone services, generated what we might call Snow's Law: "The difficulties of a service increase roughly by the square of the number of people using it." (See diagram.)

That growing numbers of people are deleterious can be seen in the increased costs of supplying them with goods.

How problems can multiply at a faster rate than people

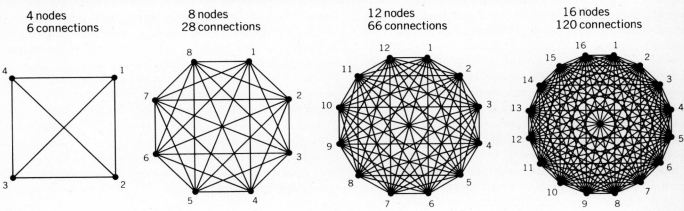

4 nodes
6 connections

8 nodes
28 connections

12 nodes
66 connections

16 nodes
120 connections

English writer C.P. Snow saw that when the number of people using a service grows, the number of service problems grows at an even faster rate, proportional to the number of possible connections between users. The diagram above illustrates this principle. For example, when the number of nodes doubles, from four to eight, the number of connections between them more than quadruples. This law of connections helps explain many of the problems of an expanding, ever more complex, global society. From 1930 to 1976 the number of people on earth doubled from two billion to four billion — and the number of possible connections quadrupled. Based on current growth rates, the population could be expected to double again to eight billion by 2020, bringing another quadrupling of possible connections — and of Snow's "service difficulties."

Informational graphics in population articles: Michael Yanoff

If today's population were of constant size, it still would have to run continually, like the Red Queen, to stay in place. Technology would have to be constantly improved to compensate for depleted supplies of fuels, declining quality of ores, and deterioration of soils, and to protect people from the environmental consequences of using both old and new technologies. But today's population is growing. Each additional person, on average, must be cared for by using lower-quality resources that must be transported further, and by food grown on more marginal land. Supplying the additional energy needed for these tasks creates both economic and environmental problems.

Although population growth rates are highest in poor countries, overpopulation and continuing population growth in rich countries are the prime threat to global resources and environmental systems, simply because the average individual in rich nations has a large impact on resources and environment. One of the best available measures of that impact is a nation's per capita use of commercially produced energy. In the United States, the average person uses the commercial energy equivalent of about 10,000 kilograms of coal annually. In contrast, an average South American uses about 1,000 kilograms, an Asian about 600, and an African about 425. By this measure, the birth of an average baby in the United States will be about 200 times as disastrous for the world as the birth of an average Bangladeshi, who will consume the commercial equivalent of some 45 kilograms of coal annually.

This does not mean that population growth in poor nations is harmless. But, as a great oversimplification, one can say that population growth in less developed countries primarily harms those countries, while that in industrialized nations harms the world as a whole. Consider what it means for poor countries such as Bangladesh or Kenya to double their populations in 25 and 17 years, respectively, as they are currently doing. If those nations are to maintain their present standards of living, low as they are, they will have to duplicate every amenity for the support of human beings in that time period. Among other things, that means doubling their food production, their supply of teachers, doctors, engineers, and scientists, and the capacity of housing, hospitals, road and rail systems, and manufacturing plants. It would be a daunting task for a rich nation such as the United States, with abundant capital, incredibly rich soils, good supplies of most other resources, fine transport and communications systems, vast industrial capacity, and a largely educated and literate population. Most poor nations have none of these things.

Indeed, even without further growth, most less developed countries face massive problems in the near future just because of their age structures. In Mexico, for instance, about half of the labor force is unemployed or underemployed. That is bad enough, but in the next 15 years Mexico will have to find jobs for perhaps 10 million people if the unemployment rate is not to increase, and 20 million people if the number of unemployed is not to rise. Those numbers are not based on future population growth, but on people

already born. Today there are over 30 million Mexicans under the age of 15.

IN THE FACE of all these factors, why is the population component so often ignored in discussions of the human predicament? One possible reason is evolutionary. Evolutionary success has meant, and means today, outbreeding your friends and neighbors.

Animal populations have often increased in size to the point where they exceeded the carrying capacity of their environments, which, in turn, led to catastrophic population declines or extinctions. *Homo sapiens* is the first of billions of species that has developed the ability both to detect the relation of its population size to carrying capacity and, through birth control, to adjust its numbers to fit within that capacity.

Yet, problems of overpopulation were rarely present or perceived over most of the millions of years that human beings were evolving into the most successful of the great apes. Throughout our evolutionary history, the emphasis has been on successful reproduction. People therefore have great difficulty facing the fact that either humanity must consciously halt population growth and then gradually reduce its numbers, or nature will end the explosion of human numbers with a catastrophic population crash. Decreasing the death rate goes with the evolutionary grain, and in the past century our species has been extraordinarily successful at it. Avoiding disaster by humanely causing a compensating decline in the birth rate, unfortunately for humanity, goes against long-evolved prejudices.

A second evolutionary reason is that the human nervous system developed little capacity to detect gradual trends in the environment—trends that only created significant changes after decades or centuries. Over most of our history, our ancestors did not need to respond adaptively to such changes, but to sudden or rapid alterations in their environments. So our nervous systems evolved a high capacity to detect the sudden appearance of tigers and the crack of falling trees and a low capacity to respond to growth in human numbers, nuclear arms, or environmental damage from acid rains.

The inability of people to register change over long periods often leads them to ignore history, to believe that the world has not and will not change. It allows economists to think that there have always been and always will be high rates of economic growth. Even those who admit that there must be ultimate limits to growth struggle to keep their world constant by assuming, against all the evidence, that those limits are so far in the future that they can be safely ignored.

There is no longer any substitute for analysis; the world can no longer afford to believe such myths as that economic growth can be infinite, contraception is immoral, and more nuclear weapons will prevent nuclear war. All the educational mechanisms that society can deploy must be focused on teaching people to understand the physical and biological constraints on human activities, to recognize and deal with long-term trends, and to perceive the connections between various aspects of the human predicament. Then everyone will recognize that problems as seemingly disconnected as Star Wars and the population bomb are simply different aspects of a general problem of social organization. But while that global enlightenment is awaited, action must be taken to end both the arms race and the baby race, for either is likely to result in the premature deaths of billions of human beings.

The world at a glance:
Population, growth, military spending, "quality of life," and abortion policies

Africa	Population, 1980 (millions)	Annual population growth (percent)[a]	Gross national product per capita (1980 U.S.$)	Military share of state budget (percent)[b]	Life expectancy (men/women)[a]	Adult illiteracy (percent)[b]	Abortion restrictions[c]
Algeria	19.6	3.3	2,140	7.6	56/58	74	H
Angola	7.8	2.6	393	——	40/43	97	?
Benin	3.6	2.8	320	15.3	44/48	72	Li
Botswana	0.9	4.8	1,010	6.4	47/50	59	?
Burundi	4.2	2.4	230	22.2	40/43	73	I
Cameroon	8.7	1.9	880	9.1	44/48	60	H,J
Central African Republic	2.4	1.6	320	8.9	44/48	67	I
Chad	4.5	2.3	110	——	43/45	94	Li
Congo	1.7	2.6	1,110	14.4	44/48	84	H
Djibouti	0.4	2.1	480	——	n/a	——	?
Egypt	43.3	1.7	650	16.0	54/56	62	I
Ethiopia	31.8	2.7	140	42.0	38/41	87	H
Gabon	0.7	1.6	3,810	1.1	42/45	88	?
Gambia	0.6	0.9	370	——	39/43	80	?
Ghana	11.8	3.2	400	3.7	47/50	70	H
Guinea	5.6	2.3	300	——	42/45	91	H
Guinea-Bissau	0.8	2.2	190	6.6	39/43	81	?
Ivory Coast	8.5	3.6	1,200	3.6	44/48	65	Li
Kenya	17.4	4.1	420	12.9	54/58	53	H
Lesotho	1.4	2.6	540	0	49/51	41	Li
Liberia	1.9	3.2	520	5.1	46/44	79	H,E,J
Libya	3.1	4.0	8,450	4.5	54/57	——	Li
Madagascar	9.0	2.6	330	13.5	44/48	——	Li
Malawi	6.2	6.3	200	6.0	41/44	78	Li
Mali	6.9	2.5	190	20.5	41/44	91	I
Mauritania	1.6	2.0	460	25.9	41/44	——	I
Mauritius	1.0	1.2	1,270	0.6	61/65	21	?
Morocco	20.9	3.3	860	17.8	54/57	79	H
Mozambique	12.5	3.2	208	——	44/48	67	Li
Namibia	1.0	2.8	1,960	——	50/53	62	H,E,J
Niger	5.7	2.8	330	4.8	41/44	90	I
Nigeria	87.6	3.4	870	9.3	46/49	66	Li
Reunion	0.5	3.8	3,840	——	56/62	——	?
Rwanda	5.3	4.1	250	15.3	44/48	50	I
Senegal	5.9	2.7	430	11.4	41/44	94	Li
Sierra Leone	3.6	1.8	320	3.8	44/48	93	H
Somalia	4.4	4.5	280	18.4	41/45	94	I
South Africa	29.5	1.9	2,770	13.9	59/62	43	H,E,J
Sudan	19.2	2.2	380	12.2	46/48	85	Li
Swaziland	0.6	3.4	760	6.2	44/48	45	?
Tanzania	19.0	2.3	280	16.3	49/52	27	H
Togo	2.7	2.1	380	7.0	44/48	84	Li
Tunisia	6.5	1.9	1,420	4.5	56/58	62	
Uganda	13.0	3.5	220	20.6	51/54	48	H
Upper Volta	6.3	1.2	240	18.2	41/44	91	I
Zaire	29.8	n/a	210	12.4	44/48	46	I
Zambia	5.8	2.3	600	8.6	48/50	53	H,E,S
Zimbabwe	7.2	2.7	870	25.9	52/55	31	H,E,J

7. THE ENVIRONMENT

Asia	Population, 1980 (millions)	Annual population growth (percent)[a]	Gross national product per capita (1980 U.S.$)	Military share of state budget (percent)[b]	Life expectancy (men/women)[a]	Adult illiteracy (percent)[b]	Abortion restrictions[c]
Afghanistan	16.3	2.6	160	——	42/43	80	Li
Bangladesh	90.7	2.2	140	6.4	46/47	74	Li
Bhutan	1.3	2.0	80	——	44/43	——	?
Burma	34.1	n/a	190	22.1	56/60	34	Li
China	991.3	1.2	300	——	62/66	34	
Hong Kong	5.2	1.8	5,100	——	68/74	23	H
India	690.2	n/a	260	16.0	46/45	66	H,E,J,S
Indonesia	149.5	1.8	530	10.2	49/51	43	I
Japan	117.6	0.7	10,080	5.0	73/79	2	H,E,J,S
Kampuchea	7.1	2.5	150	——	47/50	64	?
Korea, North	18.7	2.5	1,055	10.4	61/65	——	Li,H,E,J,S
Korea, South	38.9	1.6	1,700	28.4	63/69	12	H,E,J
Laos	3.5	2.6	80	——	42/45	56	Li
Malaysia	14.2	2.3	1,840	13.5	67/72	47	Li,E,J
Mongolia	1.7	3.2	744	——	61/65	5	I
Nepal	15.0	n/a	150	6.4	44/43	81	H
Pakistan	84.5	3.0	350	23.2	52/62	79	Li
Philippines	49.6	2.6	790	16.0	59/62	17	I
Singapore	2.4	1.2	5,240	20.6	69/74	8	
Sri Lanka	15.0	1.5	300	1.7	62/65	22	Li
Taiwan	18.5	n/a	2,543	——	70/75	——	I
Thailand	19.0	2.1	770	16.9	58/63	21	H,J
Vietnam	55.7	2.1	n/a	——	46/49	24	

Central America and Caribbean

	Population, 1980 (millions)	Annual population growth (percent)[a]	Gross national product per capita (1980 U.S.$)	Military share of state budget (percent)[b]	Life expectancy (men/women)[a]	Adult illiteracy (percent)[b]	Abortion restrictions[c]
Costa Rica	2.3	2.7	1,430	——	66/70	12	H
Cuba	9.7	0.6	1,435	——	69/72	5	
Dominican Republic	5.6	3.1	1,260	9.1	58/62	33	I
El Salvador	4.7	3.3	650	13.7	60/65	38	Li,E,J
Guatemala	7.5	3.0	1,140	5.8	54/56	49	Li
Haiti	5.1	n/a	300	7.7	49/52	79	I
Honduras	3.8	3.5	600	7.9	55/59	43	H
Jamaica	2.2	1.3	1,180	1.7	68/73	4	H
Nicaragua	2.8	3.8	860	9.1	54/57	43	H
Panama	1.9	2.2	1,910	2.9	64/68	15	I
Puerto Rico	3.7	1.6	3,350	0	70/76	12	
Trinidad and Tobago	1.2	1.6	5,670	0.9	64/68	8	H

Europe

	Population, 1980 (millions)	Annual population growth (percent)[a]	Gross national product per capita (1980 U.S.$)	Military share of state budget (percent)[b]	Life expectancy (men/women)[a]	Adult illiteracy (percent)[b]	Abortion restrictions[c]
Albania	2.8	2.1	895	——	65/71	——	H
Austria	7.8	0.2	10,210	6.0	69/76	——	
Belgium	9.9	0.0	11,920	10.5	69/75	3	Li
Bulgaria	8.9	0.3	3,820	33.3	69/74	9	H,E,J,S
Czechoslovakia	15.3	0.2	5,490	15.5	67/74	——	H,E,J,S
Denmark	5.1	−0.1	13,120	6.0	71/77	——	
Finland	4.8	0.6	10,680	7.0	69/78	——	H
France	54.0	0.6	12,190	19.0	70/78	4	H,E,J,S
Germany, East	16.7	−0.1	6,766	11.1	69/75	——	
Germany, West	61.7	−0.1	13,450	22.5	69/76	——	
Greece	9.7	0.7	4,420	23.5	70/74	16	H,E,J
Hungary	10.7	−0.1	2,100	8.7	67/74	2	H,E,J,S
Ireland	3.4	1.0	5,230	3.3	69/74	——	Li
Italy	56.2	n/a	6,960	10.1	70/76	6	
Netherlands	14.2	0.5	11,790	9.2	72/79	——	H
Norway	4.1	0.3	14,060	6.7	72/79	——	
Poland	35.9	0.9	4,345	13.8	66/74	11	H,J,S
Portugal	9.8	0.7	2,520	12.9	65/73	29	I
Romania	22.5	0.5	2,540	11.2	67/72	11	H,E,J,S

	Population, 1980 (millions)	Annual population growth (percent)[a]	Gross national product per capita (1980 U.S.$)	Military share of state budget (percent)[b]	Life expectancy (men/women)[a]	Adult illiteracy (percent)[b]	Abortion restrictions[c]
Spain	38.0	0.7	5,640	10.1	70/76	8	I
Sweden	8.3	0.1	14,870	6.8	73/79	——	
Switzerland	6.5	0.6	17,430	22.1	70/76	——	H
Turkey	45.5	2.1	1,540	19.1	60/62	40	Li,E,J
United Kingdom	56.0	−0.2	9,110	12.5	68/74	——	H,E,S
USSR	265.0	0.9	4,155	48.3	64/74	0.2	
Yugoslavia	22.5	0.8	2,790	20.8	65/70	17	
Middle East							
Cyprus	0.6	1.4	3,740	7.8	72/75	24	?
Iran	40.1	2.8	797	——	58/57	76	Li
Iraq	13.5	3.5	2,254	——	54/57	76	Li
Israel	4.0	1.9	5,160	34.2	72/76	12	H,E,J
Jordan	3.4	3.6	1,620	23.4	62/66	32	H,J
Kuwait	1.5	6.8	20,900	14.1	66/72	40	Li
Lebanon	2.7	−0.4	n/a	——	63/67	——	Li
Oman	0.9	5.0	5,920	44.0	46/48	——	?
Qatar	0.2	4.5	27,720	20.1	n/a	——	?
Saudi Arabia	9.3	4.1	12,600	26.7	47/49	75	Li
Syria	9.3	n/a	1,570	35.4	63/65	60	Li
United Arab Emirates	1.1	7.2	24,660	41.4	n/a	44	?
Yemen, North	7.3	2.3[d]	460	30.0	39/40	91	Li
Yemen, South	2.0	2.3[d]	460	45.7	43/45	73	?
North America							
Canada	24.2	1.2	11,400	7.8	70/77	——	H
Mexico	71.2	2.7	2,250	2.4	63/67	17	Li,J
United States	229.8	0.9	12,820	24.9	70/78	1	
Oceania							
Australia	14.9	1.5	11,080	9.1	71/78	——	H,E,S
Fiji	0.6	1.9	2,000	3.0	70/73	21	?
New Zealand	3.3	1.0	7,700	4.8	69/75	——	H,E,J
Papua New Guinea	3.1	2.1	840	3.5	50/50	68	H
South America							
Argentina	28.2	1.6	2,560	12.8	65/71	7	H,J
Bolivia	5.7	2.8	600	——	48/53	37	H,J
Brazil	120.5	2.3	2,220	9.3	58/61	24	Li,J
Chile	11.3	1.7	2,560	9.8	61/68	11	Li
Colombia	26.4	n/a	1,360	10.3	61/64	19	Li
Ecuador	8.6	3.5	1,180	10.2	60/62	26	Li,J
Guyana	0.8	2.0	720	5.9	67/72	8	?
Paraguay	3.1	3.1	1,630	11.4	62/65	20	Li
Peru	17.0	2.7	1,170	24.4	55/58	28	Li
Uruguay	2.9	0.7	2,820	9.9	66/73	6	Li,J,S
Venezuela	15.4	3.0	4,220	5.0	65/70	——	Li

The following low-populated countries are not included in the table: Antigua and Barbuda, Bahamas, Bahrain, Barbados, Belize, Bermuda, Brunei, Cape Verde, Comoros, Dominica, Equatorial Guinea, French Guiana, Greenland, Grenada, Guadeloupe, Iceland, Kiribati, Luxembourg, Macao, Maldives, Malta, Martinique, Santa Lucia, St. Vincent and the Grenadines, São Tomé and Principe, Seychelles, Surinam, Western Samoa.

[a] n/a: information unavailable.

[b] ——: negligible amount, or information unavailable.

[c] Key to national restrictions on abortion: I: illegal (no exceptions, although abortions to save woman's life may be authorized under general principles of criminal law); Li: permitted only if woman's life would be endangered by full-term pregnancy; H: permitted if the pregnancy will endanger the woman's health; E: permitted on eugenic grounds (fetal birth defects); J: permitted on juridical grounds (pregnancies resulting from rape or incest); S: permitted on social or social-medical grounds (woman's age, marital and economic status, size of family); ?: information not available. Countries without listings provide legal abortion on demand, although restrictions are often placed on second- or third-trimester abortions.

[d] Separate listing not available.

Sources: 1982 Statistical Yearbook, 33rd edition (New York: United Nations, 1985); Christopher Tietze, *Induced Abortion: A World View,* 4th edition (New York: The Population Council, 1981); Pete Ayrton, Tom Engelhardt, and Vron Ware, eds., *World View* (New York: Pantheon, 1984).

19th ENVIRONMENTAL QUALITY INDEX

A Nation Troubled By Toxics

EACH YEAR, the National Wildlife Federation looks back over the environmental record of the past 12 months. Each year, much of the news is grim. And 1986 was no exception, as a witch's brew of chemicals continued to foul the nation's air and water.

Researchers learned, for example, that the 37 million people who live near the Great Lakes tend to have significantly higher levels of toxic chemicals in their bodies than Americans elsewhere, the result of increasing pollution in the lakes. Toxic chemicals are also contaminating one of every five National Wildlife Refuges, in some cases turning sanctuaries into lifeless graveyards. Levels of air pollutants are rising, acid rain is worsening and forests are mysteriously dying.

Despite successful efforts to slow the flow of sewage and industrial waste into lakes and rivers, thousands of waterways are being damaged by an uncontrolled form of pollution—chemical-laden runoff from farms, mines, cities and suburbs. Authorities have also identified nearly 900 leaking hazardous waste sites around the country that demand immediate attention, and thousands more that may be threatening underground water supplies.

But 1986 also brought good news as well as bad. The nation took a historic step in the battle against soil erosion, taking some nine million acres of fragile farmland out of cultivation. The outlook for many birds improved when the Environmental Protection Agency banned the use of two deadly pesticides, and the U.S. Fish and Wildlife Service approved a plan to phase out lead shot. Numbers of such endangered species as bald eagles and alligators rose significantly. Americans continued to use energy more efficiently. And at the eleventh hour, the 99th Congress passed a strengthened Superfund bill that should improve clean-up efforts at many toxic dumps.

"The success stories," said Jay D. Hair, executive vice president of the National Wildlife Federation, "should remind Americans that environmental progress is still possible despite the formidable problems we face, and despite the increased pressure on our diminishing resources."

On the following pages, *National Wildlife*'s nineteenth annual EQ Index takes a closer look at last year's gains and losses.

ILLUSTRATIONS BY MARSHALL AND RICHIE MOSELEY

Some Gains, But New Concerns About Poisoned Species

THERE were signs last year that the federal government was finally getting serious about attacking the sources of toxic threats to wildlife. But the task is a daunting one. Poisons continue to kill many plants and animals throughout the nation—including the inhabitants of a shocking number of wildlife refuges.

The bald eagle is one creature that already has benefited from efforts to detoxify wildlife habitat. The bird's numbers have risen from 500 nesting pairs during the early 1960s to more than 1,700 in 1986. And the eagle received a further boost last summer, when the U.S. Department of Interior responded to a National Wildlife Federation lawsuit by announcing it would phase out the use of lead shot for waterfowl hunting by 1991. In the last two decades, at least 124 bald eagles have died after eating lead-contaminated prey, and currently, more than two million waterfowl fall victim to lead

poisoning each year. Although the Federation had sought an immediate halt, Executive Vice President Jay D. Hair said the decision still "represents a victory for concerned conservationists and responsible hunters."

The chemical threat to birds was further relieved last year when, for the first time, the Environmental Protection Agency restricted the use of two pesticides solely on the basis of their harmful effects on wildlife. The agency outlawed the pesticide dicofol, which has been found to jeopardize the West Coast populations of peregrine falcons. Dicofol is contaminated with DDT, a compound that nearly wiped out the bald eagle in the lower 48 states two decades ago. Another pesticide, diazinon, was the object of a partial EPA ban forbidding its use on golf courses and sod farms. Scientists have linked the chemical to at least 60 mass poisonings of birds in 18 states.

Meanwhile, ducks have begun to recover from their disastrous decline of 1985, the worst year for waterfowl numbers on record. Population estimates for the ten most common U.S. species increased 14 percent in 1986 from the previous year—but they still remained 12 percent below the average for the last 30 years. This represented the second-lowest count on record.

Unfortunately, even the most encouraging numbers are marred by the continuing nationwide loss of habitat. Federal officials caution that wetland destruction could prevent a complete recovery of ducks. Of the nation's original 215 million acres of swamps, bogs and marshes, more than half have disappeared in the last 200 years—and the loss continues at the pace of roughly 400,000 acres a year.

Despite such statistics, the U.S. Army Corps of Engineers—now headed by a political appointee who apparently rejects the idea that Congress intended to protect wetlands through the Clean Water Act—continued to issue permits in 1986 for dozens of dredge-and-fill projects. Last spring, the EPA, exercising its rarely used veto power over Corps permits, had to step in to save a 32-acre red maple swamp in Massachusetts from being turned into a shopping center.

The U.S. Department of Agriculture, meanwhile, proposed regulations that would significantly weaken the impact of "swampbuster" provisions in the 1985 Farm Act—a law that denies federal subsidies to farmers who drain wetlands for conversion to crops. If the Administration has its way, regulatory loopholes would allow many farmers who destroy their wetlands to continue to receive subsidies. "The proposed rules are likely to lead to very weak enforcement of the swampbuster provisions," said Scott Feierabend, a National Wildlife Federation legislative representative.

Wetlands are not the only wildlife areas

under siege. Throughout the country, habitat in general is becoming tainted by increasing amounts of toxic chemicals. Interior Department investigators suspect that such toxins may be damaging one in five National Wildlife Refuges. Among the potential problem areas is the Aransas National Wildlife Refuge in Texas—crucial winter habitat for the endangered whooping crane—where agricultural pesticides and wastes from nearby oil and metal refining industries are suspected of fouling marshes and thickets. "You can't dig a moat around the refuges,"

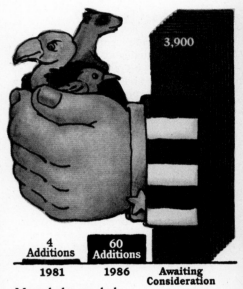

More help needed
Last year, the government added 60 plants and animals to the list of endangered and threatened species. But 3,900 candidates await consideration.

observed W. Alan Wentz, director of the Federation's fisheries and wildlife program. "They're all going to be affected by what goes on around them."

Clearly, the same can be said for some fish populations, which declined dramatically in 1986 in the face of habitat loss and overharvesting. Among the most severely depleted: striped bass and king mackerel off the East Coast, redfish in the Gulf of Mexico, and the Sacramento River's king salmon—a subspecies that scientists fear is now threatened with extinction.

The outlook for beleaguered species became even bleaker toward the end of 1986, when Congress failed to renew the Endangered Species Act. A now-defunct amendment to the law would have provided federal protection to plants and animals not yet officially listed as threatened or endangered. This is no minor matter: the total of such "candidate" species now amounts to some 3,900.

Ducks make a comeback
Duck populations have sprung back from the 1985 totals — the lowest on record. But continued wetland losses may prevent the birds from regaining their historic levels.

Are We Losing Ground in The Clean Air Battle?

AMERICA has made signficant progress in improving the quality of its air in recent years. But after more than a decade of steady decreases under the Clean Air Act, the levels of some pollutants have begun to edge upward, suggesting that the nation's battle for purer air may be losing ground.

According to the Environmental Protection Agency, the amount of sulfur dioxide, nitrogen oxides and particulates in the air increased 2 percent each in 1984 over the previous year (the latest figures available). One reason: domestic coal use increased by 7 percent between 1983 and 1984.

EPA officials cautioned that it is too early to know if the latest increases represent a trend, but the figures added a note of urgency to efforts by environmentalists to

Troubling trend?
After years of steady decreases, the levels of some dangerous air pollutants increased in America between 1983 and 1984 — the most recent figures available.

clamp tighter controls on sulfur and nitrogen emissions. These are essentially automobile and industrial pollutants released by coal and other fossil fuel-burning boilers and power plants. They are known to be hazardous to humans, particularly those people with respiratory diseases. They are also the air pollutants identified in scientific

studies as the key precursors to acid rain.

Efforts to control them got a significant boost last year when, for the first time, President Reagan formally acknowledged that action on acid rain is necessary. "We can't keep studying this thing to death," said Drew Lewis, Reagan's hand-picked representative on a U.S.-Canadian study team. "We have got to do something about it."

That fact became even more clear in 1986 as new evidence surfaced. Last summer, for instance, the National Wildlife Federation used the Freedom of Information Act to obtain unreleased data from the EPA's first systematic study of lakes in the East. The subsequent analysis of the survey showed that at least 75 percent of the lakes in New Hampshire and Rhode Island, and at least 60 percent of those in Massachusetts and Maine, will be seriously damaged if acid rainfall continues.

The situation may be just as severe in the Upper Midwest. Last spring, EPA researchers reported that 43 percent of that region's lakes could become devoid of all life as a result of high acidity. What's more, according to a Brookhaven National Laboratory study, acid rain can cause accelerated weathering on homes, buildings and other structures. As a result, some people may have to repaint, replace rusted fences and do other outside maintenance more frequently. Such maintenance, the study noted, can cost each resident in Chicago alone an estimated $45 annually.

Unfortunately, the President's tentative step forward on acid rain was one of only a few positive actions last year in the fight for clean air. Once again, it was evident in 1986 that the scope of the country's air pollution problems is much broader than federal legislators envisioned when the Clean Air Act was passed in 1970.

Scientists are particularly concerned about a spate of dangerous chemical air pollutants that have been largely ignored by federal regulators. In one investigation, a congressional panel found that more than 62 million pounds of highly toxic chemicals were being released annually into the air from some 300 manufacturing plants in 34 states. The list included more than 8 million pounds of methylene chloride and some 3 million pounds each of benzene and butadiene. All are known or suspected cancer-causing substances.

The Clean Air Act gives the federal government authority to control all airborne toxic chemicals, but to date only six out of hundreds of substances have been regulated. The others spew largely uncontrolled into the environment. The task of regulating them remains one of the nation's most difficult challenges.

Another air pollutant, ozone, has been under regulation since 1970, and its levels declined nationally by about 10 percent between 1983 and 1984. Nevertheless, last

summer, EPA Administrator Lee M. Thomas announced that more than a third of the 84 metropolitan areas being monitored will fail to meet the law's deadline next year for reducing ozone pollution to a "safe" level.

Formed when gasoline vapors and other hydrocarbons and nitrogen oxides react in the air with sunlight, ozone is considered a

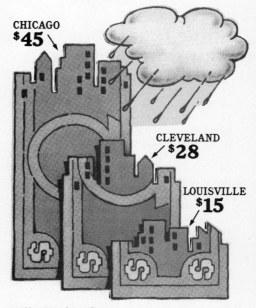

Pollution hits home
One study found that acid rain damage to homes and other structures can cost each resident in these cities the above amounts annually in added maintenance.

serious health threat. Currently, nearly 80 million Americans are breathing unhealthful levels. Unfortunately, recent studies indicate that even if cities meet federal requirements, the current ozone standard does not provide sufficient protection.

New research has also found that ozone can damage crops. In one study, a University of Wisconsin researcher discovered that typical summer ozone concentrations can reduce the growth rate of soybeans and wheat by as much as 25 percent. What's more, the study noted, ozone can slow tree growth anywhere from 1 to 10 percent a year, depending on the severity of pollution.

Meanwhile, in Boston, Harvard scientist Lance Wallace had some equally disturbing news. His five-year study, conducted in seven U.S. cities for the EPA, found levels of toxic pollutants as much as five times higher inside homes than outdoors. Such pollutants did not depend on outside air quality. "We had to conclude that the major sources of exposure were indoors," said Wallace. Among the culprits: paint, plastics, cigarettes, building materials and other consumer products.

Ground Water Pollution Remains An Urgent Problem

LITTLE more than a decade ago, Lake Erie was choking in pollution and scientists were predicting that it would soon be devoid of all life. By last summer, however, Ohio Governor Richard Celeste was reporting that the lake was cleaner now than at any time in the past 20 years. Meanwhile, in Boston, authorities were making similar claims about the Charles River, where, after two decades

Sites unseen
The Environmental Protection Agency estimates that more than one-third of the country's underground storage tanks may be leaking toxic substances into groundwater.

of cleanup efforts, sportsmen can now safely eat the fish they catch.

Both bodies of water have benefited from a national cleanup campaign that has brought major improvements. Yet both clearly illustrate how the country's water problems, far from being solved, are more pervasive and entrenched than anyone could have known when the Clean Water Act was passed 15 years ago.

Though the levels of many pollutants in

Lake Erie have declined, it still contains dangerous amounts of toxic chemicals—legacies of the region's industrial heritage that only recently have been detected. And while fish in the Charles are now edible, experts believe that the river will not be safe for swimming until they can control the pollution flowing into it from city streets and agricultural areas.

Scientists call such pollution "nonpoint;" its diffuse sources are extremely difficult to control, and generally, they have not been regulated by the Clean Water Act. A survey conducted by state water officials found that nearly one in every four miles of U.S. rivers and one out of every five lakes are being spoiled or threatened by pollutants cascading from farms, mines and urban areas.

Federal legislators finally appeared ready in 1986 to address the situation. Last fall, just prior to its adjournment, the 99th Congress passed an improved Clean Water Act that, among other things, required some controls of nonpoint pollution. Unfortunately, after Congress adjourned, President Reagan vetoed the new act, forcing legislators to start all over again this year.

Meanwhile, environmentalists warned that even tougher laws will not succeed in the face of lax enforcement. To illustrate the point, *The Washington Post* examined records last year for 124 major industrial and sewage treatment plants that discharge into the Chesapeake Bay and found that each of them dumped more pollutants than permits allowed. Despite a well-publicized "Save the Bay" campaign endorsed by the governors of states surrounding the Chesapeake, penalties were rare—even when permit levels were exceeded by as much as 2,000 percent!

The nation's inability to effectively control surface water pollution has exacted a heavy toll, not just on waterways and the wildlife they support, but also on people. According to U.S. and Canadian researchers, for instance, the 37 million people who live around the Great Lakes generally have 20 percent higher levels of toxic chemicals in their bodies than other North Americans. One reason: the Great Lakes, which supply fish and water for regional residents, are so contaminated with hazardous chemicals that current pollution control efforts cannot adequately protect human health.

Equally alarming, to communities elsewhere, is the growing knowledge of what years of unwitting abuse have done to the country's aquifers—the invisible groundwater supplies that provide the drinking water for one of every two U.S. families. Last year, the nation pumped about 100 billion gallons of water per day from below the ground, a 12 percent increase over 1980 figures. Yet new studies on the purity of that water were disconcerting.

In California, health officials discovered that one-fifth of the state's largest drinking

water wells are contaminated at levels above legal safety limits. In Iowa, authorities detected pesticides in at least half of the state's city wells. Meanwhile, based on a random survey, the EPA estimated last year that more than one-third of the country's 800,000 underground storage tanks are leaking motor fuels and chemical solvents into groundwater. "Not everyone lives next door to a toxic waste site," noted Larry Sil-

Chemical reactions
A U.S.-Canada study found that many Great Lakes residents have about 20 percent higher levels of toxic chemicals in their bodies than other North Americans.

verman of the Environmental Task Force, "but almost everyone has a fuel tank within a stone's throw."

Congress took an important step in 1986 toward controlling groundwater pollution by revising the 12-year-old Safe Drinking Water Act. The new law provides money to help protect aquifers that are the sole source of drinking water for an area. It also requires states to develop plans to safeguard other public water supplies, and it gives the EPA five years to set maximum levels for at least 110 of the hundreds of contaminants currently found in drinking water. In the past 12 years, the agency has set only about two dozen such standards.

While conservationists pointed out that further measures at both federal and state levels are still needed to adequately protect groundwater, they were buoyed by another congressional action: a five year, nine billion dollar extension of the federal Superfund program, which provides vital funds for cleaning up toxic dump sites. "It is a significant step forward in our efforts to clean up hazardous wastes and leaking petroleum tanks," observed Norman Dean, director of the National Wildlife Federation's pollution and toxics program.

ENERGY

| WORSE | SAME | BETTER |

Will America's Energy Conservation Gains Continue?

THE BIG story in energy last year was a combination of an oil glut, the collapse of OPEC and a dramatic decline in oil prices. By late summer, a barrel of crude sold for $12, and gasoline cost less than 80 cents per gallon in many parts of the country. The lower prices helped to keep inflation low and put more money in Americans' pockets.

The oil industry was seriously hurt as prices fell and domestic drilling dropped to the lowest level since 1971. Conservationists, however, pointed out that less exploration and development meant fewer adverse impacts on wildlife and fisheries.

The cheap oil also failed to increase U.S. energy consumption in 1986, as many environmentalists had predicted. The reason: the conservation measures implemented in the last decade are now so woven into the fabric of American life that they are not easily removed. As Howard Geller of the American Council for an Energy-Efficient Economy said: "People are not going to rip the insulation out of their walls."

Much of the credit for a more energy-efficient America belongs to automobiles. Because old gas guzzlers are being replaced by new, more efficient models, American drivers used roughly the same amount of gasoline last year as they did in 1979, even though they drove 21 percent more miles.

Even more dramatic gains have occurred in industry in the United States. In 1973, 27,000 BTUs (a BTU is a measure of energy) were expended for each dollar of Gross National Product. By 1986, only 20,500 BTUs were needed for the same amount of GNP.

Conservationists, however, thought that the nation should do better. They worried that cheap oil was reducing incentives to conserve, thus slowing the drive towards an even more energy efficient economy. In 1985, for example, the Transportation Department rolled back the fuel economy standard for 1986 model cars from 27.5 to 26 miles per gallon. And in 1986, the National Academy of Sciences warned that declines in government support for energy conservation research could prevent future gains.

"Conservation is one of the remarkable success stories of the last decade," said NWF Executive Vice President Jay D. Hair. "But we risk losing many of those gains in this time of cheap oil unless we continue to fight for even greater energy conservation."

Meanwhile, disputes over the use of federal lands simmered anew. In a defeat for environmentalists, the Interior Department agreed to hand over 82,000 acres of prime grazing and recreation land in Colorado to an oil consortium for a token $2.50 per acre. The award was based on decades-old shale-mining claims and set a damaging precedent for 28 million additional acres of federal land under other mining claims. "We are concerned that this giveaway will trigger other opportunists to acquire land for a song," said NWF energy expert Karl Gawell.

Environmentalists also battled against abusive mining practices. In a suit filed against the federal Office of Surface Mining, NWF charged that hundreds of coal mines are spoiling streams with acid discharges because the government had failed to enforce strip-mining regulations. Another suit accused Kentucky of failing to enforce regulations at its strip mines and neglecting to collect fees from mine operators.

Interior Secretary Donald Hodel's response, in a speech to the National Coal Council, was to suggest that environmental restrictions on strip mining are too strict.

Last year also brought renewed concern about the "greenhouse effect," a warming of the Earth caused by the burning of fossil fuels. In Senate hearings and in several new reports, most scientists agreed that carbon dioxide given off into the atmosphere when oil, gas, coal and wood are burned will (along with other gases) trap enough heat from the sun to cause everything from rising sea levels to potentially disastrous changes in climate. "Global warming is inevitable—it's only a question of magnitude and time," said Robert Watson of NASA.

The accident at the Soviet nuclear plant at Chernobyl, meanwhile, demonstrated that fossil fuel was not the only energy source that carried risks. Beyond the horror of the immediate radiation deaths, abandoned farms and tainted foods, there will be additional long-term costs in both human health and environmental damage.

Environmentalists continued to hope that renewable energy could replace significant amounts of oil, coal and nuclear power. In general, however, 1986 was a bad year for alternative energy. Low oil prices and the phasing out of tax credits made many projects using wind, solar or geothermal power

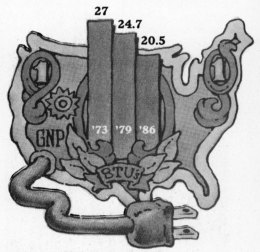

27   24.7   20.5   '73 '79 '86   GNP   BTUs

A more efficient economy?
Over the last decade, the number of BTUs (British Thermal Units) needed to produce a dollar of the country's Gross National Product has dropped dramatically.

uneconomical—and caused a number of companies to go out of business.

Still, renewable energy made significant technological strides and reached some milestones. In California, wind generators produced more than twice as much electricity as in the previous years. And by 1990, the California Energy Commission estimated, wind and hydro power will be the two cheapest energy sources in the state.

In April, a Stanford University solar cell set a new record for efficiency of conversion of sunlight into electricity. In June, Alabama Power Company officials dedicated a 100-kilowatt solar power plant, the first large-scale facility to use the thin-film technology that now powers solar calculators and watches. The plant cost more than conventional fossil-fuel plants, but analysts predicted that solar energy will be competitive with other sources of power for utilities within ten years.

$32 (1981)

$7 (1975)

$12 (1986)

More oil for the money
A combination of an oil glut and the collapse of OPEC caused gasoline and oil prices to plummet. And as a result, the lower prices helped keep inflation low.

For The Nation's Public Forests, The Heat Is On

THESE are troubled times for the nation's "woodbaskets"—the rich forests of the Northwest and South. Forest-products companies, having depleted virtually all of the valuable virgin forests on western private land, are fighting to harvest portions of the three million acres of this "old growth" that remains on public forests.

Often, the industry succeeds. Moments before a U.S. District Court judge was to convene a hearing last April on whether a 56-acre tract of 700- to 900-year-old Douglas fir trees in Oregon should be spared the woodsman's ax, conservationists withdrew their lawsuit. The reason: the trees had already been felled by Willamette Industries, Inc. "What may have been the oldest living things in Oregon are dead," said National Wildlife Federation attorney Terence Thatcher afterward.

The loss of the massive trees near Sweet Home, Oregon, illustrates the increasing struggle over a dwindling western resource. Clearcut stands of "old-growth" trees require centuries to regain their former stature; recently they have been exploited so rapidly that most cutover forests in the region have not yet reached harvestable size. That is one reason why the forest industry increasingly is relying on the South, where favorable climate and geography allow foresters to grow genetically refined "supertrees" that produce wood in record time.

But the South's rise in forestry was clouded last summer by a disturbing report that trees in the region are growing more slowly than previously—and that air pollution may be a factor. For the first time, the U.S. Forest Service's periodic study of southeastern pine forests found that the rate of growth in many species had declined. Yellow pines—the most important species for commercial uses—showed as much as 30 percent less growth between 1972 and 1982 than in the prior decade.

The study also found sharply higher numbers of tree deaths—15 percent among yellow pines compared to 9 percent a decade ago. While some of the mortality may be attributable to recurring drought and infestations of pine bark beetles, the Forest Ser-

vice said it could not rule out the impacts of pollution. The troubling symptoms of decline are similar to those noted as much as ten years ago in West Germany and other European countries, where scientists blame air pollutants for catastrophic damage to forests.

Even without the growth decline, the South's future as a center of forestry is coming under question. The Forest Service reported last year that some 85 percent of cutover forestlands in the region are not being reforested.

Concern over the future of forest resources added heat to the battle over the 191 million acres of national forest land last year,

30% LESS GROWTH

1961–1972   1972–1982

Stunted growth
Pine trees in the South grew less and died more often during the last decade than they did in the 1960s. Researchers now cite air pollution as one suspect.

as the U.S. Forest Service reached the halfway point in a planning process that will decide the fate of 155 national forests in 44 states and two territories.

The agency has tempered some of its early drafts, which almost universally called for massive increases in timber harvesting, even in forests that contain low-value lumber but are heavily used for recreation. Reacting to a prompt and angry response from conservationists and the general public, the government is now contemplating some reductions in timber cutting—a 20-percent decrease in Oregon and Washington, for example.

Still, of the first 32 forest plans completed, 22 prompted 139 separate appeals—the majority of them protesting overly generous allocations to timber firms. A proposal to increase clearcutting 300 percent and to increase government-funded road building in California's Tahoe National Forest drew more than 6,000 written objections from local citizens who opposed the

expensive destruction of 30,000 acres of old-growth forest and wildlife habitat for more than 100 species.

Conservationists charged that federal costs in building roads to harvest such timber frequently exceed the value of the wood. "We're not the economic conscience of the country—we object to a lot of profitable timber sales, too," said NWF regional executive Thomas Dougherty. "But when they plunder the forest and taxpayers have to shell out for it, there's a rat in the woodpile."

A Historic Step Forward in The War Against Soil Erosion

THE NATION took a historic step forward in the fight against soil erosion last year as the United States Department of Agriculture (USDA) began implementing key provisions of a new farm bill. The law, which passed late in 1985, authorized the government to create a "conservation reserve" by paying farmers for each acre of highly erodible land that they take out of crop production and replant to soil-saving vegetation. By November, nearly nine million acres of land had been placed in the reserve.

The new law also contains a "sodbuster" provision that penalizes farmers who plow and plant fragile grasslands without approved plans to control wind and water erosion. What's more, the farm bill says that by 1990, all farmers must have soil-conservation plans for their entire farms or they will lose crop subsidies and other farm assistance. "The government is through financing soil erosion," says Peter C. Myers, deputy secretary of USDA. "We're saying, if you want to abuse your soil, do it on your own."

The actions were greeted enthusiastically by conservationists, who have long argued that much of the blame for the current high rate of soil erosion rests with the federal government. For years, federal subsidies have encouraged farmers and speculators to plow up windbreaks and shelter belts, and to expand crop production onto marginal land, thus accelerating the loss of soil. Through the conservation reserve, the government hopes to retire as much as 45 million acres—about 10 percent of U.S. farmland—by 1990.

Fragile lands
Forty-one million acres of highly erodible farmland (10 percent of the nation's total cropland) are responsible for nearly half of the soil erosion in the United States.

In addition to reducing erosion, the conservation reserve program also promises to benefit wildlife. In contrast to farmland that is retired for one season in order to regulate harvests, land enrolled in the conservation reserve must be kept in vegetation for at least ten years, and cannot be grazed or cut for hay. The resulting meadows and groves could restore crucial wildlife habitat.

"The rejuvenated habitat will help restore numerous game and nongame birds to areas where agriculture had eliminated them," said W. Alan Wentz, director of the National Wildlife Federation's Fisheries and Wildlife Division. "It might be possible to reintroduce ring-necked pheasant in some areas."

Despite the high expectations, however, the soil conservation reserve was not as popular as UDSA officials had hoped. It took three rounds of bidding instead of the originally planned single round to push the total amount of land enrolled in the program up to the nine million acre level. The reason was cold cash: despite an average payment of $45 per acre per year from the USDA for land placed in the reserve, farmers could get even more money in crop supports by continuing to cultivate their erodible land. "We're fighting our own farm programs," admitted Myers. "The deficiency payments [crop subsidies], even on marginal corn land in places like Missouri, are higher than what we can pay in the reserve."

USDA officials, however, believe the program will do better in future years. In 1990, they explain, farmers will lose the lucrative crop subsidies on their erodible land unless they comply with conservation plans that could require the construction of expensive terraces and waterways.

Meanwhile, soil erosion from the nation's farmland continued at an estimated rate of 2.7 billion tons of soil each year. Nearly half of it came from just 41 million acres of highly erodible cropland in western Kentucky and Tennessee, the Delta region of Mississippi, eastern Washington, and sections of Iowa, Kansas and Nebraska. Yet, according to a National Academy of Sciences (NAS) report released in May, most of the money spent on such conservation practices as reduced tillage or strip-cropping has been used to finance projects in regions with much lower erosion rates.

Experts have estimated the annual losses in crop yields due to soil erosion at $500 to $600 million per year. But a number of studies have concluded that erosion causes

The high cost of erosion
Annual damages off the farm, caused when silt and pesticide residues are carried into rivers and lakes, are far greater than the losses due to lower crop yields.

far more damage off the farm, primarily by carrying silt and residues from pesticides and fertilizers into streams, rivers and lakes. One survey by state officials, for example, discovered that agriculture was the chief culprit in 57 percent of lakes polluted by runoff from the surrounding land. In addition, a number of rivers and reservoirs must be dredged frequently because they keep filling up with silt. Experts estimate that the total damage caused by erosion off the farm is between $4 and $16 billion per year.

Still, 1986 was a year for cautious optimism. The soil conservation reserve and the "sodbuster" provision promised future gains in the battle against erosion, and several experts suggested that the worst is behind us. "Soil erosion," concluded M. Gordon Wolman, the Johns Hopkins geographer who chaired the NAS study, "is not an imminent disaster."

Americans Are More Informed— And More Worried

THROUGHOUT 1986, there were sobering signs that the nation's environmental problems were striking closer to home, threatening the quality of life of many Americans who have always believed that "it can't happen here."

During one period last winter, for instance, the nation's highest levels of carbon monoxide were detected in Phoenix, Arizona, a city long considered by doctors to be a safe haven for people with asthma. Residents with respiratory problems were urged to stay indoors.

In Short Hills, New Jersey, where the average annual household income exceeds $91,000, homeowners discovered that wealth provides little protection against the threat of pollution. A Council on Economic Priorities study found that the community lies close to about 25 abandoned hazardous waste dump sites.

Meanwhile, in rural Gallatin County, Montana, residents feared for both their livelihoods and the condition of the area's pristine mountain streams. Reason: despite a study which found that trout fishing and other angling brings more money into the local economy than logging, a federal plan has targeted parts of the region for timber, oil and gas development.

"None of us can bury our heads in the sand," observes Jay D. Hair, executive vice president of the National Wildlife Federation. "We all drink the water. We all breathe the air. We must all be concerned with the quality of our environment." Clearly, many

What are your concerns?

To find out how people rate the quality of their environment, and what problems they are most concerned about, National Wildlife *posed these questions to readers.*

Americans are concerned, and they are willing to do whatever is necessary to clean up and protect the environment in the United States in years to come.

In a cross-section survey of some 1,300 readers of *National Wildlife*, 91 percent of the respondents said that they would rather pay higher taxes than have the federal government reduce budget deficits by cutting back on important pollution clean-up programs. Even with those programs, 48 percent of the readers believed that their environment would be less livable five years from now than it is today. Only 11 percent believed it would be better. (In a similar poll in 1985, 17 percent of the respondents thought it would be better.)

The survey, conducted late last year, found that drinking water contamination is considered the country's greatest environmental threat. (Leaking hazardous waste sites, a major cause of groundwater pollution, ranked second.) As with the previous year's poll, readers ranked the threat posed by indoor air pollution last, perhaps reflecting the continued lack of information on this dangerous problem. Acid rain, a subject that has received widespread attention, was considered a serious threat by more than half the respondents.

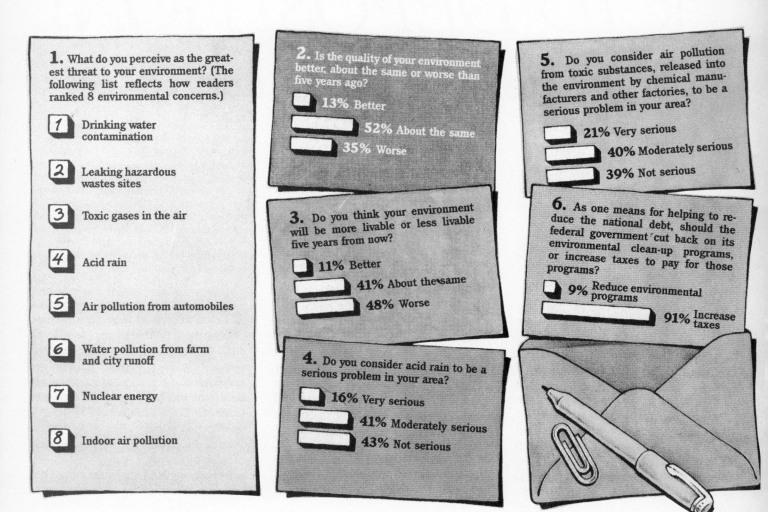

1. What do you perceive as the greatest threat to your environment? (The following list reflects how readers ranked 8 environmental concerns.)

1. Drinking water contamination
2. Leaking hazardous wastes sites
3. Toxic gases in the air
4. Acid rain
5. Air pollution from automobiles
6. Water pollution from farm and city runoff
7. Nuclear energy
8. Indoor air pollution

2. Is the quality of your environment better, about the same or worse than five years ago?
- 13% Better
- 52% About the same
- 35% Worse

3. Do you think your environment will be more livable or less livable five years from now?
- 11% Better
- 41% About the same
- 48% Worse

4. Do you consider acid rain to be a serious problem in your area?
- 16% Very serious
- 41% Moderately serious
- 43% Not serious

5. Do you consider air pollution from toxic substances, released into the environment by chemical manufacturers and other factories, to be a serious problem in your area?
- 21% Very serious
- 40% Moderately serious
- 39% Not serious

6. As one means for helping to reduce the national debt, should the federal government cut back on its environmental clean-up programs, or increase taxes to pay for those programs?
- 9% Reduce environmental programs
- 91% Increase taxes

ACID DEPOSITION:

Trends, Relationships, and Effects

Arthur H. Johnson

ARTHUR H. JOHNSON, a soil scientist and professor at the University of Pennsylvania, has served on three National Research Council committees and panels involved with acid deposition. This article is based on the most recent National Research Council report, *Acid Deposition: Long-Term Trends* (Washington, D.C.: National Academy Press, 1986). The figures in the article are adapted from the report.

Over the past decade, acid deposition has grown into a major, widely publicized environmental issue for Canada and the United States. "Acid rain" has come to symbolize the spread of air pollution to large areas of natural systems thought to be vulnerable to adverse effects, and the costs of controlling it are perceived to be far-reaching. Uncertainty over the effects of acid deposition and related pollutants has stimulated research directed, ultimately, toward determining national policy on emission levels of nitrogen and sulfur oxides, the gaseous precursors of acid deposition.

The causes of acid deposition are now well known. In eastern North America, fossil-fuel combustion is responsible for more than 90 percent of the sulfur and nitrogen oxides (SO_x and NO_x) emitted to the atmosphere.[1] When fossil fuels that contain sulfur are burned, sulfur oxides are formed. Nitrogen oxides are produced from atmospheric nitrogen during high-temperature combustion of

fossil fuel.[2] Sulfur dioxide (SO_2) and nitrogen oxides are converted in the atmosphere to sulfuric and nitric acids (H_2SO_4, HNO_3), which are removed from the atmosphere by wet and dry deposition processes (see Figure 1 on page 8). (See also Steven L. Rhodes and Paulette Middleton, "The Complex Challenge of Controlling Acid Rain," and Gregory S. Wetstone and Sarah A. Foster, "Acid Precipitation: What Is It Doing to Our Forests?" *Environment*, May 1983.)

The effects of acid deposition have been studied and debated since the mid-1960s, when Europeans began to ask whether the deterioration of water quality in Scandinavian lakes was related to atmospheric acids. Many claims have been made about the effects of acid deposition, and the scientific and popular literature have been forums for sharp debates. From time to time, scientists have searched for areas of agreement on what acid deposition is actually doing, and have tried to identify studies needed to clarify unresolved

issues. As a step toward consensus on a sufficient number of issues to formulate policy, the Committee on Monitoring and Assessment of Trends in Acid Deposition was formed under the auspices of the Environmental Studies Board of the U.S. National Research Council. This 14-member committee reflected a broad range of views on several key issues regarding the effects of acid deposition.

Before reaching its conclusions, the committee both reviewed the literature and examined data sets selected on the basis of the following three primary criteria:

- A postulated direct or indirect relationship to acid deposition.
- Availability of published data or, preferably, of original data with sufficient documentation to make peer review possible.
- Availability of data representative of broad geographical regions and/or temporal data with unambiguous dating.

From *Environment*, May 1986, pp. 6-11, 34-39. Reprinted with permission of the Helen Dwight Reid Educational Foundation. Published by Heldref Publications, 4000 Albemarle St., N.W., Washington, D.C. 20016. Copyright © 1986.

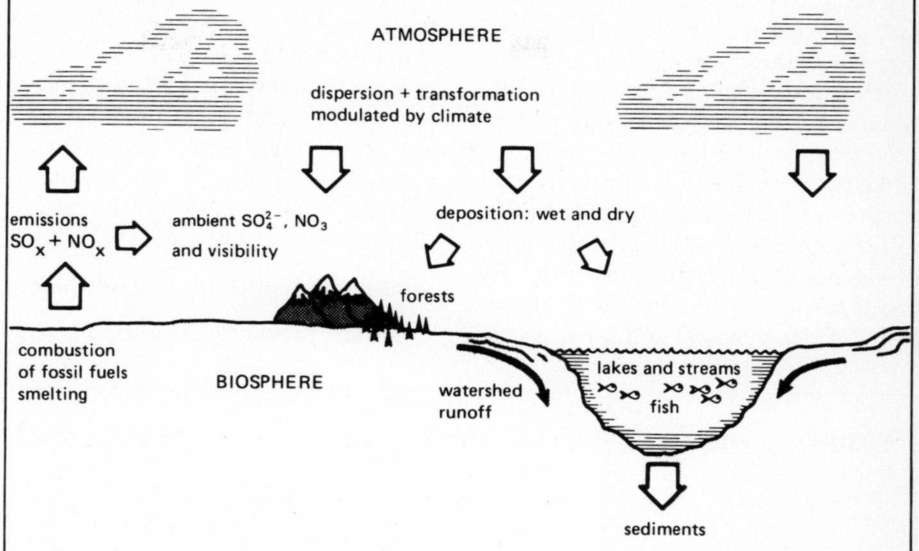

By studying trends in a variety of phenomena thought to be related to acid deposition, the committee discovered some interesting patterns that formed the basis for its conclusions (see box on page 189). The resulting report[3] covers eight topics: emissions of acid precursors (particularly sulfur dioxide); atmospheric sulfates and visibility; precipitation chemistry; acid status of surface waters; fish populations; the biological and chemical record contained in lake sediments; patterns in tree rings; and the role of climatic fluctuations.

In the ongoing debate about the desirability of controls on emissions, the establishment of cause and effect has received considerable emphasis. In essentially every case in which acid deposition has been suspected to be a factor in causing a change in the environment, an alternative explanation based on other human activities or natural phenomena has been proposed. Because many factors complicate our understanding of the mechanisms governing the effects of acid deposition, it is extremely difficult to establish cause and effect rigorously; also, agreement on the necessary criteria is difficult to obtain in an emotionally and politically charged issue.

In trying to establish cause and effect, the committee's first step was to determine whether an association existed between temporal trends and spatial patterns in acid deposition and the postulated effect. Two variables can be considered to be associated if their values are paired in a related way across a population. For example, sulfate and nitrate in wet deposition are associated across eastern North America: regions with high wet sulfate deposition also tend to have high wet nitrate deposition, and vice versa. In this case, the variables are correlated, but neither variable is the cause of the other. Thus, a statistical association between variables is a necessary, but not sufficient, condition for inferring causality.

The committee recognized three criteria—consistency, responsiveness, and mechanism—for establishing cause and effect when two variables are associated.[4] At least two of these criteria are needed to support causation. Consistency implies that the relationship be-

FIGURE 1. Acid deposition: diagram of sources and affected ecosystems.

tween variables holds in each data set across populations in direction, or perhaps even in amount. Responsiveness implies that if the causal variable is manipulated, the dependent variable will change in an appropriate and predictable direction. A test of responsiveness involves experimentation and, in the case of regional-scale phenomena involving complex ecosystems, there are many problems in establishing satisfactory experiments. Mechanism is established if there is a step-by-step pathway between the cause and the effect and if the processes are known at each step.

The criteria of mechanism and consistency were used to suggest whether a cause-and-effect relationship exists between acid deposition and the other environmental phenomena investigated. The report contains not only a review of the literature, but also the results of extensive new analyses of existing data. The findings were reviewed by more than 30 anonymous reviewers, thus expanding the level of consensus considerably beyond the members of the committee. The committee's findings are explained below with a summary of the key evidence that led to its conclusions.

Acid Precursors

Sulfur and nitrogen oxides and their

transformation products are deposited from the atmosphere onto the Earth's surface as acids or acidifying substances. Because acid deposition involves both the gaseous precursors and the acidic reaction products removed from the atmosphere by wet and dry deposition, an understanding of trends in acid deposition starts with an understanding of trends in emissions of sulfur and nitrogen. Using a sulfur-flow accounting scheme, the committee calculated sulfur emissions from coal combustion. Most of the current emissions —an estimated 11 million to 15 million tons of sulfur per year in eastern North America—are the result of coal combustion. Sulfur emissions from the burning of oil and from copper and zinc smelting were calculated using a similar flow-accounting scheme, and these estimates were added to the coal-related emissions to give trends of sulfur emissions for five regions of eastern North America (see Figure 2).

Sulfur emissions increased most rapidly from the end of the last century until about 1910, as indicated in Figure 2. Since then the roughly 50 percent increase in sulfur emissions has been characterized by peaks and dips. Economic factors and technology changes resulted in peak periods in the 1920s, early 1940s, and mid-1960s to early 1970s.

7. THE ENVIRONMENT

Until about 1970, trends in the five regions were similar. Strong differences in regional trends have since appeared as a result of varying patterns of industrial development. While in regions A and B (southeastern Canada and the northeastern United States), sulfur emissions have decreased, they have risen sharply in region C (southeastern United States). In the midwestern and north central regions (regions D and E), sulfur emissions show little change. This divergence of trends after 1970 is a key factor in some of the analyses discussed below. Since trends of sulfur emissions

diverged, there should have been different directions of change in environmental attributes affected by acid deposition.

Nitrogen oxides are related more to the nature of the combustion process than to the properties of fossil fuel. The estimated trends in emissions of nitrogen oxides, determined from inventories of emission factors[5] and fuel consumption estimates, show a monotonic increase in all regions since the turn of the century.

Atmospheric Sulfate and Visibility

Sulfate, present in the atmosphere as

fine particles, can scatter light and reduce visibility. The exact proportion of visibility reduction attributable to sulfate is unknown and variable, since many other particles such as road dust and water droplets also reduce visibility. Some investigators have estimated that atmospheric sulfate is responsible, on average, for about 50 percent of the reduction in visibility in the eastern United States.

Spatial patterns of visibility range are consistent with patterns of sulfate distribution (see Figures 3 and 4). The best air quality occurs in the mountain and

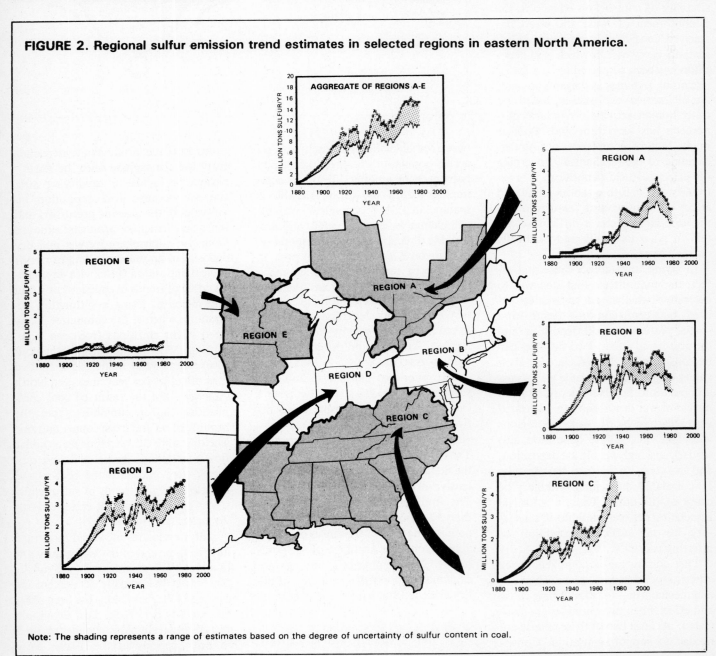

FIGURE 2. Regional sulfur emission trend estimates in selected regions in eastern North America.

Note: The shading represents a range of estimates based on the degree of uncertainty of sulfur content in coal.

HIGHLIGHTS FROM THE CONCLUSIONS OF THE COMMITTEE ON THE MONITORING AND ASSESSMENT OF TRENDS IN ACID DEPOSITION

National Research Council

1.   In eastern North America a causal relationship exists between anthropogenic emissions of sulfur dioxide (SO_2) and the presence of sulfate aerosol, reduced visibility, wet deposition of sulfate, and sulfate in small watersheds that do not have large internal sources of sulfur. Magnitudes of SO_2 emissions and deposition of sulfur oxides are highest in a region spanning the midwestern and northeastern United States.

2.   Acid precursors, particularly SO_2, have been emitted in substantial quantities into the atmosphere over eastern North America since the early 1900s. In particular, SO_2 emissions in the northeastern quadrant of the United States have fluctuated near current levels since the 1920s.

3.   Substantial differences in temporal trends in SO_2 emissions among regions of the United States have emerged since about 1970. The southeastern United States has experienced the greatest rates of increase in SO_2 emissions and related parameters; the midwestern United States has experienced lesser rates of increase; the northeastern United States has experienced modest decreases.

4.   Changes in stream sulfate flux appear to determine changes in stream water alkalinity and base cation concentrations in drainage basins that have acid soils and low-alkalinity waters, with other factors involved in some cases. The sulfate output from lakes in northeastern North America is, in general, proportional to sulfate inputs in wet deposition. Dry deposition is suggested as an important contributor to sulfate inputs, particularly near major source regions.

5.   Changes in lake-water alkalinity greater in magnitude than 150 $\mu eq/L$ can occur over time periods of about 50 years. Changes of this magnitude are too large to be caused by acid deposition alone and may result from other human activities or natural causes.

6.   Based on diatom stratigraphy, six out of the ten Adirondack Mountain lakes for which there are data became increasingly acidic between 1930 and 1970. Acid deposition is the most probable causal agent.

7.   Over the last five decades, on average, lakes sampled in Wisconsin have increased in alkalinity and pH. The New Hampshire lakes on average show no overall change in alkalinity and a small increase in pH. The weight of the chemical and biological evidence indicates that atmospheric deposition of sulfate has caused some lakes in the Adirondacks to decrease in alkalinity. We cannot quantify the number of New York lakes that have been affected by acid deposition.

8.   Emissions of oxides of nitrogen (NO_x) are estimated to have increased steadily since the early 1900s with an accelerated rate of increase in the Southeast since about 1950. Reliable data do not exist to determine long-term trends of nitrate concentrations in the atmosphere, precipitation, and surface waters.

9.   Available data indicate that fish populations decline concurrently with acidification. The strongest evidence comes from the Adirondack region.

10.   Geographically widespread reductions in tree-ring width and increased mortality of red spruce in high-elevation forests of the eastern United States began in the early 1960s and have continued to the present. The roles of competition, climatic and biotic stress, and acid deposition and other pollutants cannot be adequately evaluated with current data.

—J.S.

desert areas of the Southwest; the worst quality occurs east of the Mississippi River and south of the Great Lakes. In the rural eastern United States, both light extinction (the inverse of visibility) and sulfur emissions decreased from the 1940s to the 1950s, increased until about 1970, then leveled off (see Figure 5). The southeastern United States experienced the greatest deterioration in visibility from the 1950s to the 1970s, also consistent with the trends in sulfur emissions.

Precipitation Chemistry

Sulfur and nitrogen oxides and their reaction products are returned to the Earth's surface by a variety of mechanisms. The best known and most studied mechanism is deposition in rain and snow. The chemistry of precipitation has been measured with modern techniques for three decades, but the methods of collection, processing, and analysis have varied widely, and have evolved through time. These differences in procedures are critical, and make comparisons of precipitation chemistry over time difficult. Based on analyses of many data sets, the following regional-level conclusions were reached by the committee:[6]

• The eastern half of the United States and southeastern Canada south of James Bay experience concentrations of sulfate and nitrate in precipitation that are, in general, greater by at least a factor of five than those in remote areas of the world, indicating that levels have increased by this amount in northeastern North America since some time before 1950 (see Figure 6).

• The northeastern quadrant of the United States (region B, the northernmost states of region C, and most of region D) is the area most heavily affected by acidic species (the ions of hydrogen, sulfate, and nitrate) in precipitation.

• Precipitation is currently more acidic in parts of the eastern United States than it was in the mid-1950s or mid-1960s; however, the amount of change and the mechanisms for the change are in dispute.

• Precipitation sulfate concentra-

FIGURE 3. Nationwide distribution of annual concentrations of particulate sulfate (in micrograms per cubic meter, $\mu g/m^3$), 1970–1972.

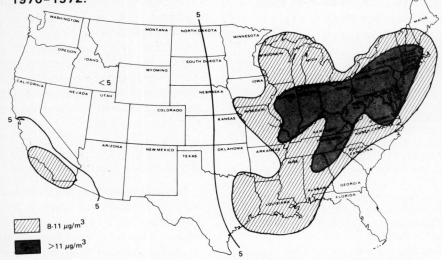

8-11 $\mu g/m^3$

>11 $\mu g/m^3$

SOURCE: N. Frank and N. Possiel, "Seasonality and Regional Trends in Atmospheric Sulfates" (Paper presented at the meeting of the American Chemical Society, San Francisco, California, August 30 to September 3, 1976).

FIGURE 4. Median annual visual range (in miles) at suburban/nonurban locations in the United States, 1974–1976.

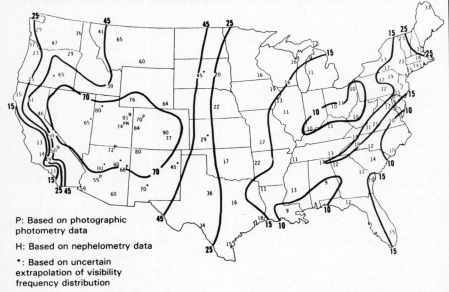

P: Based on photographic photometry data

H: Based on nephelometry data

*: Based on uncertain extrapolation of visibility frequency distribution

SOURCE: J. Trijonis, "Existing and Natural Background Levels of Visibility and Fine Particles in the Rural East," *Atmospheric Environment* 16(1982):2431–2445.

tions and possibly acidity have increased in the southeastern United States (region C) since the mid-1950s.

• Trends in precipitation sulfate at three stations in the Northeast suggest that sulfate has decreased there since the mid-1960s, but data from other sampling sites in the Northeast do not show such a trend.

• Regional data on precipitation chemistry are suitable for comparison with spatial patterns in emissions and stream chemistry, and the spatial association between sulfate in precipitation and sulfur emissions established previously[7] is supported.

Surface-Water Acidity

One of the most important concerns about acid deposition is the acidification of surface water. There has been much confusion both about the balance of natural and anthropogenic sources of surface-water acidity and about the mechanisms that cause surface waters to acidify. The committee conducted some new analyses to test the validity of the "mobile anion" (anions are negatively charged ions) model as a mechanistic link between acid deposition and surface-water acidification, and to determine if trends in surface-water acidity have changed.

The mobile anion hypothesis implicates sulfate as a major factor in surface-water acidification in areas with thin, acidic soils.[8] The model suggests that if sulfur deposition increases:

• sulfate concentration in surface waters increases;

• concentrations of acid cations (hydrogen and aluminum species) and base cations (calcium, magnesium, sodium, and potassium) increase to balance the increase in sulfate; and

• the increase in acid cations leads to a decrease in surface-water alkalinity.

This model is based on the requirement that a charge balance must be maintained in all aqueous solutions. Thin acidic soils are expected to counter an increase in sulfate flux with mostly acid cations, while less acidic soils should counter the additional sulfate with base cations. Thus the effect of sulfate deposition and leaching on surface-water alkalinity is dependent on watershed characteristics. Sulfate is not mobile in all soils. Many sources and sinks in the terrestrial part of the watershed cause its cycling to be quite complex, hence the sulfate in stream water might not be proportional to sulfate inputs in some cases.

Overall, the acidification of surface waters by sulfate deposition takes place when: sulfate moves readily through the soils; the terrestrial system supplies acid cations, rather than base cations, to match the increased sulfate flux; and in-stream or in-lake processes do not increase alkalinity by chemical reduction of the incoming sulfate. Although nitric acid from snow-melt run-off has been implicated in acidification, nitrate is normally retained by the biota of the terrestrial portion of the system and is generally not a mobile anion.

Surface-water chemistry may be affected by several factors, the sources of which may be difficult to sort out with limited data. Disturbances within the watershed (for instance, those caused by fire or logging), forest regrowth, and fluctuations in water flow may affect surface-water chemistry. Many of the naturally occurring acidifying processes would be expected to result in a reduction in base cation concentrations as surface waters acidify, while acidification caused by increased sulfate flux would be expected to result in increased base cation flux and increased acidity.

In testing the relationship between sulfate fluxes into lakes from atmospheric deposition and sulfate output fluxes, data from 626 lakes in the Northeast and in eastern Canada showed that the relationship between wet sulfate input and sulfate output from the lakes is linear when examined on the basis of contiguous regions (southern New England, New York, northern New England, Quebec, New-foundland, and Labrador). Lakes in regions nearest the sources of sulfate (southern New England and New York) had higher ratios of sulfate output to wet sulfate input than did regions remote from atmospheric sulfur sources. This suggests that wet inputs are not the only sources of sulfate and that dry deposition, or perhaps water-shed sources of sulfur, are important sources of lake-water sulfate. Overall, this analysis is consistent with the idea that sulfate deposited from the atmosphere is finding its way into surface waters in the Northeast in most types of watersheds.

Next, it was important to test the idea that changes in sulfate flux are countered by changes in alkalinity and in base cation flux. The data base selected for this analysis was a nationwide set of small streams, the chemistry of which

has been carefully monitored by the U.S. Geological Survey over the last 20 years. Analysis of the chemistry of the 16 low-alkalinity streams in the bench-mark stream network showed that on a nationwide basis, changes in sulfate flux were accompanied by changes in both alkalinity and base cation flux where there were no dominating internal sources of sulfate. For the 16 streams with average alkalinities of less than 500 microequivalents per liter (μeq/L), alkalinity changes balanced from zero to 78 percent of the sulfate changes, with an average of about 30 percent. These findings support the contention that changes in mobile anion flux affect surface-water alkalinity in areas of acid soils.

The third part of the surface-water analysis was designed to determine where long-term changes in surface-water alkalinity had occurred. For this purpose, the committee examined the 15- to 20-year records of chemistry at the U.S. Geological Survey benchmark streams, and analyzed recent and historical water-quality measurements from lakes in the Adirondacks, New Hampshire, and Wisconsin. After con-

siderable discussion, the committee agreed on the following findings:

Bench-Mark Streams

Trends in sulfate are discernible in Bench-Mark streams over a period extending from the mid-1960s to 1983. On a regional basis, those trends in stream sulfate are consistent with trends in SO$_2$ [sulfur dioxide] emissions. . . . In general in the northeastern United States (Region B) since the mid-1960s, the Bench-Mark streams have experienced no change or decreases in sulfate concentrations and no change or increases in alkalinity. In the Southeast (Region C), streams have shown increases in sulfate concentrations and have either shown no changes or have decreased in alkalinity. . . . Alkalinity and nonprotolytic cation [that is, base cations calcium, magnesium, sodium, and potassium] changes in low-alkalinity Bench-Mark streams over the 15- to 20-year period of record are consistent with changes caused by a changing flux of strong-acid anions [that is, sulfate and nitrate] from the atmosphere, but there is also evidence suggesting that internal watershed processes or in-stream processes govern trends in nonprotolytic cations and alkalinity. At many eastern low-alkalinity stations changes in strong-acid anions are sufficient to account for alkalinity and nonprotolytic cation changes, but at some western stations that receive low rates of SO$_4^{2-}$ deposition, sub-

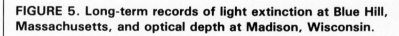

FIGURE 5. Long-term records of light extinction at Blue Hill, Massachusetts, and optical depth at Madison, Wisconsin.

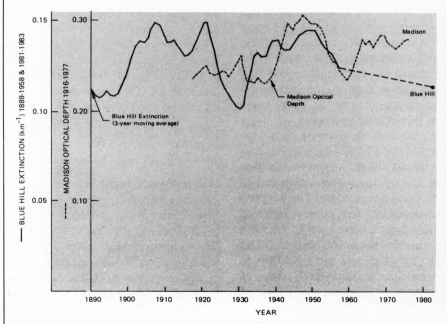

SOURCE: R. B. Husar, J. M. Holloway, and D. E. Patterson, "Spatial and Temporal Patterns of Eastern U.S. Haziness: A Summary," *Atmospheric Environment* 15(1981):1919–1928.

stantial reductions in alkalinity have occurred that have a large component not accounted for by changes in strong acid anions. . . . [These] may result primarily from changes in internal-terrestrial or instream processes.

Lakes in Wisconsin, New York and New Hampshire

Data from historical surveys of lake water parameters in Wisconsin, New York [Adirondack lakes], and New Hampshire are of varying quality, but sufficient data and documentation are available to test for internal consistency. Applying the criteria of Kramer and Tessier,[9] we obtained consistent sets of pH and alkalinity data for approximately 300 lakes for which recent data also exist. Comparing historical with recent data, it is possible to estimate pH and alkalinity changes over the past half-century.

This method is not without problems. The use of MO [methyl orange] indicator in the historical studies called for titration to the "faintest pink" endpoint. Estimates in the literature of the pH of this endpoint range between about 4.0 to higher than 4.3. The color of the endpoint in this range has been variously described as "faint pink," "pink," and "salmon pink." The Wisconsin survey provides the best documentation of the pH of the MO endpoint, citing a range of pH values from 4.06 to 4.18 for the "fainter" pink color, and recommending that a value of 4.18 be used in calculating titrations.

Our analysis shows that changes in alkalinity and pH in New York lakes are sensitive to the assumptions made about the pH of the MO endpoint. Assuming an endpoint pH of 4.2 (or greater), New York lakes, on average, appear to have increased substantially in acidic status over the past 50 years, as reflected in reductions in both alkalinity and pH. Alternatively, if a value of the [MO] pH endpoint close to 4.0 is assumed and if the historical data are compared with the 1984 data set, then there appears to have been little change, on average, in the acidic status of New York lakes over the past 50 years. However, if the historical data are compared with the 1980 New York data set, there may have been an overall decline in alkalinity [regardless of the endpoint assumption]. New Hampshire lakes do not appear to be so sensitive to the choice of the endpoint pH of MO. If an endpoint pH close to 4.2 is assumed, these lakes, on average, show little or no decrease in alkalinity and show a slight increase in pH. . . . Wisconsin lakes, on average, appear to show a significant increase in alkalinity (38 μeq/L) and pH (+0.5 unit) if an MO endpoint of 4.2 is assumed.[10]

While this summary suggests some ambiguity about whether Adirondack lakes have acidified over the past 50 years, other data led to the conclusion that lake acidification did indeed occur. Those findings follow.

Lake Sediments

Siliceous remains of diatoms (single-celled algae) and chrysophytes (plankton with silica scales) that are preserved in the bottom sediments of lakes provide additional evidence of trends in surface-water acidity. The distributions of diatom and chrysophyte taxa are closely related to water chemistry and are strongly correlated with lake-water pH. Consequently, changes in these faunal assemblages can be used to detect changes in water quality and thus surface-water acidification can be inferred from analysis of cores taken from lake-bottom sediments.

Diatom and chrysophyte data were analyzed for 31 lakes for the committee's report. Shifts in species distributions toward those that are found in acidic lakes were evident. The data sets were available for poorly buffered lakes, which are expected to be most susceptible to acidification. The evidence of recent acidification is strongest for the Adirondack lakes—of the ten clear-water lakes studied, six of the seven with a current pH at or below 5.0 show evidence of recent acidification. None of the lakes with a current pH above 5.0 shows any evidence of substantial pH decline. The most rapid

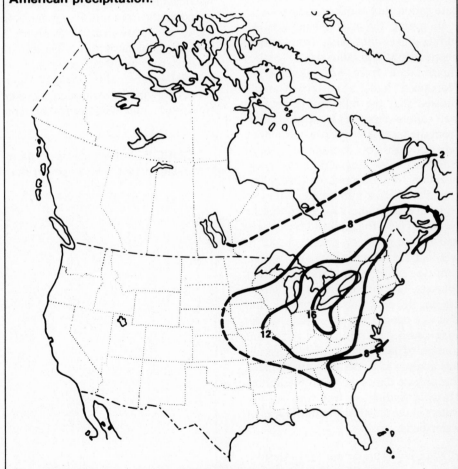

FIGURE 6. Enrichment factors for sulfate concentrations in North American precipitation.

Note: Numbers signify the ratio of sulfate concentration in precipitation in eastern North American to sulfate concentration in precipitation in remote areas of the world.

SOURCE: J. N. Galloway, G. E. Likens, and M. E. Hawley, "Acid Precipitation: Natural Versus Anthropogenic Components," *Science* 226(1984):829–831.

pH changes (decrease of 0.5 to 1.0 pH unit) occurred between 1930 and 1970. While several other factors involving changes in land use might contribute to lake-water acidification, analysis of the available land-use data for these Adirondack watersheds led to the conclusion that the recent rapid declines in lake-water pH are related to increased acid deposition rather than to watershed disturbances.

These findings are consistent with the trend in water-chemistry data that show on average a reduction in pH and alkalinity in Adirondack lakes over the past 50 years. In addition, the diatom data for lakes in Vermont, New Hampshire, and Maine are consistent with the New Hampshire water-chemistry comparisons—no substantial pH reduction was detected.

The committee also analyzed trends in heavy metals and other products of smelting and fossil fuel combustion (for instance, polycyclic aromatic hydrocarbons) that are also present in lake sediment cores. These data are generally consistent with the historical trends in fossil fuel and smelting emissions, indicating that, starting in the late 1800s, the products of smelting and fossil fuel combustion were deposited in lakes of the Adirondacks and New England.

Changes in Fish Populations

Much of the concern over acid deposition results from the perception that fish populations are dwindling or being eliminated in lakes and streams of sensitive regions. Changes in fish populations, however, can have many causes unrelated to acid deposition. Because of the lack of records of changing land use and changing stocking practices, it is often difficult to determine confidently why fish populations changed. In the cases cited below, the available evidence suggested that factors other than acidification were probably unimportant in causing the changes observed in fish populations. The committee used a number of high-quality data sets on changes in fish populations and found:

• The strongest evidence that increased surface-water acidity has reduced or eliminated fish populations

comes from data collected over the past 20 to 40 years in Adirondack lakes and in Nova Scotia rivers. Both data sets clearly show declines in acid-sensitive species over that period. The limited water-chemistry data suggest that these lakes and streams are more acidic than they were previously, and that fish population status is related to present pH. A similar result was obtained from analyzing fish and stream-chemistry data from Massachusetts.

• Data from Pennsylvania, Vermont, and Maine show that fish population status is related to pH. Limited temporal data indicate that some of the waters that used to support fish populations currently have summer pH values in the range of 5.0 to 5.5 and are now fishless. No old pH data are available for comparison.

• Experimental data indicate that waters at pH 5.0 to pH 5.5 will not support introduced fish populations, and that if surface waters are acidified to that level fish populations will be eliminated.

Overall, spatial and temporal trends in surface-water chemistry and in fish populations are consistent with the trends in sulfur emissions. In addition, the mobile anion hypothesis is supported as a major mechanism linking acid deposition to surface-water acidification. Thus there are grounds for inferring cause and effect in the chain of relationships beginning with sulfur emissions and ending in the acidification of surface waters and the elimination of fish populations. The impact on aquatic systems is reasonably clear. Much less clear at this time, however, are the nature and severity of impacts on terrestrial ecosystems.

Tree Rings and Forest Decline

In the case of forests, tree-ring data and historical reports of stand composition and mortality give some insight into changes that have occurred in the past few decades. Whether acid deposition and/or associated pollutants like ozone play a role in recent changes in North American forests is the least understood and probably most controversial topic studied by the committee.

The recent episode of forest decline in Germany and the early hypotheses that implicated soil acidification and acid deposition as major factors have sparked much speculation and concern about the impact of acid deposition on forest health in North America. The deterioration of red spruce populations on the mountains of the Adirondacks and New England has become one focus of this concern.

The committee examined quantitative studies that have been carried out over the past 25 years. These studies found reductions in density (the number of individuals per unit area) and basal area (the total cross-sectional area of the trees at breast height) of 50 to 70 percent in red spruce at some sites. Analyses of tree rings showed a pronounced reduction both in annual increment and in summer-wood density beginning in the early to mid-1960s, and several field reports indicate that substantial spruce mortality was noticeable about that time. Although the deterioration of red spruce—evaluated in terms of individual tree deaths and substantial top dieback—is greatest at high elevation where acid deposition is greatest, other natural stresses also increase with elevation. Much work is needed to evaluate the role of natural agents, particularly pathogens, before confidently judging the role of airborne chemicals.

In an analysis of the relationship between tree-ring width and climate, it appears that red spruce did not react to climate in a predictable way after the mid- to late 1960s and entered a period of decline. The committee suggested that the string of cold winters that began in the late 1950s may have had an important role in triggering the decline, as red spruce has a demonstrated sensitivity to cold winters.

According to the committee's criteria, the case of spruce decline cannot be linked to acid deposition because there is no documented mechanism. In addition, because a similar response of decreased annual ring width (but without abnormal mortality) has been observed in areas receiving relatively low rates of acid deposition, consistency between deposition rates and symptoms is not established. Ultimately, many possible mechanisms will be tested, and as the natural factors contributing to spruce decline are identified, it will be possible to rigorously evaluate whether acid deposition combines with those factors in a detrimental way.

Some Personal Comments

The findings of the Committee on Monitoring and Assessment of Trends in Acid Deposition are in agreement with the larger range of available data that suggest acid deposition has adverse effects on visibility, surface-water quality, and fish populations. While the conclusions are based on a narrow range of data, those data were subjected to rigorous quality checks. Under its own guidelines the committee was not reticent to make judgments on cause and effect.

The three-year study shows that there is no longer reasonable doubt that aquatic ecosystems are affected by sulfur emissions. The extent of the adverse effects, however, is not clear. Although it is reasonable to conclude that lakes in the Adirondacks have been affected, it does not appear that acid deposition has had a marked effect elsewhere in New England.

At present, there is no clear evidence that links acid deposition to changes in forests. The studies of recent changes in some forests in eastern North America are in an early stage, and the effects of acid deposition and airborne chemicals will receive additional attention.

I believe that the message is clear: acid deposition does cause changes in the environment. We do not know all of the long-term consequences of the changes. For instance, we do not know to what extent soils acidify under the influence of acid deposition, although data now emerging from Europe suggest that over half a century, some soils may have acidified to a greater than expected degree because of acid deposition. We do not know to what extent a reduction in sulfur emissions will reverse surface-water acidification, and we cannot yet predict the trends in water quality in regions like the southeastern United States, where surface-water acidification is delayed by the sulfur-retention characteristics of the soils. We are not certain that the products of fossil fuel combustion are assimilated benignly by forests, and we cannot predict the overall impact of many decades or centuries of airborne chemical deposition on the complex of processes that regulate terrestrial ecosystems.

In balancing the pros and cons of instituting controls, it seems to me that the issue now boils down to one of values more than to information that science can provide. Those who believe that the pride in practicing good stewardship of the environment is worth the monetary costs can demonstrate the reality of adverse environmental effects. Those who insist that more needs to be known before controls are instituted do so because they are willing to take risks with the future environment in order to avoid near-term economic costs. The understanding we have achieved about surface-water acidifica- tion has taken 20 years, and details are still lacking on some key questions. It is unlikely that there will be a clear consensus on the questions concerning terrestrial systems in the next 5 or perhaps even 10 years. The clear consensus, however, is that waiting for more answers from scientific investigations will be accompanied by demonstrable environmental costs.

NOTES

1. E. Robinson, "Natural Emission Sources," in A. P. Altshuller and R. A. Linthurst, eds., *The Acidic Deposition Phenomenon and Its Effects. Critical Assessment Review Papers,* vol. I, *Atmospheric Sciences* (Washington, D.C.: Environmental Protection Agency, 1984).

2. U.S. National Research Council, *Acid Deposition: Atmospheric Processes in Eastern North America* (Washington, D.C.: National Academy Press, 1983).

3. U.S. National Research Council, *Acid Deposition: Long-Term Trends* (Washington, D.C.: National Academy Press, 1986). Copies are available from National Academy Press, 2101 Constitution Avenue, N.W., Washington, D.C. 20418.

4. The model used followed the work of F. Mosteller and J. W. Tukey, *Data Analysis and Regression: A Second Course in Statistics* (Reading, Mass.: Addison-Wesley, 1977).

5. U.S. Environmental Protection Agency, *Compilation of Air Pollutant Emission Factors*, AP–42, 3rd ed. (NTIS PB-275525), Supplements 1–7 and 8–14 (Springfield, Va.: National Technical Information Service, 1977).

6. The first four bulleted items represent the language of the report, the last two are paraphrases by the author.

7. U.S. National Research Council, note 2 above.

8. U.S. National Research Council, *Acid Deposition: Processes of Lake Acidification* (Washington, D.C.: National Academy Press, 1984).

9. J. R. Kramer and A. Tessier, "Acidification of Aquatic Systems: A Critique of Chemical Approaches," *Environmental Science & Technology* 16(1982):606A–615A.

10. U.S. National Research Council, note 2 above, 287–289.

How Much Are the Rainforests Worth?

Philip M. Fearnside

Do the consequences of deforestation outweigh the immediate economic gain? Almost certainly, contends Philip Fearnside in his recent study, when one realizes what those consequences are. But many obstacles, he notes, stand in the way of winning that case. The forces that propel deforestation are a complex mix of direct self-interests and broader economic processes. Policies, once implemented, can set up vicious cycles from which it is difficult to escape—new and better roads encourage the influx of immigrants, the presence of migrants stimulates new efforts at road-building. Trade-offs must be weighed: large-scale sugar plantations, to ease Brazil's energy crisis through alcohol production, drove many small farmers off the land and into the virgin forests. "Nothing short of a comprehensive programme of government action based on conscious decisions," Fearnside argues, can be expected to halt deforestation before the Amazonian forests are lost. Dr. Fearnside is a resident researcher at the National Research Institute for Amazonia in Manaus, Brazil.
—Editor

The present rate and probable future course of forest clearing in Brazilian Amazonia is closely linked to the human use systems that replace the forest. These systems, including the social forces leading to particular land use transformation, are at the root of the present accelerated pattern of deforestation and must be a key focus of any set of policies designed to contain the clearing process.

Cattle pasture is by far the dominant land use in cleared portions of the *terra firme* (unflooded uplands), not only in areas of large cattle ranches, such as southern Pará and northern Mato Grosso, but also in areas initially felled by smallholders for slash-and-burn cultivation and annual crops, such as the Transamazon Highway colonization areas in

Pará. Pasture is even dominant in areas like Rondônia where government programmes have intensively promoted and financed cacao and other perennial crops. The forces leading to continued increase in pasture area, despite the low productivity and poor prospects for sustainability of this use system, are those that most closely affect the present rate of deforestation.

The extent and rate of deforestation in Brazil's Amazon rainforest is a subject of profound disagreement among both scholars and policy-makers in Brazil and elsewhere. Equally controversial is the question of whether or not the potential future consequences of deforestation are sufficient to justify the immediate financial, social, and political costs of taking measures to contain the process. The lack of effective policies to control deforestation in the Amazon today speaks for both the preference among decision-makers for minimizing such concerns and the strength of forces driving the deforestation process. But I would argue that deforestation is rapid and its potential impact severe, amply justifying the substantial costs of speedy government action needed to slow—and at some point to stop—forest clearing.

EXTENT AND RATE OF DEFORESTATION

The vast areas of as yet undisturbed forest in the Brazilian Amazon frequently lead visitors, researchers and government officials to the mistaken conclusion that deforestation is a minor concern unlikely to reach environmentally significant proportions within the "foreseeable" fu-

ture. Such conclusions are unwarranted; they also have the dangerous effect of decreasing the likelihood that timely policy decisions will be made with a view of slowing and limiting the process of deforestation. Not only is better monitoring information needed for accompanying the process, but also better understanding of the underlying causes of deforestation. Such understanding would allow more realistic projections of future trends under present and alternative policy regimes, and permit identification of effective measures to control the process.

The most recent available survey of deforestation covering the entire Brazilian Amazon was made by Brazil's Institute for Space Research (INPE) based on LANDSAT satellite images taken in 1978. The survey's finding that only 1.55 per cent of the area legally defined as Amazonia had been deforested up to that time contributed to the popular portrayal in Brazil of deforestation as an issue raised only by "alarmists". The INPE figure underestimates clearing due to inability of the technique to detect "very small" clearings and to the difficulty of distinguishing secondary growth from virgin forest.

For example, a 30,000 square kilometres (km²) region surrounding the town of Bragança in northeastern Pará that has been deforested since the early years of this century, is larger than the area indicated by 1975 images analysed in the INPE study to have been deforested in Brazil's entire Legal Amazon, and almost four times the area indicated as cleared in the state of Pará. Regardless of any underestimation due to image interpretation limitations, the conclusion that the area cleared through 1978 was small in relation to the 4,975,527 km² Legal Amazon is quite correct.

Unfortunately, the small area cleared by 1978 is a far less important finding than another less publicized one apparent from the same data set: the explosive rate of clearing implied by comparing values for cleared areas at the two image dates analysed, 1975 and 1978. If the growth pattern was exponential over the region as a whole during this period, the observed increase in cleared area from 28,595.25 to 77,171.75 km² implies an exponential growth rate of better than 33 per cent, and a doubling time of only a little over two years. Deforestation rates vary widely among different parts of the region, being highest in southern Pará, northern Mato Grosso, and in Rondônia and Acre. Comparisons of cleared areas for 1973, 1975, 1976, and 1978 in two areas of government-sponsored colonization by farmers with 100-hectare lots, and in two areas dominated by 3,000-hectare cattle ranches, indicate that deforestation in these areas may have been progressing in an exponential fashion during the period, although data are too few for firm conclusions.

LANDSAT image interpretation by the Brazilian Government for the state of Rondônia as a whole indicates that cleared areas rose from 1,216.5 km² in 1975 to 4,184.5 km² in 1978 to 7,579.3 km² in 1980. The cleared area therefore increased from 0.5 per cent to 3.1 per cent of Rondônia's total area in only five years, but it should be remembered that limitations of the image interpretation methodology mean that the true cleared areas were probably larger than these numbers imply. Even with this limitation, the clearing estimates reveal not only that deforestation proceeded rapidly throughout the period, but that it showed no signs of slowing as of 1980.

FORCES BEHIND DEFORESTATION

Some of the forces behind deforestation are linked to positive feedback processes, which can be expected to produce exponential changes. In Rondônia the population has been growing even more rapidly than in other parts of the region due to the flood of new immigrants from southern Brazil. Projections of unchanging exponential rates for deforestation, even in deforestation foci like Rondônia, are hazardous due to the many other factors affecting the process. As the relative importance of different factors shifts in future years, some of the changes will serve to increase deforestation rates, while others will slow them. Within completely occupied blocks of colonist lots, for example, clearing of virgin forest proceeds roughly linearly for about six years, after which a plateau is reached. The rate at which an individual lot is cleared is increased by such events as the arrival of road access and turnover of the lot's occupants.

At present, regional scale clearing statistics appear to be dominated by immigration, along with other forces such as the positive effect of improved road access on market availability and land value appreciation leading to accelerating deforestation. In the future, the behaviour of the population already established in the region should gain in relative importance. Other reasons for an eventual slowing (but not halting) of clearing include poorer soil quality and inaccessibility of remaining unoccupied land, the finite capacity of source areas to supply immigrants at ever increasing rates, decreased relative attractiveness of Amazonia after this frontier of unclaimed land "closes", and limits to available capital, petroleum and other inputs that would be necessary if rates of felling should greatly increase.

The accelerating course of deforestation cannot be adequately represented in any simple algebraic formula such as the exponential equation, nor can its eventual slowing be expected to follow a smooth and symmetrical trajectory such as a logistic growth path. The complex interacting factors bearing on the process are more appropriate for analysis with the aid of computer simulation. An idea can be gained of the relationships of the factors involved by examining more closely some of the causes of deforestation in Amazonia.

CAUSES OF DEFORESTATION

Present causes of deforestation can be divided, somewhat artificially, into proximal causes and underlying causes. Proximal causes motivate land owners and claimants to direct their efforts to clearing forest as quickly as possible. The underlying causes link wider processes in Brazil's economy either to the proximal motivations of each individual deforester, or to increases in numbers of deforesters present in the region.

Some of the principal motives for deforestation, especially those motives connected to government incentive programmes, apply most forcefully to large landholders. These represent forces relatively easily controlled by governmental actions, as has already occurred to a small degree. Deforestation is also linked to long-standing economic patterns in Brazil, such as high inflation rates, which have shown themselves to be particularly resistant to government control.

Changes in agricultural patterns in southern Brazil have had heavy impacts. The increase in soybean production has lead to the displacement of an estimated 11 agricultural workers for every one finding employment in the new extensive production systems. Sugar-cane plantations, encouraged by the Government to enter into alcohol production, have likewise expelled smallholders. Replacement of labour-intensive coffee plantations by mechanized farms raising wheat and other crops, a trend driven by lethal frosts and relatively unfavourable prices, has further swollen the ranks of Amazonian immigrants.

Within Amazonia, most evident are the forces of land speculation, the magnifying effect of cattle pasture on the impact of population, and the positive feedback relationship between road-building and population increases.

Profits from sale of agricultural production are added to speculative gains, tax incentives and other forms of government subsidy to make clearing financially attractive. Small farmers often come to the region intent on making their fortunes as commercial farmers, but they gradually see the higher profits to be made from speculation as their neighbours sell their plots of land for prices that dwarf the returns realized from years of hard labour. Agriculture then becomes a means of meeting living expenses while awaiting the opportunity of a profitable land sale and a move to a more distant frontier. Although individual variability is high, most aspire to produce enough to live well by the standards of their own pasts while awaiting an eventual sale. Farmers usually see such sales as providing the reward for "improvements" made on the land during their tenure, rather than as speculation. Larger operators are more likely to begin their activities in the region with speculation in mind but are likewise always careful to describe themselves as "producers" rather than speculators.

Subsistence production is always a contributor to forest clearing, although it is not precisely the major factor that it is in many other rainforest areas, as in Africa. The speculative and commercial motives for clearing in Amazonia mean that the relationship of commodity prices to clearing is positive for most of the farmers involved. In areas of the tropics where cash crops are grown primarily for supplying subsistence needs, the relationship can be the reverse: a positive feedback loop exists whereby falling prices for a product mean that larger areas must be planted for the farmer to obtain the same subsistence level of cash income, while the resulting increased supply of the product further drives prices down. For most Amazonian farmers, however, desire for cash so greatly exceeds the income-producing capacity of the farms that only the restraints of available labour and capital limit the areas cleared and planted.

Future deforestation trends should reflect changes in the balance of many forces—population growth, land speculation, road-building, export potential, subsistence needs, and a host of other factors. Future trends can also be expected to show the effects of projected major developments. As timber export, presently a negligible factor, becomes more important, outright deforestation will be supplemented by the often heavy disturbances following selective felling that presently characterize much of the forest conversion in Asia and Africa. Charcoal production, especially that derived from native forest, is foreseen as a major factor in the southeastern portion of the region in the coming decades.

Large firms, such as lumber companies requiring marketable timber, or steel manufacturing industries requiring a large charcoal supply, pose the additional problem of playing more active and forceful roles in seeing that environmental conflicts of interest are resolved in their favour. Chances are higher, as compared to the case of relatively small investors, that concessions will be made at the expense of previous governmental commitments to reserves of untouched forest. This recently occurred in the case of timber concessions operating in the area now flooded by the Tucuruf hydroelectric dam. Despite not having fulfilled its role in removing forest from areas to be flooded, the concessionaire was reportedly granted logging rights to 93,000 hectares in two nearby Amerindian reservations when commercially valuable tree species proved less common than anticipated in the reservoir area, according to the head of the firm involved.

Future deforestation appears likely to proceed at a rapid rate. Although limited availability of fossil fuel, capital, and other resources should eventually force a slowdown, this cannot be counted on to prevent loss of large areas of forest. Even at rates slower than those of the recent past, the forest could be reduced to remnants within a short span of years. The deforestation process is subject to control and influence at many points. Decisions affecting rates of clearing must be based on an understanding of the causes of deforestation. Such decisions are taken, either actively or by default, for all areas undergoing agricultural or other development, as well as in defining reserves where such development will be excluded. Making timely choices of this kind depends on decision-makers' conceptions of the likely course of deforestation. Understanding the system of forces driving the process is also essential for evaluating the probable effectiveness of any changes contemplated.

POLICY IMPLICATIONS

The negative consequences of deforestation should give pause to planners intent on promoting forms of development requiring large areas of cleared rainforest. Nevertheless, such plans continue to be proposed and realized. Part of the problem is a lack of awareness among decision-makers of the magnitude of the eventual costs implied by these actions. But such lack of knowledge explains only a part of the reluctance to take effective actions to contain and slow deforestation. At least as important is the distribution of the costs and benefits, both in time and space. Most of the costs of deforestation will be paid only in the future while the benefits are immediate. Many of the costs are also distributed over society at large while the benefits accrue to a select few. In the many cases where land is controlled by absentee investors, negative consequences within the region have even less reason to enter individual decisions. In other cases the costs are highly concentrated, as for indigenous groups deprived of their resource base, while the perhaps meagre benefits of clearing are enjoyed by a constituency that is both wider and more influential.

Brazil's national Government has the task of balancing the interests of different generations and interest groups. At the same time, the Amazon has long suffered from exploitation as a colony whose products serve mainly to benefit other parts of the globe, most recently and importantly the industrialized regions of Brazil's Central-South. The unsustainable land uses resulting from this kind of "endocolonialism" require that decision-making procedures guarantee the interests of the Amazon's residents when conflicts arise with more influential regions of the country. Clear definitions of the development objectives are essential as a prerequisite for any planning. Development alternatives should be evaluated on the basis of benefits to the residents of the Amazon region and their descendants. Coherent policies must include the maintenance of the human population below carrying capacity, the implantation of agronomically and socially sustainable agro-ecosystems, and limitations on total consumption and on the concentration of resources. The inclusion of future generations of local residents in any considerations means that greater weight must be accorded the delayed costs implied as hydrological changes, degradation of agricultural resources, and sacrifice of as yet untappable benefits from the rainforest. The folly of present trends toward rapid conversion of rainforest to low-yielding and short-lived cattle pasture is evident, at least when decisions are based on the long-term interests of Amazonia's residents.

Beyond the green revolution

New strategies are needed to increase food production for more than one billion people bypassed by agricultural technology

Laura Tangley

Twenty years ago, when famine seemed inevitable throughout much of the developing world, plant breeders at two international research centers produced what is now considered one of the most significant technological achievements in agriculture. Working in Mexico and the Philippines, these scientists had developed new varieties of high-yielding wheat and rice that were introduced—along with fertilizers, pesticides, and modern farm equipment—into many Third World countries. This effort, now known as the green revolution, dramatically increased global food production over the next two decades. According to one recent estimate, the improved wheat and rice varieties are directly responsible for providing 50 million tons of grain annually, or enough to feed 500 million people.[1]

In some countries, the gains have been particularly impressive. For instance, India, which once faced chronic food shortages, today can boast of substantial grain reserves. Indonesia, formerly the world's leading rice importer, is now not only self-sufficient, but also a major exporter of rice.

But in other parts of the world, millions of people have enjoyed no benefits from the green revolution. The high-yielding varieties were adopted primarily by farmers who had access to markets and irrigation and who could afford the costly chemical inputs required for peak performance of these crops. According to a report released last fall by the Worldwatch Institute in Washington, DC, green-revolution crop varieties today are grown on less than a third of the Third World's 423 million hectares planted in cereal grains. On most of the remaining 300 million acres—primarily marginal, rainfed lands that support subsistence agriculture—yields have not increased measurably over the last two decades.

The report, *Beyond the Green Revolution: New Approaches for Third World Agriculture,* says that these marginal lands provide food for 230 million rural households—or nearly 1.4 billion people. "Because their agriculture remains unproductive and vulnerable to crop failure, drought, and natural catastrophe, these rural people remain among the poorest in their societies," the report says. "Failure to address their needs has slowed economic progress in dozens of countries."

The green revolution's impact varies widely by region. According to the report, 36% of grain-producing lands in Asia and the Middle East are now growing high-yielding wheat and rice varieties, while 22% of Latin America's grain lands and just 1% of Africa's are planted with these crops.

Africa, where per capita food production has actually declined over the last two decades, gained the least from green-revolution innovations. One reason is that, more than any other continent, Africa is faced with poor soils and insufficient rainfall. Its subsistence farmers cannot afford irrigation or costly fertilizers and pesticides. In addition, rice and wheat are not important crops in most parts of Africa, where people rely more heavily on maize, sorghum, millet, cassava, sweet potatoes, and yams.

Green revolution technology did not focus on these African staples in part because of a perception that food needs were far greater elsewhere, says Edward C. Wolf, author of the Worldwatch report. Wolf says that in the middle-1960s Africa as a whole was still exporting food, in contrast to Asia where poor rainfall over several consecutive seasons had threatened to cause famine. He adds, however, that because Africa's colonial leaders focused on export rather than subsistence crops, we do not know whether rural Africa's food demands were even then being met.

Wolf says the arguments for boosting global food production today are at least as compelling as they were 20 years ago. "The world's population is projected to grow from today's 5 billion to 6.2 billion over the next 13 years, though little new land will be

Transplanting rice shoots in Bangladesh, one of several countries in southern Asia that quickly adopted the green revolution's high-yielding crop varieties; in contrast, these varieties are grown on just one percent of Africa's grain lands. Photo: Tomas Sennett, courtesy the World Bank.

brought under cultivation by then," he says. "Just to maintain food consumption levels will require a 26% increase over the 1985 average grain yield. And by 2020, feeding the projected population of 7.8 billion will require grain yields 56% higher."

A study conducted by the International Food Policy Research Institute in Washington, DC, confirms this need. Released last summer, *Food in the Third World: Past Trends and Projections to 2000,* looks at trends in production, consumption, and trade of basic food staples in 105 developing countries and makes projections for the next 14 years. It finds that although Asia could end up with a food surplus of 50 million metric tons, most parts of the Third World will come up short—North Africa and the Middle East by 60 million metric tons, Sub-Saharan Africa by

50 million metric tons, and Latin America by 10 million metric tons. Overall, the report concludes that global production of basic food staples will fall short of demand by 70 million metric tons by the year 2000 if present trends continue.

Biotechnology meets tradition

The Worldwatch paper emphasizes that production gains in the nations that need them most will depend on finding creative new strategies rather than simply extending the green-revolution approach. Farmers in these countries work under different conditions than those who adopted high-yielding wheat and rice 20 years ago. Rather than producing monocultures, they generally grow a variety of different crops, often mixing several species in the same field, and they cannot

afford chemical fertilizers and pesticides or the fuel to run irrigation systems and, sometimes, even farm machinery. For many of these farmers, decreasing crop vulnerability is more important than increasing yields. The report says: "Unlike past yield increases achieved under favorable cropping conditions, future improvements in average yields must come from raising the productivity of traditional farmers who cultivate low-yield crops under marginal conditions."

The report says that "(t)wo sets of technical opportunities, already stirring the agricultural research community, promise rapid progress." The first is a reappraisal of traditional farming practices, and the second is the application of biotechnology to Third World staple crops. According to the report, "joining biotechnologies with the ecological insights of traditional farming promises innovative solutions to agriculture's economic and environmental problems."

Farming practices that have succeeded for thousands of years, including intercropping, agroforestry, and shifting cultivation, use resources more efficiently than do modern techniques, the report notes, because they mimic natural ecological processes. These ancient practices have much to offer modern farmers who want to increase both the productivity and long-term sustainability of their crops (see Oldfield and Alcorn, p. 199, this issue).

As an example of traditional farming's potential contributions, the report describes agroforestry systems used in West Africa's Sahel region. Sahelian farmers traditionally plant sorghum and millet in fields that also include a permanent intercrop of *Acacia albida* trees. Because *Acacia* fixes nitrogen and improves soil quality—by returning organic matter to the topsoil, drawing nutrients from deep soil layers to the surface, and altering soil texture so that rain can infiltrate more readily—grain yields are exceptionally high in these mixed crop systems.

Traditional practices can also be combined with modern techniques to magnify the impact of each. The report describes research in Burkina Faso in western Africa demonstrating that a combination of organic fertilizer and a small amount of chemical fertilizer produces far higher sorghum yields than does either treatment alone. In one experiment, organic fertilizers enhanced the efficiency of artificial nitrogen by 20–30%. This finding could be significant in regions where farmers cannot afford much artificial fertilizer or where fertilizers are damaging the environment.

Despite these important contributions, "until recently, a myopia has kept the research community from recognizing the opportunities for agricultural innovations that lie in traditional practices," says Wolf. He notes that in West Africa, 70–80% of all farmland is planted simultaneously with more than one crop. Cowpeas, one of the region's widely grown staples, is always planted as an intercrop. Yet only about 20% of agricultural research in sub-Saharan Africa focuses on intercropping.

Workers stacking harvested sorghum in Burkina Faso in western Africa. Sorghum is an important staple crop throughout Africa. Photo: Ray Witlin, courtesy the World Bank.

Similarly, the techniques of biotechnology, although widely used in developed countries' agricultural research, have not yet been applied to most crops that are important in the developing world. Yet "biotechnologies may offer cheaper and quicker ways to improve Third World staples than the costly innovations of the mechanical and chemical eras," says the report. "It took decades of work to produce high-yielding varieties of wheat and rice. With biotechnology, comparable improvements in millet, sorghum, cassava, or tropical legumes could come more quickly." Specifically, techniques like tissue culture and recombinant DNA may hasten the development of new crop varieties that are not only more productive, but also more resistant to disease, drought, insects, and other environmental stresses common in the Third World.

Because there is little commercial incentive either to apply biotechnology to low-cost, locally grown staple crops or to analyze traditional farming practices, this research must become a priority of national and international agricultural research programs, says Wolf. According to his report, some of these organizations have already begun to make significant progress.

Moving in the right direction

In addition to creating new crop varieties, the green revolution spawned a system of international agricultural research centers, which included the two centers that had developed high-yielding wheat and rice. Paid for initially by the Ford Foundation and Rockefeller Foundation, this network today comprises 13 centers funded through the Washington, DC-based Consultative Group on International Agricultural Research (CGIAR). Dozens of donors, such as the United Nations, the World Bank, and several national governments, contribute to the system's annual budget of about $170 million. CGIAR activities include research on: 21 food crops, animal husbandry, livestock diseases, conservation of crop genetic resources (see Kloppenburg, p. 190, this issue), and policy issues related to food production.

According to CGIAR Scientific Advisor Donald L. Plucknett, research on traditional farming and biotechnology applied to Third World staples has been emphasized at some centers for many years and is likely to become increasingly important in the future. In particular, "there has been a great deal of effort to understand both the strengths and weaknesses of traditional agricultural systems," says Plucknett.

Wolf says he is encouraged by recent changes at some centers. For example, the Philippines-based International Rice Research Institute (IRRI), headquarters to the scientists who developed green revolution rice varieties, recently shifted its focus from high-yield, irrigated rice to varieties suited to a wide range of growing conditions. In addition, IRRI researchers have begun investigating farm-grown nutrient sources that could substitute for expensive, artificial fertilizers. One promising approach employs *Azolla* ferns, which support populations of nitrogen-fix-

Checking the millet crop on a farm in Nigeria. Millet is also one of Africa's most important subsistence crops. Photo: Yosef Hadar, courtesy the World Bank.

ing blue-green algae. In experiments in the Philippines, farmers who planted *Azolla* in their rice paddies were able to reduce the use of artificial fertilizers by 50% without lowering their yields.

Last spring, CGIAR's technical advisory committee broadened the focus of all 13 centers when it adopted a resolution stating that the success of future research will be judged by its contributions not only to agricultural yields, but also to sustainability and environmental protection. Sustainability in agriculture refers to practices that promote the preservation of natural resources and thus long-term productivity. "This was considered to be a fairly significant policy change," says one CGIAR staff member.

A change in CGIAR focus could have an impact far beyond its own research programs. According to Wolf, the centers contribute only about five percent of agricultural research funding in the Third World.

Most funding comes from individual governments. But because these governments look to the CGIAR for guidance, the system's influence on research directions is "disproportionate" to the amount of money it spends, says Wolf.

Decisions at the World Bank, a major contributor to CGIAR research and to agricultural development projects worldwide, also "help shape the attitudes and approaches of national governments around the world," says Wolf. Although long criticized for insensitivity to the environmental consequences of its projects, the bank recently has begun to make changes in this area (See *BioScience* 36: 712–715 and p. 186, this issue).

Another important sponsor of Third World agricultural research is the US Agency for International Development (AID), which provides about 25% of the CGIAR system's funding. According to Wolf, AID has begun to emphasize more research on small-scale, sustainable agriculture, although the agency is "still in the consciousness-raising stage." The most significant change is a new policy mandating the integration of natural resource and agricultural projects to strengthen the latter's resource-protection components. In the past, staff members from these two parts of AID had little to do with one another even when they were working in the same region.

At last year's meeting of the American Association for the Advancement of Science in Philadelphia, Pennsylvania, AID's senior assistant administrator for science and technology, Nyle C. Brady, emphasized the agency's commitment to biotechnology research in the Third World. He cited several ongoing projects, including research using recombinant DNA to alter nitrogen-fixing *Rhizobium* bacteria that could reduce artificial fertilizer use. According to Brady and other agency officials, biotechnology and low-input, sustainable agriculture will receive even greater emphasis in the future.

National governments and independent foundations are also beginning to make progress. The Rockefeller Foundation, for example, is funding projects to make rice more resistant to environmental stress and to apply the techniques of biotechnology to sorghum, millet, and other

staple crops receiving little attention from the commercial sector. Wolf hopes his report will help stimulate more of these kinds of efforts; toward this goal, the Worldwatch Institute has widely distributed the document to AID project officers, CGIAR center directors, and heads of national agricultural research programs.

Funding problems: national and international

Although national and international programs are pursuing much important research neglected in the private sector, Wolf worries that competition from private companies may hurt these public sector efforts. Specifically, he is concerned that international centers will be unable to compete with corporations for the most talented scientists because they cannot match the salaries and facilities that private companies offer. In addition, "the full exchange of scientific information that is essential to the international centers may be curtailed if it appears to compromise proprietary corporate research," the report warns.

A more serious problem for national agricultural research programs is insufficient funding and scientific expertise. Even research focusing on low-input, traditional farming costs more than many people assume, says CGIAR's Plucknett. Because the complexities of the resource base must be understood, such research demands a wide range of experts on soil, climate, water, and the many different crops grown in such farming systems. "It's much easier and often less expensive to organize research around a single commodity," says Plucknett.

In contrast to wheat and rice, which are grown under relatively homogeneous conditions, Third World staples are planted in a variety of different situations. Moreover, while crops improved by the green revolution had already been studied for decades, "for most staple crops, research has only begun in the last ten years," says Wolf. He adds, however, that this research is now "reaching a critical mass" so that major improvements in crop performance should be attainable soon.

Threatening to hamper this progress is the possibility of decreased funding for the international agricul-

tural research system. After growing steadily since the CGIAR was officially created in 1971, the system's budget remained stable for the first time in 1986, meaning an actual decline once inflation is considered. One reason funding has stagnated may be that many countries that faced food shortages 20 years ago are now self-sufficient and some developed countries are plagued with problems of crop surplus. "The sense of urgency with which the green revolution was launched has largely disappeared," notes the report.

In addition, the world economy faces more serious problems than it did two decades ago. In a paper presented at a Federation of American Societies for Experimental Biology symposium entitled Biomedical Aspects of World Famine,[2] Robert M. Goodman of Calgene in Davis, California, said that an annual economic growth rate of five percent between 1950 and 1973 helped both attract funding for agricultural research and magnify the impact this research had on the world's living standards. Today, however, economic growth has slowed considerably, while population growth is unacceptably high in many of the world's poorest nations. Goodman believes that overpopulation and global environmental deterioration may cancel any progress that new technology can bring to Third World agriculture.

Although Wolf acknowledges these many serious problems, he remains optimistic that agricultural research, if properly focused, will once again be able to raise living standards for millions of people worldwide. According to his report, "opportunities have never been greater for moving agriculture toward sustainability and reaching the quarter of the world's people—and quarter of the world's cropland—left out of the green revolution."

Notes

[1] Warren C. Baum, 1986, *Partners Against Hunger*. The World Bank, Washington, DC.

[2] Robert M. Goodman, 1986, *New technology and its role in enhancing global food production*. Pages 2432–2437 in *Federation Proceedings*. Vol. 45 no. 10. Federation of American Societies for Experimental Biology, Bethesda, MD.

Mimicking nature

The sustainability of many traditional farming practices lies in the ecological models they follow.

Edward C. Wolf

Edward C. Wolf *is a senior researcher for the Worldwatch Institute in Washington, DC. This article has been adapted from his study* Beyond the Green Revolution: New Approaches for Third World Agriculture, *recently published by the Worldwatch Institute.*

Over the next 13 years, the world's population will increase from today's 5 billion to over 6 billion. Few analysts expect a significant increase in cultivated land by then. Merely maintaining current consumption levels will require a 26 per cent increase in the world's average grain yields. And by 2020, feeding the projected population of 7.8 billion will require grain yields 56 per cent higher than 1985 levels. Unlike past spectacular yield increases achieved under favourable cropping conditions, future improvements in average yields must come from raising the productivity of traditional farmers who cultivate unimproved crops under marginal conditions—perhaps the most demanding challenge that national governments and the international development community have faced.[1]

Small farmers growing food primarily for subsistence face extraordinary diversity in ecological conditions, ranging from the rain-soaked volcanic archipelagos of Southeast Asia to the arid sub-Saharan savannahs and the high altiplano of Latin America. Farming methods, and staple foods, vary enormously as well. In Southeast Asia, for example, where one-third of the farms are less than half a hectare in size, most farmers depend exclusively on manual labour and draught power supplied by water buffalos to cultivate their rice. On Africa's small farms, cultivated more with hoes than with ploughs, families grow root and tuber crops, including cassava and yams, as their primary staples. Despite such variety, subsistence farms around the world have certain common features. Farmers often mix different crops in the same field to reduce the risk if a particular crop fails; they grow a variety of staple crops and vegetables to meet family food needs; and they rarely purchase artificial fertilizers or pesticides.[2] Per-

haps it is not surprising that high-yielding varieties of wheat and rice have been introduced to less than a third of the 423 million hectares planted to cereal grains in the Third World. For members of the 230 million rural households in Africa, Asia, and Latin America who use farming methods little different from those of their ancestors, green-revolution approaches will only be part of the answer. One reason is energy. Past advances have come from increasing the energy intensity of farming: fuel to run machinery, fossil-fuel-based artificial fertilizers, and diesel fuel or electricity to run irrigation pumps. Few of the rural poor can afford these costly materials and services. Even if they have the income to purchase such inputs, many farmers are not served by roads or markets that can reliably supply them.

In addition, subsistence farmers grow crops that have received comparatively little research attention. There is as yet no research base for achieving high yields in many staple crops.

Reprinted from *CERES*, The FAO Review, Vol. 20, No. 1, January-February 1987, pp. 20-24. *CERES*, the FAO Review, published bimonthly by the Food and Agriculture Organization of the United Nations.

7. THE ENVIRONMENT

Agricultural research has been needlessly hindered for two decades by some disparaging attitudes toward traditional farming. Some scientists assumed that because peasant farmers produced low grain yields, their practices had little relevance to twentieth-century agriculture. Until recently, few researchers recognized the ecological and agronomic strengths of traditional practices that had allowed farmers over the centuries to maintain their land's fertility. In pursuit of higher productivity, many agricultural scientists overlooked the need for long-term sustainability.

Economic analysis of traditional farming reinforced the belief that traditional practices had little to offer in solving contemporary agricultural problems. In *Transforming Traditional Agriculture*, published in 1964, University of Chicago economist Theodore Schultz argued that peasant farmers were rational and efficient individuals who had reached the limits of their technologies. He concluded that no significant increase in harvests could be achieved using only the resources and methods that traditional farmers had at their command. Schultz advocated investments in agricultural research, new technologies, and rural education that would allow traditional farmers to choose innovations to increase their productivity.[3]

Many scientists and policy-makers, however, saw traditional methods as an obstacle to be eradicated rather than a basis for introducing new seeds and farming methods. The food crisis in India and throughout Asia in the late 1960s lent a sense of urgency to efforts to promote the green revolution. The strengths of traditional practices and the reasons for their persistence were swept aside. A report by US President Lyndon Johnson's Science Advisory Committee warned in 1966 that "the very fabric of traditional societies must be rewoven if the situation is to change permanently."[4]

Overdue reappraisal. Agricultural scientists have recently begun to recognize that many farming systems that have persisted for millennia exemplify careful management of soil, water, and nutrients, precisely the methods required to make high-input farming practices sustainable. This overdue reappraisal stems in part from the need to use inputs more efficiently, and in part from the growing interest in biological technologies. The complex challenge of Africa's food crisis in the early 1980s forced scientists to reexamine what peasant farmers were already doing. Many researchers today seek to "improve existing farming systems rather than attempting to transform them in a major way," according to William Liebhardt, Director of Research at the Rodale Research Center.[5]

Traditional farming systems face real agronomic limits and can rarely compete ton for harvested ton with high-input modern methods. It is important to recognize these limitations, for they determine both how traditional practices can be modified and what such practices can contribute to the effort to raise agricultural productivity.

First, most traditional crop varieties have limited genetic potential for high grain yields. They are often large-leaved and tall, for example. These traits help farmers meet non-food needs, supplying thatch, fuel, and fodder as well as food to farm households. Traditional varieties respond poorly to the two elements of agronomic management that make high grain yields possible: dense planting and artificial fertilizer. Despite these limitations, traditional varieties also contain genetic diversity that is invaluable to breeders in search of genes for disease- and pest-resistance and for other traits. [6]

Second, peasant farmers often have to plant in soils with serious nutrient deficiencies, where crop combinations and rotations are needed to help offset the limitations. Many tropical soils, for instance, lack sufficient nitrogen to sustain a robust crop. Soils in vast areas of semi-arid Africa are deficient in phosphorus. High-yielding varieties, more efficient in converting available nutrients into edible grain, can rapidly deplete soil nutrients if they are planted by peasant farmers who cannot purchase supplemental fertilizers.[7]

Traditional agriculture practised under biological and physical limitations often breaks down under growing population pressure. As rural populations grow, farmers try to squeeze more production from existing fields, often accelerating the loss of fertility. Or they may cultivate new, often marginal or sloping, land that is vulnerable to soil erosion and unsuited to farming.

Nonetheless, traditional methods can make an important contribution to efforts to raise agricultural productivity. They offer what Gerald Marten of the East-West Center in Hawaii calls "principles of permanence". They use few external inputs, accumulate and cycle natural nutrients effectively, protect soils, and rely on genetic diversity. "Neither modern Western agriculture nor indigenous traditional agriculture, in their present forms, are exactly what will be needed by most small-scale farmers," says Marten. "The challenge for agricultural research is to improve agriculture in ways that retain the strengths of traditional agriculture while meeting the needs of changing times."[8]

Farming methods like the traditional agroforestry systems of West Africa's Sahel region offer improvements in water-use efficiency and soil fertility that subsistence farmers can afford. Sahelian farmers traditionally planted their sorghum and millet crops in fields interspersed with a permanent intercrop of *Acacia albida* trees. Acacia trees fix nitrogen and improve the soil. In the Sahel, grain yields are often highest under an acacia's crown.[9]

More productive and profitable. Fields that include acacia trees produce more grain, support more and require shorter fallow periods between crops than fields sown to grain only. *Acacia albida* naturally enhances productivity by returning organic matter to the topsoil, drawing nutrients from deep soil layers to the surface, and changing soil texture so that rainwater infiltrates the topsoil more readily. All of these benefits make farming on marginal lands more productive and profitable without requiring the farmer to purchase fertilizers year after year. [10]

Equally important, improvements in soil structure, organic matter content, water-holding capacity, and biological nitrogen fixation allow the most productive application of conventional fertilizers. Programmes promoting acacia-based agroforestry could complement fertilizer extension in semi-arid countries—with agroforestry playing a role analogous to irrigation. Governments that have modest fertilizer-promotion programmes may find that they can maximize the benefits from fertilizer by promoting agroforestry as well. [11]

Legume-based crop rotations and traditional intercropping systems share the advantage of husbanding organic material and nutrients much more carefully than modern monoculture practices do. While organic manures and composts contribute significant amounts of nutrients in their own right, they can, like agroforestry, make small amounts of artificial fertilizers effective.

Research in Burkina Faso illustrates the complementary effect (see Table). This study looked at the contributions of straw, manure, and compost to sorghum yields with and without the addition of small amounts of artificial nitrogen. The results show that the most productive organic method, applying compost, can increase sorghum yields from 1.8 tons to 2.5 tons per hectare. Artificial fertilizer alone produced grain yields slightly higher than any of the organic practices. But the best result was achieved by combining compost with artificial fertilizer; this raised sorghum yields to 3.7 tons per hectare. The three organic practices increased the efficiency of nitrogen application by 20 to 30 per cent. Given responsive crop varieties and small amounts of artificial fertilizer, traditional practices that cycle organic materials effectively would raise yields in the same manner. [12]

Some conventional analysts looking at the study would argue that fertilizer outperforms the organic

Table
Complementary effect of artificial and organic fertilizer on sorghum yields in Burkina Faso, 1981

Sorghum yield (metric tons per hectare)

Treatment [1]	Without artifial nitrogen	With 60 kg/ha nitrogen
No organic treatment	1.8	2.8
Sorghum straw	1.6	3.4
Manure	2.4	3.6
Compost	2.5	3.7

[1] *All organic materials applied at a rate of 10 tons per hectare.*
Source: M. Sedogo, "Contribution à la valorisation des residus culturaux en sol ferrugineux et sous climats semi-aride" (doctoral thesis, Nancy, France: ENSAIA, 1981), quoted in Herbert W. Ohm and Joseph G. Nagy, eds., Appropriate Technologies for Farmers in Semi-Arid West Africa (West Lafayette, Ind.: Purdue University International Programs in Agriculture, 1985).

practices. Yet exclusive reliance on fertilizer would sacrifice a significant part of the additional harvest. As French researcher Christian Pieri, who has worked in West Africa, points out, "Fertilization is a prime technique for increasing agricultural productivity in this part of the world, but in order to obtain a greater and lasting production it is indispensable to combine the effects of mineral fertilizers, the recycling of organic residues and biological nitrogen fixation, and also to optimize the use of local mineral resources such as natural phosphates." [13] Neglecting the local internal resources can undermine a farmer's investments in conventional inputs.

Mimic natural processes. Intercropping, agroforestry, shifting cultivation, and other traditional farming methods mimic natural ecological processes, and the sustainability of many traditional practices lies in the ecological models they follow. This use of natural analogies suggests principles for the design of agricultural systems to make the most of sunlight, soil nutrients, and rainfall.

Shifting cultivation practices, such as bush-fallow methods in Africa, demonstrate how farmers can harness the land's natural regeneration. Farmers using bush-fallow systems clear fields by burning off the shrubs and woody vegetation. Ashes fertilize the first crop. After a couple of seasons, as nutrients are depleted, harvests begin to decline, so farmers

abandon the field and move on to clear new land. Natural regeneration takes over; shrubs and trees gradually reseed the field, returning nutrients to the topsoil and restoring the land's inherent fertility. After 15 to 20 years, the land can be burned and cultivated again. [14]

The bush-fallow system has obvious limitations. It requires enormous amounts of land, and when population growth pushes farmers to return too quickly to abandoned fields, serious environmental deterioration can result. Declining land productivity in crowded countries like Rwanda is testimony to this danger. But even disintegrating systems offer a basis for designing productive and sustainable farming practices.

Researchers at the Nigeria-based International Institute of Tropical Agriculture, for instance, have adapted the principles of natural regeneration in bush-fallow systems to a continuous-cultivation agroforestry system called alley cropping. Field crops are grown between rows of nitrogen-fixing trees; foliage from the trees enhances the soil organic matter, while nitrogen fixed in root nodules increases soil fertility. A high level of crop production is possible without a fallow interval. Traditional shifting cultivation provided the model for this system. [15]

Conventional research tools can also be used to overcome the agronomic constraints that have limited traditional systems to low productivity. For decades, crop breeders have tailored varieties to respond to high levels of artificial fertilizers, assured water supplies, and dense monoculture plantings. Working with the genetic diversity available in traditional crop varieties, they can apply breeding methods to produce varieties better matched to the conditions faced by subsistence farmers. At an Agency for International Development (AID) workshop on regenerative farming practices, Charles Francis of the University of Nebraska concluded, "A new generation of varieties and hybrids adapted to marginal conditions and to intercropping could be the start of a new revolution aimed at meeting the needs of the majority of limited resource farmers in the developing world."[16]

Traditional practices exemplify efficiency and the regenerative approach to agricultural development. Yet until recently, a kind of myopia has kept the research community from recognizing the opportunities for agricultural innovations that lie in traditional practices. In West Africa, for example, 70 to 80 per cent of the cultivated area is sown to combinations of crops in traditional intercropping systems. Cowpeas, one of Africa's most widely grown food staples, are always planted as an intercrop. But only about 20 per cent of the research effort devoted to agriculture in sub-Saharan Africa investigates intercropping.[17]

As the African examples described here show, researchers can use traditional principles to develop new techniques that preserve the land's stability and productivity even as populations increase. Though traditional methods have limitations, they are not archaic practices to be swept aside. Traditional farming constitutes a foundation on which scientific improvements in agriculture can build.

[1] Population projections are from Population Reference Bureau, 1985 World Population Data Sheet (Washington, DC, 1985). Grain-yield projections are by Worldwatch Institute, based on world grain utilization data from USDA, Foreign Agricultural Service, Foreign Agriculture Circular: Grains, FG-9-86 (Washington, DC, 1986).
[2] For descriptions of traditional farming, see D. B. Grigg, The Agricultural Systems of the World (Cambridge: Cambridge University Press, 1974); Gerald G. Marten (ed.), Traditional Agriculture in Southeast Asia (Boulder, Colo.: Westview Press, 1986), and US Office of Technology Assessment (OTA), Africa Tomorrow: Issues in Technology, Agriculture, and U.S. Foreign Aid (Washington, DC: U.S. Government Printing Office, December 1984).
[3] Theodore Schultz, Transforming Traditional Agriculture (New Haven, Conn.: Yale University Press, 1964).
[4] Quoted in Sterling Wortman and Ralph W. Cummings, Jr., To Feed This World (Baltimore: The Johns Hopkins University Press, 1978).
[5] W. C. Liebhardt, C. A. Francis, and M. Sands, "Research Needs for the Development of Resource Efficient Technologies", in Rodale Institute, Regenerative Farming Systems.
[6] Peter R. Jennings, "The Amplification of Agricultural Production", Scientific American, September 1976.
[7] Ibid.
[8] Gerald G. Marten, "Traditional Agriculture and Agricultural Research in Southeast Asia", in Marten, op. cit.
[9] Traditional Acacia-based systems are described in National Research Council, Board on Science and Technology for International Development, Environmental Change in the West African Sahel (Washington, DC: National Academy Press, 1983).
[10] Ibid.
[11] Michael McGahuey, Impact of Forestry Initiatives in the Sahel (Washington, DC: Chemonics, 1986).
[12] Christian Pieri, "Food Crop Fertilization and Soil Fertility: The IRAT Experience", in Herbert W. Ohm and Joseph G. Nagy (eds.), Appropriate Technologies for Farmers in Semi-Arid West Africa (West Lafayette, Ind.: Purdue University International Programs in Agriculture, 1985).
[13] Ibid.
[14] Shifting cultivation is described in Grigg, Agricultural Systems.
[15] Current research in alley cropping is described in IITA, IITA Annual Report.
[16] C. A. Francis, et al., "Resource Efficient Farming Sytems and Technologies", in Rodale Institute, Regenerative Farming Systems.
[17] Dunstan S. C. Spencer, "Agricultural Research: Lessons of the Past, Strategies for the Future", in Robert J. Berg and Jennifer Seymour Whittaker (eds.), Strategies for African Development (Berkeley: University of California Press, 1986).

Ozone Depletion's New Environmental Threat

JANET RALOFF

The seeds of future stratospheric ozone depletion have already been sown. Not only have many of the ozone-attacking chemicals — such as chlorofluorocarbons — been irretrievably dispersed into the atmosphere, but the global rate at which such chemicals are emitted continues to grow annually. As a result, atmospheric chemists no longer talk about whether stratospheric ozone depletion will occur, but rather how much will be lost. In fact, the first apparent evidence that measurable depletion has already occurred was reported earlier this year (SN: 6/28/86, p. 404).

Until recently, most predictions of the environmental hazards posed by this ozone depletion focused on the direct effects — such as human skin cancers — of bathing the earth's living things in a more intense field of solar ultraviolet radiation, or on the climate change that might occur as many of the ozone-destroying "greenhouse gases" initiated a global warming (SN: 5/18/85, p.308).

But the recent findings of a pair of atmospheric chemists have added two new dimensions. Declines in stratospheric ozone, they say, could exacerbate not only smog but also the acid rain with which urban areas may have to contend. Details of their preliminary estimates of such effects — based on air quality data from Philadelphia, Nashville and New York — were circulated in a discussion paper last week in Amsterdam at a meeting of the United Nations Coordinating Committee on the Ozone Layer.

One of the chief benefits of earth's stratospheric ozone layer is its ability to filter out much of the sun's biologically harmful ultraviolet radiation. But as air pollutants such as the chlorofluorocarbons CFC-11 and -12 reach the stratosphere and begin destroying that ozone, increased ultraviolet levels will penetrate to the lower atmosphere, the troposphere. There, the higher ultraviolet levels will begin driving subtle perturbations in tropospheric chemistry. A computer model to simulate those tropospheric chemistry changes and their potential impacts on the human environment is being developed by Gary Z. Whitten and Michael Gery of Systems Applications Inc. in San Rafael, Calif.

"We see about a 2 percent increase in smog ozone from a 1 percent decrease in stratospheric ozone," says Whitten. Ultraviolet increases the rate of ozone formation — a process that occurs only during daylight hours. Whitten says this means not only that there is the potential for an overall increase in how much smog ozone is produced from reactions involving combustion pollutants such as hydrocarbons and nitrogen oxides (NO_x), but also that ozone production will peak earlier in the day.

The latter could have serious implications for human exposure to ozone — the primary irritant in smog. Both smog production and human activity tend to be concentrated around industrial urban centers. If smog production peaked after many of the urban workers had commuted home to the suburbs, relatively few individuals would be exposed to the most intense smog. However, if, as the new Whitten-Gery simulations suggest, smog peaks in early to mid-afternoon, far more people could be exposed to serious smog-ozone pollution than most future projections would indicate.

But smog ozone is not the only hazard. Another potentially serious effect of more efficient smog ozone production is a dramatic increase in the production of

At right is a damaged leaf grown in an environment rich in sulfur dioxide and ozone. At left is a healthier leaf that grew in carbon-filtered air.

hydrogen peroxide, a key chemical precursor to acid rain. The generation of hydrogen peroxide — also a product of ultraviolet-driven reactions between derivatives of hydrocarbons and NO_x — only occurs after ozone production shuts down, explains Whitten. "If the ozone process is still going when the sun goes down, you won't make any hydrogen peroxide. But if the ozone process finishes at noon," he says, "you have the whole rest of the day to make hydrogen peroxide."

Based on preliminary analyses of their data for Nashville, Gery says it appears there could be "about an 80 percent increase in hydrogen peroxide production for each 1 percent decrease in stratospheric ozone." Moreover, in contrast to smog ozone formation, the ultraviolet-in-

Solid marble caryatid figure outside the Field Museum of Natural History in Chicago shows decades of damage caused by acid rain. Photo dates from left to right: circa 1920 (pre-installation), January 1967, June 1981.

Photos: Courtesy, Field Museum of Natural History, Chicago

duced changes in hydrogen peroxide are nonlinear — for each successive unit of stratospheric ozone depletion there is a disproportionately larger increase in hydrogen peroxide generation.

In their computer modeling calculations, Whitten and Gery tried to account for whether the city being studied would be enacting major controls on the emission of hydrocarbons and NO_x in the future to limit smog and acid rain production. But when they accounted for the increased tropospheric ultraviolet levels that would correspond to an 8 percent depletion of stratospheric ozone, Gery says, these cities all but lost the benefit of the expensive hydrocarbon- and NO_x-control measures in controlling acid rain precursors. Their analyses indicate that levels of hydrogen peroxide would increase dramatically — roughly to levels that match those forming under today's lower tropospheric-ultraviolet levels and no emissions controls.

"The implications of this for people who do smog simulation studies," he says, "is that if they use today's stratospheric ozone figures they may end up dramatically underestimating not only the future amount of smog ozone that is going to form, but also the cost of [hydrocarbon and NO_x] control measures they're going to need" to meet federal pollution standards.

Humans are not the only victims in this scenario. The Agriculture Department's National Crop Loss Assessment Network estimates that ozone damage already costs U.S. farmers $2 billion annually in reduced crop yields. Some commercially important trees also appear sensitive to ozone.

For example, according to a new field study by researchers at the Yale School of Forestry and Michigan Technological University's forestry department, a 16 to 19 percent stunting was measured in the growth (dry mass) of young hybrid poplars, cottonwoods and black locusts exposed to outdoor ozone levels that were generally well within the current federal air-quality limit. Ironically, write the authors in the November ENVIRONMENTAL SCIENCE AND TECHNOLOGY, although the effect was serious, it was generally masked by the fact that there were no pathological symptoms in the ozone-affected trees; even their stunted growth was virtually invisible.

Explains Deane Wang, one of the authors (and now working at the University of Washington's Center for Urban Horticulture in Seattle), it wasn't that the ozone-stunted trees were much shorter, so much as that they had fewer branches and somewhat fewer leaves on their branches. Wang described this subtle change in the "architecture" of the tree as being a possible sign that the tree's energy was being severely sapped — perhaps so severely that it might not have enough energy to fight additional stressors, like pests. Because the ozone produced no overt signs of ill health, the study's authors report, "it is probably prudent to conclude that ozone effects on forest ecosystems are more widespread and intensive than indicated by visible symptomology alone."

The reduced plant productivity indicated by this study, taken together with ozone's current toll on crops, suggests that all who depend on forestry and agriculture may bear a much higher cost if the emissions of pollutants that destroy stratospheric ozone are not regulated soon, says Daniel J. Dudek, senior economist for the Environmental Defense Fund in New York City.

And, adds David Doniger, a senior attorney with the Natural Resources Defense Council in Washington, D.C., ozone-depletion-initiated "increases in smog production threaten to make it harder for cities to meet the ambient ozone standard [SN: 6/28/86, p.405]." He says, "We've mentioned this to as many . . . state air-pollution-control officials as we can. We've been telling them they ought to become concerned about fluorocarbon emissions because they're going to make the smog problem worse down here."

Donald Stedman, an atmospheric chemist at the University of Denver, describes the Whitten-Gery link between stratospheric ozone depletion and tropospheric environmental chemistry as "extremely exciting and very interesting. It puts together a couple of fields of researchers who haven't been talking to each other — those who work on photochemical smog and those who work on stratospheric ozone."

Jack G. Calvert of the Boulder, Colo.-based National Center for Atmospheric Research and chairman of the National Academy of Sciences panel that authored "Acid Deposition: Atmospheric Processes in Eastern North America" (SN:

7/2/83, p.7), has a note of caution about attempts to interpret the new findings. "I would be a little skeptical about its numbers," he says — in part because the computer model used to simulate the atmospheric chemistry is, of necessity, rather simplistic. Moreover, he adds, the quantity of hydrogen peroxide present is not the only factor in determining whether acid rain production will increase.

Inside cloud water, hydrogen peroxide reacts with sulfur dioxide to form sulfuric acid. Maximum acid production occurs when there are roughly equal quantities of each precursor. Calvert points out that if there is already enough hydrogen peroxide to react with all of the sulfur dioxide, adding more hydrogen peroxide to the atmosphere will not increase acid rain production.

However, Stedman notes, because there is usually a relative shortage of hydrogen peroxide in urban air during the winter, this season would likely be susceptible to the most dramatic increases in acid rain production as tropospheric levels of ultraviolet radiation increase.

But even if there were no summer shortage of hydrogen peroxide, Stedman believes more prodigious hydrogen peroxide generation in cities could still have a profound effect on the environment. First, he says, if there were enough of it present to oxidize all of the sulfur dioxide — before that airborne sulfur dioxide got a chance to drift away from cities — more of the production and deposition of acid would occur closer to pollution sources. Under this scenario, he says, the city that generates the acid precursors would suffer more from acid rain than it does today, and the forests downwind would suffer relatively less.

Second, he points out that it's not yet clear whether ozone and acid rain are the primary actors accounting for forest losses downwind of industrial centers. One of the better of several new, competing theories, Stedman says, it that hydrogen peroxide may by itself be directly responsible for injuring the leaves of trees.

To date the Whitten-Gery projections are based on only several days' worth of air pollution data collected at three cities. Within the next six months, the researchers expect to run simulations based on data from nine additional cities and more than 100 days of air pollution readings. By that time, Gery also expects to have completed a three-day simulation of an air mass traveling from West Virginia into the Adirondacks. It will contain an acid rain model he's developed to account for how changes in the ultraviolet-light field might affect acid formation.

In the end, Gery says, the precise numbers of ozone and hydrogen peroxide increases they've calculated thus far may not hold up. But the researchers do not expect the type of effects they're reporting today to change much, Gery says, because "what we see is a definite trend in the data" toward increased urban ozone and "dramatically increased" acid rain precursors. Considering the strength of this trend, he says, there's no reason to believe it won't continue.

Index

abortion, policies on, in different countries, 175-177

acetylcholine: and Alzheimer's disease, 106; and Huntington's chorea, 116, 117; and Parkinson's disease, 115

acid deposition, 186-194

acid rain: and acid deposition, 189, 190; and air pollution, 178, 180, 185; and ozone depletion, 207

acquired immune deficiency syndrome, see AIDS

acupuncture, and endogenous opiates in pain relief, 145, 147

adenosine deaminase disease (ADA), and gene therapy, 154

aging, effect of, on body, 120-125

agroforestry, 200, 204, 205

AIDS, 130, 162-167

air, origin of, on Earth, 18-19

air pollution, 178, 180, 183, 185-194

allergies, 148-153

alpha factor, 46, 47

altruism, and natural selection theory, 13

Alzheimer's disease, drug research on memory component of, 106

amphibian egg, study of, after fertilization, 59-60

antiallergy treatment, 150-153

APUD theory, criticism of, 84

autism, and limbic system of brain, 107

autogen theory, of life's origins, 24

autopoiesis, 24

axonal transport technique, 106

axons, 84, 126-128

bacteria: 50; gene splicing with, 39-40; vs. yeast in recombinant DNA research, 44-47

B cells, 71, 129-130

beer brewers, and genetic engineers, 44-47

behavior, influence of, on health, 70-73

behavioral therapy, and immune system's response to disease, 71

behaviorists, vs. cognitivists and how the brain works, 104-108

bewitchment, and ergot fungus, 135-138

biofeedback, and controlling pain, 147

biology, limitations to possibilities in, 25-31

biomass, as alternative to fossil fuels, 170, 171

biotechnology: advances in, 48-52; future of, 48; and gene therapy, 154-161; and green revolution in Third World, 199, 200; and use of yeast as bacteria in recombinant DNA research, 44-47

body, effects of aging on, 120-125

body cycles, and chronobiology, 77-81

Boyer, Herbert W., 41, 45, 49, 50

brain: and aging of the body, 173; and dreaming, 109-113; injury to, and regrowth of nerve cells, 126-128; and pain, 146; structures of, involved in learning and memory, 104-108

Burkino Faso, and green revolution, 200, 205

bush-fallow system, and shifting cultivation, 205

cancer: 47; gene research involving, 56-58; and gene therapy, 155, 156, 160

cells, see specific type

cellular communication: inter-, 83, 84; in plants and animals, 83-85

central nervous system, and regrowth of nerves, 126-128

chemicals: plant-derived, 82, 83; toxic, 178-181

chlorofluorocarbons, and ozone depletion, 207

chlorpromazine, and Parkinson's disease, 115

chronobiology: 77-81; history behind, 79

circadian rhythms, 77-81

Clean Air Act, 180

Clean Water Act, 181

cognitivists, vs. behaviorists and how the brain works, 104-108

Cohen, Stanley, 41, 45, 49, 50

computers, as intelligent, 101-103

connectionism, and computers teaching themselves, 101-103

conservation reserve program, and soil erosion, 184

continuous-cultivation agroforestry, 205

convulsive ergotism, and Salem witchcraft affair, 135-138

cortex: and dreaming, 109-113; involvement in memory and learning, 104-108

cortico-limbic system, and memory, 104

critical periods: and imprinting, 87, 88; and learning, 87, 88

crops: and damage to, by ozone, 180; and soil erosion, 183, 184

cultural bias, and intelligence tests, 99

cystic fibrosis, and gene therapy, 157, 158

Darwin, Charles, 6, 8, 9, 10-13

Darwinism, influence of Mendelism on, 13-14

deforestation, of rainforests in Brazil, 195-197

depression: effect of, on immune system, 70; and tyrosine, 92

desensitization, and allergies, 149, 152

diet, effect of, on human behavior, 91, 92

dieting: and fat cells, 131-134; yo-yo, 134

differentiation, 62, 63

disease: effect of emotions on, 70-73; genetic, and gene mapping, 159-161; genetic, and gene therapy, 155-158; and stress, 70

DNA: 34-38, 55, 65-67; see also, recombinant DNA research

dopamine, and Parkinson's disease, 115, 116

dreams: content of, 110, 111; Crick-Mitchison model of, 111, 112; function of, 110, 111; interpreting, 109-113

drugs: and body cycles, 80, 81; and pain relief, 145; and placebo effect, 93; and plants, 82

Earth: factors accounting for origin of life on, 17-19; simulation of prebiotic life on, 20-24

eating disorders, and weight control, 132, 133

E. coli: 66; use of, in gene splicing, 41; recombinant DNA research using, 45-47

egg cell, research on, 59-64

embryology, and genetic research, 59-64, 141

emotions, and chemical links with health, 70-73

endangered species: 178; and

Environmental Quality Index on wildlife, 179

endocrine system: and APUD theory, 84; and cellular communication, 83, 85

endogenous opiate system, and acupuncture, 145, 147

endorphins, and pain research, 144-147

energy: alternatives to fossil fuels, 170-171; Environmental Quality Index on, 182

enkephalins, and pain research, 144-147

Environmental Quality Index, 180-184

ergot, and witchcraft affair at Salem, 135-138

Escherichia coli, see E. coli

ethics: of genetic engineering, 43, 48-51; of recombinant DNA research, 40

evolution: Darwin's theory of, 6, 8-13; as discontinuous, 15, 16; impact of genetic research on theories of, 51, 63-64; limitations to possibilities of, 25-31; modern synthesis view of, 6-10; synthetic view of, 15; tempo of, 8-9

extraterrestrial life, possibility of, 17-19

farmers, and deforestation in Brazil, 196, 197; and green revolution in Third World, 198-205; and soil erosion, 183, 184

fat, and aging of body, 121, 123

fat cells, and weight regulation, 131-134

fat eating behavior, criticism of, 132

feature detectors, 88, 89

fetal surgery: morality of, 139; risks vs. rewards of, 139-141

fetus: rights of, 141; risk and rewards to, of fetal surgery, 139-141

fish populations: and acid deposition, 193; and toxic chemicals, 179

5S gene, 62

food, effect of, on behavior, 91, 92

food allergies, 150

food-avoidance phenomenon, and learning, 89

food production: global and green revolution, 198-206; traditional methods combined with biotechnology in, 199, 200, 201, 203

forebrain, and dreaming, 110

forests: and acid deposition, 193, 194; and acid rain, 178, 208; Environmental Quality Index on, 183; effect of ozone depletion on, 208; rain forests in Brazil, 195-197

fossil fuels: and acid deposition, 186-188; alternatives to, 170-171, 182

Freud, Sigmund, and interpretation of dreams, 110, 113

GABA, 115-117

gate theory of pain, 146, 147

gene mapping, 41, 159-161

gene markers, 42, 157, 158

Genentech: 51; production of human insulin by, 41

gene replacement: 159; see also, gene transfers

genes: and behavior of animals, 88; man-made fabrication of, 41

gene sequencing, 67, 159-161

gene splicing: 39, 50-52, 155; see also, gene transfers

gene therapy, 42, 154-158

genetic code: 35; advances in deciphering, 65-67; implications of intervention in human, 51 genetic counseling, 42; see also, genetic screening

genetic determinism, and personality traits as inherited, 74-76

genetic diseases, 42, 114, 116, 154-161

genetic engineering: applications of, 42, 43; commercial uses, 48-50; and plants, 52-54; see also, recombinant DNA research

genetic programming, and connectionism concerning computers, 102

genetics, impact of Mendelian, on Darwinism, 13-14

genetic screening, 158, 159

gene transfer: 39, 41, 42, 45, 48, 49; and cancer research, 56-58; in plants, 52-54; see also, gene splicing

Genex, 39, 40, 43

genome, 41, 160, 161

geothermal energy, as alternative to fossil fuels, 171, 182

germ-cell therapy, 156

graft(s), and regrowth of nerves in peripheral nervous system (PNS), 126-128

greenhouse effect, 18, 182

green revolution, and global food production, 198-206

groundwater pollution, 181

Hardy-Weinberg equilibrium principle, 14

health, and chemical links with emotions, 70-73

heart disease, and gene therapy, 156

height, and aging of body, 122

helper T cells, 130

hemophilia, type A, and genetic testing, 158

hepatitis B vaccine, 44, 45

herbal medicine, 82

heredity: Darwin's theory of, 12; rise of Mendelian theory of, 13-14; and personality traits, 74-76

hippocampus, 105-107, 112

homeobox, 61, 62

homeotic genes, 61

honey bee, and programmed learning, 86, 87

hormone(s), as a messenger molecule, 82, 83, 84

human evolution, see evolution

Huntington's chorea: and gene therapy, 156, 158; and neurotransmitters of the brain, 114-117

hypothalamic-pituitary-adrenal axis, 72

immune system: 129-130; and allergies, 149; effect of emotions on, 70-73; and T-cells, 71

Immunoglobulin E (IgE), and allergies, 149, 150, 152

imprinting, and learning, 87, 88

instinct, and learning, 86-90

insulin, human, made by E. coli, 45, 50, 51

intelligence test, development of, 98-100

intercellular communication: 83, 84; and APUD theory, 84; between plants and animals, 83-85

intercropping, 200, 205

interferon, 44, 46, 51, 84, 160

interleukin, 2, 46, 84, 160

introns, 42

junk food diet, effect of, on behavior, 91

Kaufman Assessment Battery for Children (KABC), 98, 99, 100

life: origins of, on Earth, 17-19; search for origins of, and first synthetic cell, 20-24, 29-31

learning, and genetic programming, 86-90

Lashley, Karl, 104, 108

learning, and brain structures involved in, 104-108

limbic system, involvement in learning and memory, 104-108

L-DOPA, and Parkinson's disease, 115

lysergic and diethylamide (LSD), and ergot, 136

Lesch-Nyhan syndrome, and gene therapy, 155, 157

life expectancy, rates around world, 175-177

lakes, and acid deposition, 190, 191, 192

loss of habitat, 179, 184

macroevolution, theories on, 6-10

major histocompatibility complex, 130

mammals, and critical periods for imprinting, 87, 88

medicine: chronobiology and, 77-80; herbal, 82

memory: and the amygdala, 105; and dreams, 111, 112; structures of the brain and, 104-108

Mendel, theory of heredity by, 13-14

messenger molecules, and intercellular communication, 82, 83, 84

messenger RNA, 35, 36, 37, 38, 55, 60

metabolism, and obesity, 131, 134

mobile anion hypothesis, and surface-water acidification, 190, 193

modern synthesis, view of evolution, 6-10

molecular biology: 34-38; research of, on egg cell, 59-64

mothers' rights, vs. rights of fetus, 141

National Institute of Mental Health (NIMH), 104, 105, 141, 155

natural selection, role of, in evolution, 7-8, 11-14

Nature, limitations to possibilities in, 25-31

nature/nurture debate, and personality traits as inherited, 74-76

Neanderthal man, and herbal medicine, 82

neostriatum, 104, 116

nerve cells: and pain, 146; regrowth of, 126-128

nerve growth factor, 126

nervous system: and cellular communication, 83, 85; peripheral, and regrowth of nerve cells, 126-128

neural transmission: 106-108; and Parkinson's disease and Huntington's chorea, 114-117

neurocomputing, 101-103

neurons: and nerve regrowth, 126-128; transmission of information through, 107, 108, 114-117

neurotransmitters, 72, 83, 84, 91, 114-117

nitrogen oxides, and acid deposition, 186-194

non-REM sleep, 109, 112

non-Western cultures, and attitudes toward fat people, 132, 133

nucleotides: and AIDS, 163; and genetic code, 155, 160; see also, DNA

obesity: attitudes in other cultures concerning, 132, 133; and fat cells research, 131-134; psychogenic theory of, 131, 132

oil-eating microbe, patenting of, 50

oncogenes, and cancer research, 56-57

ontogeny, 26-29

opiate receptors, and pain, 144, 146

organ transplants, and immune system, 130

origin of life: on Earth, 17-22; and laboratory synthesis of living cell, 23-24, 29-31

Origin of Species (Darwin), 10, 11, 12, 13, 15

overpopulation, of world, 172-177

ozone: and air pollution, 180, 193; depletion of, 207-209; origin of, 18

pain research, 144-147

Parkinson's disease, and neurotransmitters of the brain, 114-117

patents, on genetically engineered life forms, 48, 50, 51

peripheral nervous system (PNS), and regrowth of nerve cells, 126-128

personality traits, as inherited, 74-76

pesticides, and wildlife, 179

phenylketonuria, genetic testing for, 158

photovoltaics, as alternative to fossil fuels, 171, 182

phylogeny, 26-29

placebo effect: 93-95; as a chemical pain-relief mechanism, 145, 147

planets, Earth as unique among, 17-19

plants, gene transfers in, 52-54

pons, and dreams, 110

population crisis, global, 172-177

positive mental states, effect of, on immune system, 70, 71

post-synaptic neuron, 114, 115

prefrontal cortex, and dreams, 109-113

prenatal diagnosis, and genetic disease, 157, 158

pre-synaptic neuron, 114, 115

programmed memorization, and birds, 87

prostaglandins, 145

protein synthesis, 35-37

pseudogenes, 42

psychogenic theory, of obesity, 131, 132

psychotherapy, and placebo effect, 93

public opinion, on risks of genetic engineering, 48

quality of life, Environmental Quality Index on, 184, 185

rainforests, destruction of, in Brazil, 195-197

RAST, and allergies, 149

recombinant DNA research: 39, 49-51; and cancer research, 56-58; using E. coli bacteria, 45-47; ethics of, 40, 43, 48-51; in plants, 52-54; safety concerns over, 48, 54; using yeast, 44-47; see also, genetic engineering

REM sleep, and dreaming, 109-113

renewable energy, as alternative to fossil fuels, 170, 171, 182

research laboratories, neuropsychological, at National Institute of Mental Health, 105-108

restriction enzymes, role of, in gene splicing, 41, 49, 50, 51, 55

rheumatoid arthritis, 130

RNA, 34-38, 55, 60
Rosvold, H.E., 104, 105

Saccharomyces cerevisiae, see
    S. cerevisiae
Safe Drinking Water Act, 181
Salem, explanation of witchcraft affair in,
    135-138
S. cerevisiae, 46, 47
schizophrenia, and Huntington's chorea,
    117
Schwann cells, 126
sequential processing, 98, 99, 100
sex, and aging of body, 123, 124
sexual selection, Darwin's theory of, 13
shifting cultivation, 200, 205
silver-stain technique, 106
single-celled organisms, 84
skin, and aging of body, 121, 124
skin cancer, and hazards of ozone
    depletion, 207
sleep, effect of diet on, 91, 92
smog ozone, 207
snow, acid, 189, 190
Snow's Law, and population crisis of
    world, 173, 174
soil erosion: 178; Environmental Quality
    Index on, 183, 184
solar energy, as alternative to fossil fuels,
    171, 182
somatic-cell therapy, 156
speciation, 6, 9
species selection, 9
species-specific songs, and programmed
    memorization in birds, 87
spinal cord, damage to, and regrowth of
    nerves, 126-128

stimulus response bonds, 105
stress: and allergies, 149; and disease,
    70, 73; and endorphins, 144;
    vulnerability to, as inherited trait, 74,
    75
strip mining, 182
subsistence farming: and deforestation in
    Brazil, 196, 197; and green revolution
    in Third World, 198, 199, 203
sugar, and hyperactivity, 92
sulfur oxides, and acid deposition,
    186-194
sun, "life zone" of, 19
suppressor T cells, 130
surface-water acidity, and acid deposition,
    190, 191
synthetic cell, laboratory creation of,
    20-24, 29-31

T-cell activating factor, 72
T cells: 71, 72; and immune system, 129,
    130; and organ transplants, 130
tensor analysis, 103
T-4 cells, 130
Third World: and biotechnology, 199-201;
    and green revolution, 198-200, 202,
    204-206
thymosins, as immunotransmitters, 73
toxic chemicals: air and water pollution
    from, 178, 180, 181; effect of, on
    wildlife, 179
traditionalism, as an inherited trait, 74
traits, personality, as inherited, 74-76
transcutaneous nerve stimulation, and
    pain, 146, 147
transfer RNA, 36, 37

transgenic pig, and gene transfer
    experiment, 48, 49
transgenic plants, 48
transposons, 42
tryptophan, effect of, on sleep, 91, 92
2-deoxyglucose method, 106
tyrosine, and depression, 92

unicellular organisms, see single-celled
    organisms
universe, possibility of extraterrestrial life
    in, 17-19

virus, cancer-causing, 56
vitamin B deficiency, effect of, on learning
    and memory, 107

Walsh descriptors, 107
water pollution: 178; and acid deposition,
    186, 190, 191; Environmental Quality
    Index on, 181
weight, and aging of body, 121
weight control, and fat cells, 131-134
White, David, experiments with laboratory
    creation of synthetic cell, 20-24
wildlife: and conservation reserve program
    for soil erosion as beneficial to, 184;
    Environmental Quality Index on, 179
wind power, as alternative to fossil fuels,
    171, 182
witchcraft, ergot as explanation for, in
    Salem, 135-138
world population, crisis in, 172-177

yeast, use of, in recombinant DNA
    research, 44-47
yo-yo dieting, 134

Credits/ Acknowledgments

Cover design by Charles Vitelli

1. Evolution
Facing overview—The New York Public Library.

2. Genetics
Facing overview—WHO/photo.

3. Behavior
Facing overview—United Nations. 80—M.E. Challinor.

4. The Brain
Facing overview—Medical World News.

5. Physiology
Facing overview—United Nations/Photo by L. Barns.

6. Medicine
Facing overview—Courtesy of Jean Bailey.

7. The Environment
Facing overview—UN photo/Viviane Holbrooke.
178-185—Marshall & Richie Moseley/Little Apple Art.

ANNUAL EDITIONS: BIOLOGY, Fifth Edition

Article Rating Form

Here is an opportunity for you to have direct input into the next revision of this volume. We would like you to rate each of the 46 articles listed below, using the following scale:

1. **Excellent: should definitely be retained**
2. **Above average: should probably be retained**
3. **Below average: should probably be deleted**
4. **Poor: should definitely be deleted**

Your ratings will play a vital part in the next revision. So please mail this prepaid form to us just as soon as you complete it.
Thanks for your help!

Annual Editions revisions depend on two major opinion sources: one is our Advisory Board, listed in the front of this volume, which works with us in scanning the thousands of articles published in the public press each year; the other is you—the person actually using the book. Please help us and the users of the next edition by completing the prepaid article rating form on this page and returning it to us. Thank you.

Rating	Article	Rating	Article
	1. The Evolution Revolution		24. Lab Team Enjoys a Cosmic Cruise Inside the Mind
	2. Influence of Darwin's Ideas on the Study of Evolution		25. Such Stuff as Dreams Are Made On
	3. Earth's Lucky Break		26. The Unbalanced Brain
	4. The Search for Life's Origins—and a First ''Synthetic Cell''		27. The Aging Body
	5. On the Wings of an Angel		28. Grow, Nerves, Grow
	6. Bits of Life		29. Immune System: Great Mystery Is Solved After Long Quest
	7. Genetics: The Edge of Creation		30. Weight Regulation May Start in Our Cells, Not Psyches
	8. Yeast at Work		31. Ergot and the Salem Witchcraft Affair
	9. Science Debates Using Tools to Redesign Life		32. Saving Babies
	10. The Gene Revolution		33. Pain: Many Causes, Fewer Cures
	11. How Genes Work		34. Racing Toward the Last Sneeze
	12. Spelling Out a Cancer Gene		35. The Gene Doctors
	13. The Fate of the Egg		36. Mapping the Genes, Inside and Out
	14. Life's Recipe		37. The Natural History of AIDS
	15. Thinking Well: The Chemical Links Between Emotions and Health		38. What Triggers AIDS?
	16. Major Personality Study Finds That Traits Are Mostly Inherited		39. Energy: A Promise Renewed
	17. The Clock Within		40. World Population Crisis
	18. Chemical Cross Talk		41. 19th Environmental Quality Index: A Nation Troubled by Toxics
	19. The Instinct to Learn		42. Acid Deposition: Trends, Relationships, and Effects
	20. Food Affects Human Behavior		43. How Much Are the Rainforests Worth?
	21. The Placebo Effect: It's Not All in Your Mind		44. Beyond the Green Revolution
	22. Intelligence Test: Sizing Up a Newcomer		45. Mimicking Nature
	23. What the Brain Builders Have in Mind		46. Ozone Depletion's New Environmental Threat

(continued on back)

ABOUT YOU

Name_____ Date_____

Are you a teacher? ☐ Or student? ☐

Your School Name _____

Department _____

Address _____

City _____ State _____ Zip _____

School Telephone # _____

YOUR COMMENTS ARE IMPORTANT TO US!

Please fill in the following information:

For which course did you use this book? _____

Did you use a text with this Annual Edition? ☐ yes ☐ no

The title of the text? _____

What are your general reactions to the Annual Editions concept?

Have you read any particular articles recently that you think should be included in the next edition?

Are there any articles you feel should be replaced in the next edition? Why?

Are there other areas that you feel would utilize an Annual Edition?

May we contact you for editorial input?

May we quote you from above?

BIOLOGY, Fifth Edition